Environmental Security and Environmental Management: The Role of Risk Assessment

NATO Security through Science Series

A Series presenting the results of scientific meetings supported under the NATO Programme for Security through Science (STS).

The Series is published by IOS Press, Amsterdam, and Springer Science and Business Media, Dordrecht, in conjunction with the NATO Public Diplomacy Division.

Sub-Series

A. Chemistry and Biology	(Springer Science and Business Media)
B. Physics and Biophysics	(Springer Science and Business Media)
C. Environmental Security	(Springer Science and Business Media)
D. Information and Communication Security	(IOS Press)
E. Human and Societal Dynamics	(IOS Press)

Meetings supported by the NATO STS Programme are in security-related priority areas of Defence Against Terrorism or Countering Other Threats to Security. The types of meeting supported are generally "Advanced Study Institutes" and "Advanced Research Workshops". The NATO STS Series collects together the results of these meetings. The meetings are co-organized by scientists from NATO countries and scientists from NATO's "Partner" or "Mediterranean Dialogue" countries. The observations and recommendations made at the meetings, as well as the contents of the volumes in the Series, reflect those of the participants in the workshop. They should not necessarily be regarded as reflecting NATO views or policy.

Advanced Study Institutes (ASI) are high-level tutorial courses to convey the latest developments in a subject to an advanced-level audience

Advanced Research Workshops (ARW) are expert meetings where an intense but informal exchange of views at the frontiers of a subject aims at identifying directions for future action

Following a transformation of the programme in 2004 the Series has been re-named and re-organised. Recent volumes on topics not related to security, which result from meetings supported under the programme earlier, may be found in the NATO Science Series

www.nato.int/science
www.springeronline.com
www.iospress.nl

Series C: Environmental Security – Vol. 5

Environmental Security and Environmental Management: The Role of Risk Assessment

edited by

Benoit Morel

Carnegie Mellon University,
Pittsburgh, PA, U.S.A.

and

Igor Linkov

Cambridge Environmental, Inc.,
Cambridge, Massachusetts, U.S.A. and
Carnegie Mellon University,
Pittsburgh, PA, U.S.A.

Springer

Published in cooperation with NATO Public Diplomacy Division

Based on papers presented at the NATO Advanced Research Workshop on
The Role of Risk Assessment in Environmental Security and Emergency Preparedness in the Mediterranean Region
Eilat, Israel
15–18 April 2004

A C.I.P. Catalogue record for this book is available from the Library of Congress.

ISBN-10 1-4020-3892-5 (PB)
ISBN-13 978-1-4020-3892-1 (PB)
ISBN-10 1-4020-3891-7 (HB)
ISBN-13 978-1-4020-3891-4 (HB)
ISBN-10 1-4020-3893-3 (e-book)
ISBN-13 978-1-4020-3893-8 (e-book)

Published by Springer,
P.O. Box 17, 3300 AA Dordrecht, The Netherlands.

www.springeronline.com

Printed on acid-free paper

Printed in the Netherlands.

TABLE OF CONTENTS

PREFACE

The concept of "environmental security" has emerged as one basis for understanding international conflicts. This phrase can mean a variety of things. It can signify security issues stemming from environmental concerns or conflicting needs, or it can mean that the environment is treated as a resource for the long term, and the question is what should be done today to preserve the quality of the environment in the future. In the same way that energy security is about ensuring access to energy for the long run, it can also mean that pressing environmental concerns create a situation where different countries and communities are forced to collaboratively design a unified response, even if cooperation is not generally in the logic of their relations.

Over the last several years, the authors of this book and their colleagues have tried to demonstrate the power of risk assessment and decision analysis as valuable tools that decision makers should use for a broad range of environmental problems, including environmental security. Risk analysis is almost more a state of mind or a way of looking at problems than it is a kind of algorithm or a set of recipes. It projects a kind of rationality on problems and forces a certain degree of quantitative rigor, as opposed to the all too common tendency of making environmental recommendations based on anecdotal evidence.

This book is based on the discussions and papers prepared for the NATO Advanced Research Workshop that took place in Eilat, Israel, in April 2004. The workshop in Eilat was the third meeting sponsored jointly by the Society for Risk Analysis and NATO. The goal of the first workshop in this series, entitled "Assessment and Management of Environmental Risks: Cost-efficient Methods and Applications[1]" (Lisbon, Portugal, October 2000), was to present risk assessment as a unified technique for providing a scientific basis for environmentally sound and cost-efficient policies, strategies, and solutions for various environmental problems. The workshop confirmed risk assessment as a viable tool for developing countries. One of the workshop suggestions was to organize a more focused topical meeting on the application of specific risk-based techniques in developing Mediterranean countries. The second workshop, "Comparative Risk Assessment and Environmental Management,[2]" was an important step in the development and application of comparative risk assessment (CRA) and other risk-based decision-analytical tools in environmental management. Comparative Risk Assessment (CRA) is a methodology applied to facilitate decision-making when various activities compete for limited resources. This application is thus especially relevant to developing countries. The workshop in Italy was one of the first

---

1 Linkov, I., Palma Oliveira, J.M., eds "Assessment and Management of Environmental Risks," Kluwer, Amsterdam, 2001.

2 Linkov, I., Ramadan, A.B., eds "Comparative Risk Assessment and Environmental Decision Making," Kluwer, Amsterdam, 2004.

to place CRA as part of the more general but as yet quite academic field of multi-criteria decision analysis (MCDA).

The organization of the book reflects sessions and discussions during the meeting in Eilat. The goal of the workshop was to expand concepts introduced in the first two meetings and explore their applications to the field of environmental security as well as to specific realities of the Middle East. One question the workshop did not answer definitively is in fact what could have been construed as the fundamental question: what is exactly "environmental security"? The introductory chapter by Belluck *et al.* provides alternative definition to this term and explores its connection with risk assessment. In certain situations, such as extreme events, risk assessment provides a unique level of analytic power, as was illustrated in this workshop in the case of the earthquakes in Turkey, for example. The individual chapters are grouped into sections which reflect broad issues discussed at the meeting.

Many of the contributions to the book do not deal directly with the region of Israel and its neighbors. Instead, they tend to apply a risk analytical approach to a variety of environmental problems. The goal was as much to inform about the problems as it was to show the power of risk analysis as a tool to discuss environmental policy. Risk assessment also points to the need for certain data. It helps to identify what data would be useful or are needed. It clarifies the discussion and provides a common language to the policy makers. It is also clear that risk analysis still needs refinements for it to be more universally accepted as a tool for environmental policy discussions.

In addition to plenary sessions, the meeting included two discussion groups which took two very different perspectives on the issues. One took a risk analysis approach to environmental security. The other originally tried to put the environmental issues pertaining to the region in the broader context of climate change and international environmental concerns. Reports from these groups are included in this book.

The workshop concluded that "environmental security" has a special connotation in the Middle East. The region defined by Israel and its neighbors is confronted with immediate environmental issues which make concerns about global warming a less urgent priority. Since the resumption of the "Intifada" in September 2000, the West Bank and Gaza have been suffering from environmental neglect. The fact that the environmental situation has been degrading is beyond debate, but the extent of the problem is not precisely known. This is happening in a location where for years water quantity (and now quality) have been a source of growing concern and also controversy. The agreements negotiated in the 90's during the Oslo process in fact did not resolve the problems associated with water allocation and completely ignored the environmental implication of letting the Jordan River without replenishment of its water. One clear message of the discussions is that this region needs immediate and international attention.

Igor Linkov and Benoit Morel
April, 2005
Cambridge, MA

ACKNOWLEDGEMENTS

The meeting took place at a time when terrorist acts and suicide bombings were relatively common in Israel and US nationals were advised to avoid visiting the region. We would like to thank workshop participants for their contribution to the meeting and the book. One unfortunate impact of the security situation was to preclude any Arab participation. We would like to thank Dr. Abou Bakr Ramadan for his enthusiastic support of the meeting and scientific collaboration in the region. We are grateful to Mr. David Letnik of the Israel Engineering Academy and Arava Institute for helping with meeting logistics. The editors would like to thank Drs. Tal, Levner, and Marquina for their help in the workshop organization. Excellent editorial and technical assistance was provided by Lori Simpson, Alexandra Morel and Elena Belinkaia. The workshop agenda was prepared in collaboration with the Society for Risk Analysis (SRA). Financial support for the workshop organization was provided mainly by NATO. Additional funding was provided by Isrotel.

Part 1.

Environmental Security

ENVIRONMENTAL SECURITY, CRITICAL INFRASTRUCTURE AND RISK ASSESSMENT: DEFINITIONS AND CURRENT TRENDS[1]

D.A. BELLUCK
Senior Transportation Toxicologist, FHWA/USDOT
400 7th Street, S.W., Washington, D.C. 20590 USA

R.N. HULL
Cantox Environmental, Inc., 1900 Minnesota Court, Suite 130
Mississauga, Ontario L5N 3C9 CANADA

S.L. BENJAMIN
National Wildlife Biologist, Farm Service Agency
U.S. Department of Agriculture, 1400 Independence Avenue, S.W.
Washington, D.C. 20501 USA

J. ALCORN
SAIC, 11251 Roger Bacon Dr.
Reston, VA 20190 USA

I. LINKOV
Cambridge Environmental Inc. 58 Charles Street,
Cambridge, MA 02141 USA

Abstract

Population growth, needed economic growth, and social pressures for improved infrastructure coupled to the need for human health and ecological protection and environmental security make systematic and transparent environmental decision-making a complex and often difficult task. Evaluating complex technical data and developing feasible risk management options requires procedural flexibility that may not be part of existing evaluative structures. Experience has demonstrated that direct transposition of risk assessment and risk management frameworks (e.g. those developed in the United States and European Union) may not work in regions whose social, legal, historical, political and economic situations are not suitable or prepared for acceptance of these methodologies. Flexible decision-making, including the use and development of acceptable or unacceptable risk levels based on the critical nature of an infrastructure

[1] The work presented by the authors was performed in their private capacities and does not reflect the policies or views of their parent institutions.

B. Morel and I. Linkov (eds.),
Environmental Security and Environmental Management: The Role of Risk Assessment, 3–16.

type, is one potential approach to assist risk managers in their decision-making. Unfortunately, the newness of the discussions on the interrelatedness of environmental security and critical infrastructure has yet to produce a unified and comprehensive treatment of the fields. As a result, this paper will describe and define these terms in order to set the stage for discussions of human health and ecological risk assessment and risk management later in the paper. This paper reviews basic concepts defined in the field of risk assessment and extends its applicability to the areas of environmental security and critical infrastructure protection.

1. Environmental Security Defined

Environmental security has emerged as an increasingly important concern of governments and their defense establishments because of several trends that have the potential to threaten stability. These potential threat issues include: world population in 2015 will be 7.2 billion, up from 6.1 billion in year 2000; water scarcities and allocation will pose challenges to certain governments; groundwater depletion; contemporary environmental problems will persist and grow; globalization will be rocky, marked by chronic financial volatility and a widening economic divide; significant degradation of arable land; loss of tropical forests; greenhouse gas emissions will increase substantially; exacerbation of biological species loss; rapid urbanization; increasingly serious urban air and water quality problems; and global climate change induced glacial ice melt backs, sea level rise, and increasing storm frequency (National Intelligence Council 2000; Schwartz and Randall 2003; SERO 2003).

"Environmental Security" is an ill-defined term (McNeil 2000) with many definitions whose two key elements are: repairing damage to the environment for human life support and for the moral value of the environment itself; and, preventing damage to the environment from attacks and other forms of human abuse (Cheremisinoff 2002). Several definitions of environmental security exist and demonstrate that after more than two decades of discussion, the concept of environmental security still has no widely agreed upon formulation (McNeil 2000). Examples include:

- "Environmental security (ecological security or a myriad of other terms) reflects the ability of a nation or a society to withstand environmental asset scarcity, environmental risks or adverse changes, or environment-related tensions or conflicts." (Chalecki No Date);
- "Science-based case studies, which meld physical science with the discipline of political economy, are a suitable vehicle for forecasting future conflicts derived in some measure from environmental degradation" (McNeil 2000);
- "[T]hose actions and policies that provide safety from environmental dangers caused by natural or human processes due to ignorance, accident, mismanagement or intentional design, and originating within or across national borders." (Cheremisinoff 2002);
- "Environmental Security is a state of the target group, either individual, collective or national, being systematically protected from environmental risks

caused by inappropriate ecological process due to ignorance, accident, mismanagement or design." (AC/UNU Millennium Project No Date);

- "Environmental security is protectedness of natural environment and vital interests of citizens, society, the state from internal and external impacts, adverse processes and trends in development that threaten human health, biodiversity and sustainable functioning of ecosystems, and survival of humankind." (AC/UNU Millennium Project No Date);
- "Environmental security is the state of protection of vital interests of the individual, society, natural environment from threats resulting from anthropogenic and natural impacts on the environment." (AC/UNU Millennium Project No Date).

The discipline of "environmental security" is neither a pure security issue nor an environmental issue (Chalecki No Date). However, environmental issues are often security concerns because, even without directly causing open conflict, they can result in environmental perturbations or triggers that can destabilize the status quo and result in a loss of regional, national, and local political, social, economic and personal security (Schwartz and Randall 2003).

Environmental security concerns can be grouped into three general categories (AC/UNU Millennium Project No Date): 1) security of the environment which is a good in itself; 2) security from environmental change that can create societal instability and conflict; and, 3) security from environmental change (e.g. water scarcity, air pollution, etc.) that would threaten the material well-being of individuals (AC/UNU Millennium Project No Date). Common elements of environmental security definitions include: public safety from environmental dangers caused by natural or human processes due to ignorance, accident, mismanagement, or design; amelioration of natural resource scarcity; maintenance of a healthy environment; amelioration of environmental degradation; and, prevention of social disorder and conflict (promotion of social stability) (Glenn et al. 1998).

Environmental security concerns include chemical/material releases to the environment. This is because, worldwide, an estimated one quarter to one third of disease burden is attributable to environmental factors (European Environment Agency 2003). Chemical or material releases to the environment or environmental alteration result in actual or perceived health risks that can result in societal conflicts between parties in support or opposition to the environmental perturbation.

2 Critical Infrastructure Defined

Critical infrastructures are complex societal systems (Vrijling et al. 2004) and also have many definitions. Examples include:

- "Civil and critical infrastructure systems such as transportation, communication, power, and financial systems have provided the foundation for modern society" (Oren 2001);

- According to the U.S. Department of Defense, critical infrastructures are "Those systems and assets essential to plan, mobilize, deploy, and sustain military operations and transition to post-conflict military operations, and whose loss or degradation jeopardize the ability of the Department of Defense to execute the National Military Strategy" (Secretary of the Navy 2002);
- "Critical infrastructure (or assets) in the highway transportation system include all of its components, including physical and cyber-based components, which are used in attaining transportation functions to serve national, regional and local objectives. Examples of these structures include the physical structures (roadways, bridges, tunnels), facilities (parking areas, toll complex), ITS (Intelligent Transportation Systems) components (signs and signals, network, control centers), and organizational components (personnel, procedures, communication)" (Haimes et al. 2004);
- In Queensland, Australia "Critical infrastructure is defined as infrastructure which, if destroyed, degraded or rendered unavailable for an extended period, will significantly impact on social or economic well-being or affect national security or defence" (Queensland Government No Date);
- The United States Patriots Act states that critical infrastructures are: "Systems and assets, whether physical or virtual, so vital to the United States that the incapacity or destruction of such systems and assets would have a debilitating impact on security, national economic security, national public health or safety, or any combination of those matters" (National Infrastructure Institute (NI^2) Center for Infrastructure Expertise and the University of New Hampshire 2004).

Regardless of the definition, it is clear that "Our society and modern way of life depend on a complex system of critical infrastructures" (The White House 2003).

The convergence of critical infrastructure, environmental security, risk assessment and risk management is a function of the perception that these fields are inextricably interrelated and the need to make complex decisions based on multiple criteria as part of the risk management process. In order to fully understand how these two issues relate, one needs to understand how the concept of acceptable risk developed, its application to risk assessment and risk management, and its relationship to environmental security and critical infrastructure. The remainder of this paper will delve into these interrelationships.

3 Risk Assessment, Risk Management and Environmental Security

In order to survive and prosper, humans must alter their environment and use environmental resources. As modern societies increase in population, they must increase their use of renewable and nonrenewable resources to provide their citizens with essential goods, services and economic security. Expanding populations and economies build farms, homes, factories and transportation networks that use and release chemicals/materials to the environment purposefully, accidentally or incidentally. The field of risk assessment is developing to address anthropogenic risk.

3.1 ORIGINS OF RISK ASSESSMENT AND RISK MANAGEMENT

Releases of chemicals and materials, along with environmental alteration, is often monitored or regulated by government agencies. Government agencies use administrative tools to evaluate chemical/material releases and environmental alteration. A favored tool is quantitative risk assessment that has been described as "An organized process used to describe and estimate the likelihood of adverse health outcomes from environmental exposures to chemicals" (Presidential/Congressional Commission on Risk Assessment and Risk Management 1997)

Quantitative risk assessment is the preferred tool in the United States and elsewhere to regulate or evaluate facilities, activities or processes that release chemicals to the environment. More than 25 years ago, the Inter-Agency Regulatory Liaison Group and the Office of Science & Technology Policy in the White House proposed an orderly set of activities under the headings of "hazard identification", "risk characterization", and "risk reduction". In 1983, the National Research Council published "Risk Assessment in the Federal Government: Managing the Process" (NRC 1983) also known as the "Red Book" (Commission on Life Sciences 1983; Omenn 2003). The four steps of the current human health and ecological risk assessment process are hazard identification, dose-response assessment, exposure assessment, and risk characterization (Presidential/Congressional Commission on Risk Assessment and Risk Management 1997). For both human health and ecological risk assessment, numerical risk assessment findings are compared to risk management and policy based acceptable risk levels (e.g. single point risk levels or risk ranges) to determine if there is a potential for significant or unacceptable risk.

Risk management is variously defined as "The process of analyzing, selecting, implementing, and evaluating actions to reduce risk" and "...the process of identifying, evaluating, selecting, and implementing actions to reduce risk to human health and to ecosystems" (Presidential/Congressional Commission on Risk Assessment and Risk Management (1997). It has also been defined as a "...decision-making process involving considerations of political, social, economic, and technical factors with relevant risk assessment information relating to a hazard so as to develop, analyze, and compare regulatory and non-regulatory options, and to select and implement the optimal response for safety from that hazard (Duffus 2001). The goal of risk management is scientifically sound, cost-effective, integrated actions that reduce or prevent risks while taking into account social, cultural, ethical, political, and legal considerations" (Presidential/Congressional Commission on Risk Assessment and Risk Management 1997).

Acceptable and unacceptable numerical risk levels (along with their synonymous terms) can vary by governmental unit or statute. The risk management paradigm allows risk managers at all levels of government to use the risk management decision-making process to allow chemical/material releases to the environment even though they have been found to exceed applicable acceptable risk levels. This type of risk management decision-making uses the numerical risk expression as a point of departure for decision-making rather than a bright line that cannot be exceeded. How this is done and why such a decision can be acceptable will be discussed in detail later in this paper. First, let us look at a very brief history of the evolution of the acceptable risk concept.

3.2 ACCEPTABLE/UNACCEPTABLE RISK CONCEPT

The assumption that public health could be protected by chemical risk management developed in the United States in the early 1900s for food additives. By 1958, an amendment to the U.S. Food, Drug, and Cosmetics Act put forth the concept that some chemicals might have no toxic threshold and prohibited the addition of any chemical that can cause cancer. It was also recognized that it was impossible to completely remove carcinogens from the food supply and, as a result, Food and Drug Administration put forth the proposal that if risks calculated under the no-threshold assumption were below some small value, the carcinogen was effectively absent in the food. A virtually safe dose (one in one hundred million or 10^{-8}) to limit cancer risk was proposed but was found to be an almost impossible burden on regulators. An alternative level for food additives was proposed at one in a million (10^{-6}), a level considered negligible by most people. This level became the criterion for acceptable risk in the United States when cancer risks from environmental exposures became recognized in the late 1960s and early 1970s. By the 1990s, it was recognized that the one-in-one-million risk level was very stringent and the idea of a lesser risk level of one-in-ten-thousand (10^{-4}) was introduced. In general, a risk above one-in-ten-thousand is considered excessive. The Clean Air Act Amendments of 1990 led to the development of a risk range of 10^{-4} to 10^{-6}. Thus, "If a cancer risk is judged to be *significant* or *unacceptable*, then it is generally expected that some action will be taken to reduce or eliminate the risk. In contrast, a *de minimis* or *essentially negligible* risk is one that is so small that no action needs to be taken. If a risk is judged to be *insignificant* or *acceptable*, however, this does not necessarily mean that it is *de minimis* or *negligible"* (Health Canada 1998).

Defining acceptable and unacceptable risks is the foundation for determining regulatory compliance or the need for risk management interventions to reduce calculated risks. There are numerous definitions of acceptable or unacceptable risks that are discussed below and presented in an appendix to this paper. Chemical releases can result in human or ecological exposures and subsequent hazards or risks. Human and ecological risk assessment methods are used to calculate risks and hazards to potentially exposed receptors. Acceptable hazard or risk levels can vary by locality. For example, acceptable risk levels for human carcinogens can be set at a specific level (e.g. one-in-one-million) or range (e.g. one-in-one-million to one-in-ten-thousand). In Canada, the federal government does not, in general, recommend acceptable risk levels (Health Canada 1998); each province has adopted either 10^{-6} or 10^{-5} as the acceptable excess cancer risk level. Acceptable hazard levels for non-carcinogenic chemicals can be different for single chemicals (e.g. Hazard Quotient of 0.2 or 1.0) or multiple chemicals (e.g. Hazard Index of 1.0). Acceptable risks for ecological risk assessment can be a single value for individual organisms (e.g. Toxicity Index less than 1) and populations (e.g. 10% chance that no more than 20% of a population will be exposed above a benchmark value).

According to Vrijling et al. (2004), "For complex societal systems as a whole, like a nation, one normally uses *individual risk* as a measure, which varies between 1 x 10^{-5} and 3.1 x 10^{-4} deaths per year for occupational, traffic and consumer risks respectively. The individual risk is then taken over the whole population at stake and a time period of one year. Although no general individual risk criteria are set for trivial risks either, one tends to measure those against the *de minimis* value of 10^{-6} or 10^{-5} deaths per year... indicating a potential low acceptable risk level for any individual [which] everybody can live with. In some cases of critical infrastructures, like high speed train links, individual risk criteria are set in the Netherlands ..." "The same is true for the zoning between hazardous chemical facilities and residential areas, at an individual risk contour of 10^{-6} deaths per year..." "For critical infrastructures sometimes also societal risks are defined. For social or group risks, the next step is to order the scenarios with increasing measure of potential consequences (mostly deaths). The cumulative probabilities (or frequencies) for exceeding a certain number of deaths are then derived from the probabilities of all scenarios contributing to that particular number of deaths." "The societal acceptable risk is judged at a national level by placing an upper-bound upon the expected number of fatalities per activity per year."

Ecological risk assessment, the study of non-human risks, is much more complex than human health risk assessment. The tremendous diversity of habitats and species makes the study of ecological risks challenging. Walker (No Date) notes: "There is no national standard for *ecological value* for ecological risk assessments. There needs to be much discussion by stakeholders as to what the "environmental value" is at each site that must be protected. The ecological value must reflect policy goals and societal values. Societal concerns can range from protection of endangered or commercially or recreationally important species to preservation of ecosystem attributes for functional reasons (e.g. flood water retention by wetlands) or aesthetic reasons (e.g. visibility in the Grand canyon)."

Deterministic methods are most often used to assess the potential for acceptable or unacceptable ecological risks. At its most basic level, actual toxicity data for the species of interest or a similar species are used to provide an ecologically protective numerical value for a given medium (e.g. surface water, sediment, soil) and ecological receptor. These calculated media-specific concentrations are associated with risks at the threshold of acceptable/unacceptable risk *to individual organisms*. When the ratio between a known or calculated contaminant level and the ecologically protective numerical value exceeds unity (or some other value established by a responsible governmental unit), then there is the potential for unacceptable risks, and either more detailed assessment is conducted or risk management is required. Conversely, not exceeding the acceptable ratio value usually equals acceptable ecological risks. Other measures of acceptable ecological risk include those that do not allow more than a certain percentage of an exposed population to exceed a given concentration based on a selected toxicological value. Probabilistic methods and statistical models are also used to derive medium and receptor specific acceptable/unacceptable threshold concentrations. An important reason for moving away from concentration-effect data for single species is that ecological risk assessment generally is not concerned with protection of individual organisms, but rather protection of populations, communities and ecosystems (Newman et al. 2000).

Risk assessment numerical findings are not meant to be accurate or precise estimates of morbidity or mortality. They are merely numerical estimates using a systematic and transparent process based on elements of science, policy, law and professional judgment. Determining when these numerical risk or hazard estimates represent acceptable or unacceptable levels is not a scientific exercise. Rather, it is a risk management determination based on established acceptable risk levels (e.g. 10^{-6} or lower excess cancer risks are acceptable) or risk ranges (e.g. 10^{-4} to 10^{-6} risk ranges) that are in policy, guidelines, or statute. Risks greater than "a bright line" or "acceptable risk range" levels are not necessarily "unacceptable" because risk management practices allow management judgment in determining when a potentially "significant risk" becomes unacceptable. It is common practice to uniformly apply "bright lines" and risk ranges to all types of processes, activities, or facilities. In practice, risk assessors and managers often judge exceedence of a bright line or risk range as unacceptable risk unless risk management decisions, based on additional factors not related to the calculated risk value(s), are made to the contrary. It seems reasonable to assert, based on these considerations, that as risk assessment techniques evolve so should risk management techniques and along with them acceptable risk level expressions.

This advancement may be particularly important when estimating ecological risks. Ecological risk assessments are conducted using a tiered approach, where the finding of acceptable or unacceptable risks determines the move to higher levels of analysis or site remediation (Owens et al. 1999). Unlike the assessment of human health risks, where it is considered more "acceptable" or "prudent" to err on the side of conservative (over-predicted) risk estimates, there is less support for overly conservative predictions of ecological risks and subsequent conservative risk management decisions. A benefit of ecological risk assessment is that it is possible to monitor, relatively easily, whether the ecological risk prediction was correct (or more importantly, incorrect), through the use of standard biomonitoring techniques. Acceptable/unacceptable ecological risk levels also may be defined less quantitatively than in human health risk assessment, and incorporate considerations such as probability of population (or community) persistence and restoration after some disturbance (Hobbs and Kristjanson 2003). In addition, the area of disturbance relative to the remaining undisturbed habitat of suitable quality for the ecological receptors of interest can be factored in to the acceptable/unacceptable risk definition (Suter et al. 1995).

Findings of acceptable or unacceptable risk drive risk reduction efforts, regulatory compliance determinations, as well as complex and expensive litigation. It is for this reason that the subject of critical vs. non-critical infrastructure based acceptable risk levels is vital to understand and explicitly define.

Finally, acceptable and unacceptable risk levels can be determined by public perception and political decision-making processes. For example, Vogel (2003) states: "The public's perception or tolerance of particular risks often differs from that of experts and in a democratic system the former's preferences – and values – often play an important role in the policy process. Thus governments can and frequently do chose to err on the side of caution, seeking to avoid or reduce particular risks that many citizens regard as unacceptable, even if the available scientific evidence does not or cannot prove evidence of harm. As Christoforou writes, "It is generally agreed that defining the level of acceptable risk is a normative decision that belongs to the democratically elected and accountable institutions of a state."

3.3 PRECAUTIONARY PRINCIPLE

The precautionary principle is often cited as an important doctrine to follow when making a risk management determination. However, this view is not universally accepted. The precautionary principle originated in Principle 15 of the Rio Declaration (1992); it states: "Where there are threats of serious or irreversible damage, a lack of full scientific certainty shall not be used as a reason for postponing cost-effective measures to prevent environmental degradation" (Scottish Executive No Date). The precautionary principle has been used to describe "[d]ecisions about the best ways to manage or reduce risks that reflect a preference for avoiding unnecessary health risks instead of unnecessary economic expenditures when information about potential risks is incomplete" (Presidential/Congressional Commission on Risk Assessment and Risk Management 1997). While the precautionary principle has no formal legal effect in the United States, no country has so fully adopted the essence of the principle in domestic law (Vogel 2003). In contrast, the European Union officially introduced the Precautionary Principle in Article 130 (the environmental section) of the 1993 Treaty of the European Union (Maastricht) and it has been referenced in 27 resolutions between 1994 and 1999 (Vogel 2003).

What happens when risk managers have insufficient scientific knowledge to make scientifically sound decisions? According to the Presidential/ Congressional Commission on Risk Assessment and Risk Management (1997), "Decision-makers must balance the value of obtaining additional information against the need for a decision, however uncertain. Sometimes a decision must be made under the precautionary principle (decisions about the best ways to manage or reduce risks that reflect a preference for avoiding unnecessary health risks instead of unnecessary economic expenditures when information about potential risks is incomplete). Risk management determinations of acceptable risk often involve judging safety. Since safety is not absolute and is immeasurable, achieving acceptable risks means decision-makers must make tradeoffs between costs, both absolute and relative, with risks"(Haimes et al. 2004).

4 Conclusions

It is clear from this discussion that environmental security is a very broad term used to encompass a wide variety of issues. For the purpose of this book, by *ensuring environmental security* we mean *guarding against environmental degradation in order to preserve or protect human, material, and natural resources at scales ranging from global to local.* The critical infrastructure concept is directly linked to environmental security. For environmental applications, critical infrastructure may be defined as *man-made structures constructed and maintained to assure human health, environmental protection, transportation networks, water supplies, clean air, food supplies and other critical elements necessary to maintain economic and national security*.

The question not yet fully addressed is *how does one accommodate both the importance of critical infrastructure and of environmental security, given the current risk assessment/risk management paradigm*? Many papers published in this volume address different aspects of this issue. For example, Belluck et al (2005) defines environmental security in terms of chemical releases, risk assessment, and risk management, and proposes consideration of a flexible risk acceptability criteria to match the critical nature of a given type of infrastructure based on the use of a systematic and transparent risk management process that matches the rigor of the risk assessment on which it is based. Nevertheless, our review and current research shows that more work is necessary to address methodology and application of risk assessment and environmental security to emerging threats in general and specifically in the Middle East. As the fields of risk assessment, risk management, critical infrastructure, and environmental security merge, additional discussions will need to occur to define these interactions and their implications for environmental protection and regulatory activity.

Appendix: Descriptions of Acceptable Risk and Risk Ranges

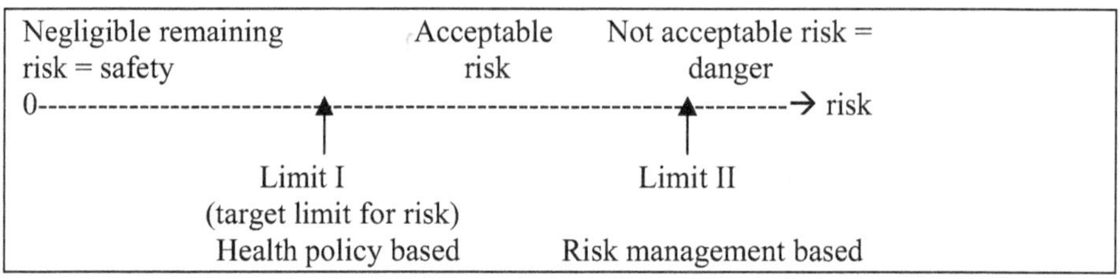

Bundesanstalt fur Arbeitsschutz und Abrbeitmedizin. 1998, slightly modified.

"Acceptable Risk Range – If the cumulative carcinogenic risk to an individual based on reasonable maximum exposure (RME) for both current and future land use is less than 1E-04 and the non-carcinogenic hazard index is less than 1, action generally is not warranted unless there are adverse environmental impacts." "*Note: The upper boundary of the risk range is not a discrete line at 1E-04, although EPA generally uses 1E-04 in making risk management decisions. A risk estimate that is greater than 1E-04 may be considered acceptable, if justified based on site-specific conditions. A risk manager may also decide that a baseline risk level less than 1E-04 is unacceptable due to site specific reasons and that remedial action is warranted."*(United States Navy 2001).

"As in the *Benzene* case, the court did not define any particular method for EPA to use in determining what risks are acceptable. On remand, the agency, after taking comment on a number of possibilities, decided that it could not use any single metric as a measure of whether a risk is acceptable. Instead, it adopted a general presumption that a lifetime excess risk of cancer of approximately one in 10,000 (10^{-4}) for the most exposed person would constitute acceptable risk and that the margin of safety should reduce the risk for the greatest possible number of persons to an individual lifetime excess risk no higher than one in 1 million (10^{-6})." (Commission on Life Sciences 1994)

"...the published acceptable risk level does not necessarily represent the "safe level" but rather a target level with the expectation that the true risk to exposure is less than the published value." (EXTOXNET 1993)

"In general terms, a risk that is so small, whose consequences are so slight or whose associated benefits (perceived or real) are so great that persons or groups in society are willing to take or be subjected to that risk. In more technical terms, an arbitrary value denoting a very low probability of occurrence of a seriously adverse effect in persons exposed daily over a lifetime. The dose associated with this risk may be considered to have an insignificant impact on human health. Synonyms: Tolerable Risk; Negligible Risk; Risk Level." (Health Canada 2000)

"It is the Agency's responsibility to determine in the first instance what it considers to be a "significant" risk. Some risks are plainly acceptable and others are plainly unacceptable. If for example, the odds are one in a billion that a person will die from cancer by taking a drink of chlorinated water, the risk clearly could not be considered significant. On the other hand, if the odds are one in a thousand that regular inhalation of gasoline vapors that are 2 percent benzene will be fatal, a reasonable person might well consider the risk significant and take the appropriate steps to decrease or eliminate it. (I.U.D. v. A.P.I., 448 U.S. 607, 655). So a risk of (1/1000) (10(-3)) is clearly significant. It represents the uppermost end of the million-fold range suggested by the Court, somewhere below which the boundary of acceptable versus unacceptable risk must fall." "...free to use conservative assumptions in interpreting the data with respect to carcinogens, risking error on the side of overprotection rather than underprotection" (448 U.S. at 655, 656)." "Further guidance for the Agency in evaluating significant risk and narrowing the million-fold range described in the "Benzene Decision" is provided by an examination of occupational risk rates, legislative intent, and the academic literature on "acceptable risk" issues. For example, in the high-risk occupations of mining and quarrying, the average risk of death from an occupational injury or an acute occupationally-related illness over a lifetime of employment (45 years) is 15.1 per 1,000 workers. The typical occupational risk of deaths for all manufacturing industries is 1.98 per 1,000. Typical lifetime occupational risk of death in an occupation of relatively low risk, like retail trade, is 0.82 per 1,000. (These rates are averages derived from 1984-1986 Bureau of Labor Statistics data for employers with 11 or more employees, adjusted to 45 years of employment, for 50 weeks per year)." (United States Department of Labor No Date)]

"Tolerable risk. Risk level below which risks would be regarded as being widely acceptable, either because they are irreducible or because they compare with other risks routinely accepted (see comparative risk assessment). Some sources distinguish tolerable risk from acceptable risk: the former being 'just acceptable' and kept under review." (U.K. Department of the Environment, Transport and the Regions 1999).

Acceptable risk is the "...type of risk such that the benefits derived by an organism, a population, or an ecological system outweigh the adverse effects that might affect them as a result of being administered or exposed to a particular agent" (Duffus 2001).
The determination of this "acceptable" or "tolerable" level of risk may have been prescribed before the risk assessment process begins - through societally determined acceptable levels of risk in the form of legislative environmental quality standards for instance, or industry derived "norms". In this case, risk management attempts to analyse which options for action based on the results of the risk assessment will produce these pre-determined risk levels. Where no acceptable risk standards exist, the risk management process will attempt to derive "acceptable" or tolerable risk on a case-by-case basis." "Decision making to determine "acceptable" or "tolerable" risk uses a number of approaches. The three major approaches to acceptable risk decisions are professional judgment where technical experts devise solutions, bootstrapping where historical precedent guides decision making and formal analyses where theory-based procedures for modeling problems and calculating the best decision are used. (European Environment Agency No Date)

"Any risk that is currently tolerated is considered to be acceptable" (Hunter and Fewtrell 2001).

"Defining an acceptable risk level gives meaning to the risk estimate generated from the risk assessment. There are few legislative, public policy, and judicial guidelines on how to define acceptable risk. Although "safe" has not been found to necessarily mean zero risk (State of Ohio v. EPA 997 F.2d 1520, 1533, D.C. Cir. 1993), the courts have not provided (1) a risk level above which risk management action must occur, (2) specific guidance as to what might be done to determine whether a risk is acceptable, or (3) workable definitions of acceptable, safe risk levels. The EPA currently "endorses" a risk range from 10^{-6} (one in a million) to 10^{-4} for one's lifetime risk from exposure to carcinogens and a hazard quotient of 1.0 for noncarcinogens. As our state survey shows, acceptable risk levels across the state regulatory agencies tend to mirror EPA guidance (Commission on Geosciences, Environment and Resources 1999).
"The level of Residual Risk that has been determined to be a reasonable level of potential loss/disruption" (National Infrastructure Institute (NI^2) Center for Infrastructure Expertise and the University of New Hampshire 2004).

5 References

1. AC/UNU Millennium Project. No Date. Emerging International Definitions, Perceptions, and Policy Considerations. Environmental Security Study. http://www.acunu.org/millennium/es-2def.html.
2. Belluck DA, Hull RN, Benjamin SL, Alcorn J and Linkov I. 2005. Are Standard Risk Acceptability Criteria Applicable to Critical Infrastructure Based on Environmental Security Needs? In: Linkov, I., Morel, B., eds. Environmental Security and Risk Assessment. Kluwer Academic Publishers (this volume).
3. Bundesanstalt fur Arbeitsschutz und Abrbeitmedizin. 1998. Guidelines on Risk Management of the Hazardous Substances Committee Concerning Decisions with Far Reaching Consequences. November 15 1998. http://www.baua.de//prax/ags/ags_guidelines_rm.pdf.

4. Chalecki, E. L. No Date. Environmental Security: A Case Study of Climate Change. Pacific Institute for Studies in Development, Environment, and Security. http://www.pacinst.org/environment_and_security/env_security_and_climate_change.pdf.
5. Cheremisinoff, N.P. 2002. Environmental Security: The Need for International Policies. Pollution Engineering. 5/1/2002. http://www.pollutionengineering.com/CDA/ ArticleInformation/features/ BNP__Features__Item/0,6649,103962,00.html.
6. Commission on Geosciences, Environment and Resources. 1999. Environmental Cleanup at Navy Facilities: Risk-Based Methods http://books.nap.edu/books/0309063434/html/37.html#page_bottom.
7. Commission on Life Sciences. 1983. Risk Assessment in the Federal Government: Managing the Process. Committee on the Institutional Means for Assessment of Risks to Public Health, National Research Council. http://books.nap.edu/nap-cgi/chaphits.cgi?term=acceptable+risk&isbn=0309033497.
8. Commission on Life Sciences. 1994. Science and Judgment in Risk Assessment. Committee on Risk Assessment of Hazardous Air Pollutants, National Research Council.
9. Duffus, J.H. 2001. Risk Assessment Terminology. Chemistry International 23(2). March 2001. http://www.iupac.org/publications/ci/2001/march/risk_assessment.html.
10. European Environment Agency. No Date. Environmental Risk Assessment - Approaches, Experiences and Information Sources Environmental issue report No 4. http://reports.eea.eu.int/GH-07-97-595-EN-C2/en/riskindex.html.
11. European Environment Agency. 2003. Europe's Environment: The Third Assessment.
12. EXTOXNET. 1993. Toxicology Information Briefs. Cooperative Extension Offices of Cornell University, Oregon State University, the University of Idaho, and the University of California at Davis and the Institute for Environmental Toxicology, Michigan State University. September 1993. http://extoxnet.orst.edu/tibs/carcino.htm.
13. Glenn, J.C., Gordon, T.J., and R. Perelet. 1998. Defining Environmental Security: Implications for the U.S. Army. Editor: Molly Landholm. AEPI-IFP-1298
14. Army Environmental Policy Institute. December 1998. http://www.aepi.army.mil/Publications/Defining%20Environmental%20Security%20-%20Implications%20for%20the%20US%20Army.pdf.
15. Haimes, Y.Y., Lambert, J.H., Horowitz, B.M., Kaplan,S., Pikus, I.M., Leung, M.F. and A.D. Mosenthal. 2004. Final Contract Report, Risk Assessment And Management Of Critical Highway Infrastructures: Executive Summary. February 2004. Center for Risk Management of Engineering Systems. University of Virginia. Virginia Transportation Research Council (A Cooperative Organization Sponsored Jointly by the Virginia Department of Transportation and the University of Virginia) Charlottesville, Virginia. VTRC 04-CR16. http://www.virginiadot.org/vtrc/main/online_reports/pdf/04-cr16.pdf.
16. Health Canada. 1998. Assessment And Management Of Cancer Risks From Radiological And Chemical Hazards. http://www.hc-sc.gc.ca/hecs-sesc/ccrpb/publication/98ehd216/ chapter4.htm.
17. Health Canada. 2000. Health Canada Decision-Making Framework for Identifying, Assessing, and Managing Health Risks. August 1, 2000. http://www.hc-sc.gc.ca/hpfb-dgpsa/ hcrisk_11_e.html.
18. Hobbs, R.J. and L.J. Kristjanson. 2003. Triage: How do we prioritize health care for landscapes? Ecological Management & Restoration 4(Suppl): S39-S45.
19. Hunter, P.R. and L. Fewtrell. 2001. Acceptable Risk. World Health Organization Water Quality Guidelines, Standards and Health. Chapter 10.
20. McNeil, F. 2000. Making Sense of Environmental Security. North-South Agenda. Paper. Thirty-Nine. http://www.miami.edu/nsc/publications/pub-ap-pdf/39AP.pdf
21. National Infrastructure Institute (NI^2) Center for Infrastructure Expertise and the University of New Hampshire. 2004. The Center for Infrastructure Expertise Library. July 22, 2004 (accessed). http://www.ni2ciel.org/Glossary/a_glossary.htm.
22. National Intelligence Council. 2000. Global Trends 2015: A Dialogue about the Future With Nongovernment Experts. December 2000. http://www.odci.gov/nic/PDF_GIF_global/globaltrend2015.pdf.
23. Newman MC, Ownby DR, Me'Zin LCA, Powell DC, Christensen TRL, Lerberg SB, and B-A Anderson. 2000. Applying Species-Sensitivity Distributions in Ecological Risk Assessment:

Assumptions of Distribution Type and Sufficient Numbers of Species. Environmental Toxicology and Chemistry, Vol. 19, No. 2, pp. 508–515.
24. NRC (National Research Council). 1983. Risk Assessment in the Federal Government: Managing the Process. Washington, D.C.: National Academy Press.
25. Omenn, G.S. 2003. Assessment of the Impact of the 1983 Red Book. W5.1. Socieity for Risk Analysis Annual Meeting 2003. W5 - Symposium: Impacts on Risk Policy and Practice, Red Book, Risk Commission, Understanding Risk. http://www.birenheide.com/sra/2003AM/program/singlesession.php3?sessid=W5.
26. Oren, S. 2001. Market Based Risk Mitigation: Risk Management vs. Risk Avoidance. University of California at Berkeley, Presented at a White House OSTP/NSFWorkshop on Critical Infrastructure Interdependencies June 14-15, Washington DC. http://www.pserc.wisc.edu/ecow/get/publicatio/2001public/marketbasdriskmitigation-v2-oren.pdf.
27. Owens, R., Broderick, P, Charters, D., Sprenger, M, and T. Bernhard. 1999. Ecological Risk Assessments. Powerpoint. http://enviro.nfesc.navy.mil/erb/erb_a/support/rits/presentations/1999-5-era.pdf.
28. Presidential/Congressional Commission on Risk Assessment and Risk Management. 1997. *Framework for Environmental Health Risk Management.* Final Report, Washington, DC. Volume 1. http://www.spea.indiana.edu:8000/dhenshel/e311/EPAJAN.pdf.
29. Queensland Government. No Date. Securing Queensland's Critical Infrastructure: Guidelines for owners/operators. Department of the Premier and Cabinet. http://www.premiers.qld.gov.au/library/pdf/infopackweb.pdf.
30. Secretary of the Navy. 2002. Department of Navy (DON) Critical Infrastructure Protection (CIP). Memorandum. June 16, 2002. http://neds.nebt.daps.mil/Directives/3501_1.pdf.
31. SERO (Southeast Region, Installation Management Agency). 2003. Sustainability 401: Part 1: Houston, we have a problem. Joint Services P2 & HWM Conference (14 August 2003). http://www.envquest.com/library-FORSCOM.asp.
32. Schwartz, P. and D. Randall. 2003. Abrupt Climate Change Scenario and Its Implications for United States National Security (October 2003). GBN: Global Business Network. http://www.gbn.com/ArticleDisplayServlet.srv?aid=26231.
33. Scottish Executive. No date. An Overview of Risk Guidance. http://www.scotland.gov.uk/library3/government/riskguidance.pdf.
34. Suter, G.W. II, B.W. Cornaby, C.T. Hadden, R.N. Hull, M. Stack and F.A. Zafran. 1995. An approach for balancing health and ecological risks at hazardous waste sites. Risk Analysis 15: 221-231.
35. The White House. 2003. National Strategy for the Physical Protection of Critical Infrastructures and Key Assets. February 2003. http://www.whitehouse.gov/pcipb/physical.html.
36. U.K. Department of the Environment, Transport and the Regions. 1999. Review of Technical Guidance on Environmental Appraisal A Report by EFTEC (Economics for the Environment Consultancy). April 30, 1999. http://www.defra.gov.uk/environment/economics/rtgea/22.htm.
37. United States Department of Labor. No Date. Occupation Exposure to 1,3-Buradiene. Regulations (Preambles to Final Rules. Section 7 – VII. Significance of Risk. http://www.osha.gov/pls/oshaweb/owadisp.show_document?p_table=PREAMBLES&p_id=748.
38. United States Navy. 2001. U.S. Navy Human Health Risk Assessment Guidance. Chapter 12 – Risk Management. December 2001. http://www-nehc.med.navy.mil/HHRA/ guidancedocuments/process/pdf/risk_mgmt.pdf.
39. Vrijling, J. K., van Gelder, P.H.A.J.M., Goossens, L.H.J., Voortman, H.G., and M. D. Pandey. 2004. A Framework for Risk criteria for critical infrastructures: Fundamentals and Case Studies in the Netherlands. *Journal of Risk Research 7:569-679.*
40. Vogel, D. 2003. The Politics of Risk Regulation in Europe and the United States. Prepared for publication in The Yearbook of European Environmental Law, Volume 3, 2003.

SECURITY AND THE ENVIRONMENT IN THE MIDDLE EAST WATER ISSUES

B. MOREL
Department of engineering and public policy Carnegie Mellon University, Pittsburgh Pa 15213, USA

Abstract

Water scarcity and quality have been a major concern for Israel and its neighbors. It has been recognized long ago that this problem has to be addressed at the regional level. Many agreements and interactions between the different nations have taken place over the years. Some contentious issues have never been solved to the satisfaction of all. But it is also clear that the nations involved will have to cooperate in the management of water and the environment for the long haul. The onset of the second Intifada in 2000 has put the peace process on hold. It has also led to an interruption of the discussions on water issues, basically at all levels. As a result despite a few small scale private initiatives the region (and in particular the West Bank and Gaza) are suffering from environmental neglect. This is creating a situation which has the potential to lead to further contamination of watershed and affect the water supply of neighboring nations.

1. Palestine and Israel and Environmental security

Environmental security does not mean the same thing to everybody. In the Middle East, there are multiple interfaces between environmental issues and security. It is common to hear that environmental issues have security implications. For example water dispute can lead to violent conflicts. This point is not uncontroversial. Environmental security could also refer to a conservationist attitude with respect to the environment and as a matter of prudence, a precautionary environmental policy. The Middle East and more precisely the occupied territories of the West Bank and Gaza are an example of another coupling between security and environmental issues. They suffer from a degradation of their environment through neglect.

Water and environmental issues in general do not respect borders. That region shares some of the environmental concerns of the rest of the world, like air quality, or the potential consequences of global warming. But those are dwarfed by the immediacy of issues such as water scarcity and water quality. There is a long history of disputes and agreements among the local protagonists on that subject. There are also disagreements among analysts on the question of whether there have been wars over water or there may be in the future[1].

B. Morel and I. Linkov (eds.),
Environmental Security and Environmental Management: The Role of Risk Assessment, 17–24.

The Middle East may be the region "with the most concentrated water scarcity on earth". Water management in that region is as complex as it gets. Even with the highest level of committed cooperation among the nations of the region, water management would be a tantalizing challenge for the long term. The political context within which those issues are approached today is definitely not conducive to that quality of management. In fact since fall 2000 and the onset of the second Intifada, the relations between Israel and its neighbors (West Bank and Gaza in particular) have deteriorated, exacerbating significantly the environmental problem. Water issues do not respect borders. The whole basin of the Jordan River is affected and with it countries like Jordan, despite the fact that it is not part of the conflict.

Despite the violence which is the hallmark of that volatile region, there is a consensus that when it comes to water, plans for the future are necessary. The rate of replenishment of water has progressively been outpaced by the growth of the demand due to increase of population. When the so-called "peace process" was progressing slowly, water was the subject of many discussions, studies, plans. There were talks of ambitious plans of desalination of water or of canals connecting the Dead Sea to the Red Sea or to the Mediterranean. The former Prime Minister of Israel Shimon Peres was quoted saying that "if roads lead to civilization, water leads to peace".

But the inter-regional dialogue between Israel and Palestine got abruptly interrupted in the fall 2000. The most visible environmental effect has been the deterioration of the situation in the West Bank and Gaza. Those regions suffer from acute "environmental neglect". The presence of the bacteria Shigella in the waters in that region is more evident than the promise of peace.

Before 2000, water quality was already a growing concern. Since, its seriousness has reached the point to make Professor Elias Salamey state in a recent testimony to the US House committee for External Relations: "It is not the **water quantity**, but its **worsening quality** that will bring us to our knees."

Every day the situation is getting worse. Little is happening to counter that trend. Integrated management through close cooperation between the local governments has been for a long time deemed the only potentially successful approach to water issues in the Jordan River Basin. In the present political climate, despite the seriousness of the situation, there is little evidence that the local political authorities feel the imperative to do so.

The resources required to significantly improve the situation and pave the way for a less bleak future, are probably significantly larger than what the nations of the area can generate. That the international community will have to be involved is a given. The question is how…

2. A few numbers

A rule of thumb is that "for a country to be considered as having sufficient water for all purposes it would be desirable to have at its disposal at least 1000 cubic meters/capita/year(m^3/C/yr)… Israel's estimated potential renewable fresh water resources are about 270 m^3/C/yr with somewhat less for the Jordan at 200 m^3/C/yr. It has also been estimated that for the Palestinians in the West Bank and Gaza, the number is about 90 $m^3/C/yr$".[2] These numbers from 2000, assume that the Jordan River basin has a total population of approximately 13 - 14 million people, 5.5 million in Jordan, 2.5 million in West Bank and Gaza, and 5.5 million in Israel. It is estimated that the population of the basin will pass 30 Millions by 2040. If as one can hope the region enjoys some economic prosperity, the demand of water by then will exceed massively the present supply of fresh water. For comparison, in the US the average water consumption is 382 liters/ capita/day. In Israel it is 170. Israel has water problems. But its neighbors are worse off. For example in Jordan the average water consumption is about 100 liters/capita/day. This is about twice more than the Palestinians.

Israel is a world-pioneer in conservation and efficient water use in agriculture and industry. For example, fresh water allocation to agriculture has been reduced from 1,200 Mm^3/year to 530 Mm^3/year. Despite the fact that Israel's water productivity is the highest in the world, Israel has been over-using the natural water resources it has access to. The average annual natural recharge in Israel's three main sources – the Coastal Aquifer, the Mountain Aquifer and Lake Tiberias (Sea of Galilee) - is 1457 Mm^3. But it varies widely from ear to year. In some drought years it has been as low as 657 Mm^3. This has to be put in comparison with the total demand for potable water, which today at 1350 Mm^3/year, is close to saturate the average replenishment rate and forecasted to rise above the replenishment rate soon.

This area is not the only place on earth suffering from water shortages. According to the US State Department, "today an estimated 1.1 billion people lack access to safe drinking water; 2.4 billion lack access to basic sanitation. Each year, over 3 billion people suffer from water related diseases resulting in 3-4 million deaths. ... The economic impact of the health related aspects related to unsafe water is estimated at $380 billion per year. ... In some countries water mismanagement and water pollution can reduce GDP by more than 2%". The Middle East is not the only area with problems, but it may be the area with the most acute ones today, also because of its entanglement with the security problems. In addition to security, the exacerbating factors include population growth and economic growth. United Nations estimates suggest that the population in the Jordan Valley, today about 16 Millions, could reach 34 Millions in 2050. Population growth may not be a completely welcome factor. It is a constraint to be met in the demand for water. Economic growth, in particular in the poorest areas, on the other hand is an imperative. Peace prospects depend on that. Population and economic growth synergize to put real demands on water management in the region.

3. The environmental situation in the West bank and Gaza

The integrated management of water resources in the West Bank and Gaza will have to be established in a politically polarized context, as the Palestinians have a "strategy [which] emphasizes the Palestinian right for sovereignty and full control over their own water resources. The strategy for the short term is to define and pursue Palestinian water rights. A first step will be the implementation and full utilization of the water allocations committed in Article 40 of the Oslo 2 Agreement. The next step will be to prepare the negotiation strategies for the final status negotiations and finally to agree upon a final water agreement between Palestine and Israel."

In the words of the Palestinian Water authority (PWA)[3]: "The sources of the water in the West Bank are those renewable fresh waters of the mountain aquifer which are estimated to be 650 Mm^3/year, in addition to the surface runoff in the Wadis, estimated to be 70 Mm^3/year. The Gaza Coastal Aquifer is essentially the sole sources for water in the Gaza Governorates, a shallow aquifer that underlies the entire Gaza Governorates and extends northward into Israel. The natural renewable recharge is in the order of 45 Mm^3/year.

The Jordan River system has a natural capacity to deliver an average annual flow of 1311 Mm^3/year.The riparian of the Jordan River are Lebanon, Syria, Palestine and "Israel" and Jordan Palestine's annual share from the Jordan River Basin has been estimated to be 20% of the total annual flow. As a result of water diversion of the upper Jordan River tributaries by the Israelis there is no fresh water down stream of Lake Tiberias, and the negligible quantity that reaches the Palestinian Riparian in the West Bank is of a deteriorated quality.

Palestinian total use form the groundwater resources in the West Bank has been estimated to be 120 Mm^3/year. About 86 Mm^3/yr. (71%) is used for irrigation. The remaining 34 Mm^3/yr. are used for domestic and industrial consumption (industry's share about 3%) with more than 40% of unaccounted for water.

In Gaza, Palestinians total use of water is about 125 mill.m3/yr. is used for irrigation. The remaining 45 Mm^3/year are used for domestic and industrial consumption (industry's share about 3% with more than 50% water is unaccounted for. The water crisis in Gaza is not limited to the deficit in quantity. However, the water quality is deteriorating and subject to continuous increase in salinity due to over-abstraction and to the percolation of sewage in the area."

The problem of the allocation of water is a major irritant in any negotiation about water in the region and it will not be resolved easily or soon. As Ismael Serageldin said as Chairman of the World Commission on Water for the 21st century, "Conventional wisdom has been that cooperation is easier in quality issues than on quantity issues." The experience with the Oslo Agreement is certainly no exceptions. Quantity and quality are important considerations. Disagreements on quantity issues have sometimes acted as obstacles to agreements on quality.

Article 40 of the Oslo 2 Agreement mentioned water quantity issues with words carefully chosen to fail to resolve the conflict. On the subject of quality or more exactly "in the sphere of Water and Sewage", on the other hand, the agreement reads nicely:

Both sides agree to coordinate the management of water and sewage resources and systems in the West Bank during the interim period, in accordance with the following principles:

- Preventing the deterioration of water quality in water resources;
- Using the water resources in a manner which will ensure sustainable use in the future, in quantity and quality;
- Treating, reusing or properly disposing of all domestic, urban, industrial, and agricultural sewage.

- Existing water and sewage systems shall be operated, maintained and developed in a coordinated manner.
- Each side shall take all necessary measures to prevent any harm to the water and sewage systems in their respective areas.

But basically none of those provisions was fulfilled… Instead, in the words of the organization Friends of the Earth Middle East (FoEME), the Mountain Aquifer, is turning into a seeping time bomb through pollution by sewage… According to that report, the largest and best quality source of water between the Jordan River and the Mediterranean Sea is progressively contaminated primarily from large quantities of untreated sewage.

FoEME is a regional organization created in 1994, made up of Jordanians, Palestinians and Israelis committed to protecting their shared environment and advancing peace between their peoples. The mere existence of FoEME is one precious evidence that at the grass root level there is a will to introduce some rationality in the environmental trans-boundary issues. In December 2003, they organized a symposium on the deterioration of the Mountain Aquifer, and put the same kind of blames both on the Israeli and Palestinian authorities: "the use of the Mountain Aquifer's pollution for propaganda against the each other is damaging, and creates distrust regarding genuine good will to find solutions. Pollution of the aquifer's recharge area originates from both Palestinian and Israeli sources, and can only be solved through maximum cooperation between all sides."

They recommend that urgent and key constructive and pro-active steps need to be taken. In the case of Israel, "the Minister of National Infrastructure [should] appoint a senior staff member to assist donor countries in implementing sewage treatment projects in the West Bank through, *inter alia removing obstacles and administrative barriers to their operations*." The recommendations to the Palestinian authorities mirror exactly the recommendations to the Israeli government. In both cases the main recommendation is to facilitate the work of donor countries… That symposium sent at least two messages: far from feeling any imperative to engage into any environmental initiatives, the local authorities are obstacles to projects from foreign donor countries. And improvements of the situation depend on the generosity of donor countries…..

Palestine is also suffering from the pernicious effect of a constantly deteriorating economic situation. According to the CIA fact book: "Real per capita GDP for the West Bank and Gaza Strip declined by about one-third between 1992 and 1996 due to the combined effect of falling aggregate incomes and rapid population growth. The most serious social effect of this downturn was rising unemployment; [which] during the 1980s was generally under 5%; by 1995 it had risen to over 20%.... In 2001, and even more severely in 2002, [violent hostilities] have resulted in the destruction of much capital plant and administrative structure, widespread business closures, and a sharp drop in GDP. […] International aid of $2 billion in 2001-02 to the West Bank and Gaza Strip has prevented the complete collapse of the economy."

The people of the West Bank and Gaza are experiencing a general degradation of their conditions of living. Few Palestinians are connected to piped water. But even the Palestinians connected to piped water receive water only intermittently. In Khursa for example, a village 18 km from Hebron connected to piped water, the inhabitants now have to draw their water from the village's ancient well which dates back to the Roman Empire. The pipe very rarely has water, and when it has it is extremely bad quality. In other cases, Palestinians have to buy their water from trucks. Water is a large fraction (more than half often) of the budget of families. One of the effects of the worsening of the quality of water is the spread of water borne diseases like Shigellosis, which apparently has become rampant.

The appropriate management of water resources should be **basin-wide.** It should incorporate the management of the **environment** together with **sound economics**. Economic prosperity needs and leads to good water as well as good water needs and leads to economic prosperity. Instead of being in that virtuous circle, the West Bank and Gaza are in the opposite vicious circle.

4.Can water issues of the Middle East be decoupled from security?

Nobody questions the fact that the environmental situation is degrading in the West Bank and Gaza. Since the interruption of discussions about environment and water in the fall 2000, a scientific and political void has been created. The experts never miss an opportunity to state their anxiety: "The environment cannot wait..., this is late but better late than never, cooperation between nations of the region is urgently needed, we need the involvement of the international community...".

At the grass root levels there are initiatives. For example, the self-appointed Center for Environmental Diplomacy (CED) is taking on itself to "solve" the problem. It has launched the West Bank Clean Up Project (WECUP)[4]. But a purely private initiative cannot be the best solution. Governments have to take their responsibility and be involved. A plan at the required scale would need more resources than the region can provide. No wonder that the recommendations of the experts tend to be to facilitate the work of the donor countries and emphasize the responsibility of donor countries...

UNESCO with its initiative PCCP (Potential Conflict to Cooperation Potential) is spearheading and effort on trans-boundary water management issues. There are no less than 300 transboundary watercourses. 40% of the population of the world lives in basin shared by more than two countries. The saliency of the problems of the West Bank and Gaza tend to be shadowed in this multiplicity.

UNEP has not staid indifferent. It produced a "Desk Study"[5] on the environmental situation in the area, detailed and exhaustive. It states: "Many of the findings in this UNEP Desk Study are alarming, and need to be addressed immediately. In the current phase of the conflict, the absence of even minimal cooperation is worsening the situation on a daily basis, with impacts not only on the environment but also on human health. [...] Given the alarming findings of the Desk Study, cooperation between the parties on acute environmental issues should be immediately revitalized. There is need for an institutional framework to negotiate these issues, especially during times of conflict."

Had the Article 40 of the Oslo 2 agreement been a separate agreement dealing exclusively with water quality and sewage management, would the situation in the West Bank be different today? The answer to this question is unclear. But the answer to the following one is not: did the interruption of the discussions on environmental issues have a harmful effect on the area? Demonstrably. Can a resumption of those interactions wait? In a rational world, the answer is emphatically no. In a rational world those discussions would never have stopped in the first place.

Part of the problem may be that the discussions on environmental and water issues were seen as an aspect of the peace process. They did not have a life of their own. Is it conceivable that that situation could be changed?

In fact local authorities and experts have tried to imagine ways to make that happen. This is where the possibility of a role for the international community becomes a consideration. Not only a consideration, in the mind of many, it may be the only realistic solution.

The role that the international community should have is not well defined. One role would be to be a facilitator to the resumption of a much needed dialogues among officials representing the nations of the area. People familiar with the diplomatic-bureaucratic process are not sure that now is a good moment.... But however difficult this may look in the present political context, this is the trivial part...

International organizations love studies. But new studies are not needed. There is plenty of expertise in the area. Over the years, hundreds of studies and projects proposals have been produced. What has not taken place is the implementation phase. Among the obstacles are lack of resources and the political reality on the ground. The international community is expected to provide a solution to both...

One problem that the international community seems to be in the unique position to solve is to provide a much needed leadership in the implementation of an environmental policy. In a region like the West Bank where sovereignty is divided in little areas like a patchwork, and where every little disagreement is allowed to grow into a major dispute, there is no local authority which can implement sensible environmental regulations. The locals seem to agree on that. Ideally, with the agreement of the locals, the international community could set-up a body with a well defined mission and enough clout to enforce a coherent environmental and water policy in the area.

This may look hopelessly unrealistic. But what about the recommendation of UNEP in its desk study?: "For these reasons, the Israelis and Palestinians, as well as the international community, should do their utmost to put an end to the conflict."

In order to successfully address the immediate water and environmental problems of the region, the hypothetical international organism would have to overcome obstacles, which are huge compared to what has been sufficient to derail much more modest international initiatives in the past. Instead of trying immediately to accomplish something unprecedented, the international community could sponsor some treaties or agreements dealing with water an/or environmental issues, and try to separate them completely from security considerations. Their immediate objectives could be at first to create conditions for the coordination of a minimal environmental and water management policy to maintain some minimum standards. Such discussions would not establish the foundations of a sustainable future. They would put a stop at the present degradation of the situation.

There is a plethora of treaties on trans-boundary water issues that provide precedents of successful international cooperation [6]. From the Baltic sea to the

Danube, the Aral sea or the Nile river, The Mekong or the Caspian sea, there is a large choice. The record shows that since 800 A.D., 3,600 treaties have been signed on different aspects of water in the world... And according to Aaron Wolf, who holds a data base on those treaties, "once cooperative water regimes are established through treaty, they turn out to be tremendously resilient over time, even between otherwise hostile riparians, and even as conflict is waged over other issues. International water regimes continued to function in the Mekong basin since 1957, despite the Vietnam War; the Jordan basin (between Israel and Jordan) since 1955, even as these riparians until only recently were in a legal state of war; and along the Indus, even through India-Pakistan warfare."[7]

Haim Shaked in a testimony to House committee for International Relations on May 4 2004 pointed to another one, less discussed in this context, which as he says may have more to offer: the International Joint Commission (IJC [8]). It is an American-Canadian institution created in 1910 as a consequence of the Boundary Waters Treaty of 1909 between two unequal neighbors very suspicious of each other: the US and Canada... In its own words: "The International Joint Commission is an independent bi-national organization established by the Boundary Waters Treaty of 1909. Its purpose is to help prevent and resolve disputes relating to the use and quality of boundary waters and to advise Canada and the United States on related questions".

The historian, John W. Holmes, 70 years after its creation, characterized the circumstances of the creation of IJC in those words: "At last Washington had reconciled itself to the existence of Canada as a large and permanent, if regrettable, fact of continental politics....As for the Canadians...they had now overcome their natural fear of joint institutions to embrace one that was imaginatively designed to protect their interests....It promised equity without interfering with national sovereignty."
IJC was a very successful institution for its impact on the water issues, but also on the relations between the US and Canada...

5. References

1. A. T. Wolf, J. H. Hammer, „Trends in Transboundary Water Disputes and Dispute Resolution", in Water for Peace in the Middle East and Southern Africa, Green Cross International, March 2000, p.55.
2. H. Shuval: "The Water Issues in the Jordan River Basin between Israel, Syria and Lebanon can be a Motivation for Peace and Regional Cooperation", in Water for Peace in the Middle East and Southern Africa, Green Cross International, March 2000, p.39.
3. Ihab Barghotti, testimony to the House Committee on International Relations, May 4, 2004.
4. http://www.wecup.org
5. http://www.unep.org/GC/GC22/Document/INF-31-WebOPT.pdf
6. Environmental Peace Making, K. Conca & G. D. Dabelko ed., Woodrow Wilson Center Press, 2004.
7. A. T. Wolf, J. H. Hammer, „Trends in Transboundary Water Disputes and Dispute Resolution", in Water for Peace in the Middle East and Southern Africa, Green Cross International, March 2000.
8. http://www.ijc.org/en/home/main_accueil.htm

POTENTIAL OF SOLAR THERMAL DESALINATION TO DEFUSE WATER AS A CONFLICT ISSUE IN THE MIDDLE EAST

Proposal for Functional Cooperation in the Gulf of Aqaba

HANS GÜNTER BRAUCH
Free University of Berlin, Otto-Suhr-Institute of Political Science
Peace Research and European Security Studies (AFES-PRESS),
Alte Bergsteige 47, 74821 Mosbach, GERMANY

1. Introduction

This author (Brauch 2005a) argued that the conceptual ideas of David Mitrany, George Marshall and Jean Monnet were instrumental for 60 years of peace in Europe and for European integration, and those of Mikhail Gorbachev for overcoming the Cold War and contributing to the reunification of the continent. Brauch (2003, 2005) contrasted different security perceptions of narrow *national security threats* with a widened security concept that includes economic, societal and environmental dimensions and other levels of analysis and referents, with a special focus on *human security* (Bogardi/ Brauch 2005).

Brauch (2005a) projected the regional impact of global environmental change and potential extreme outcomes for the Middle East until 2050 and 2100 arguing that these environmental challenges are not yet perceived by Arabs and Israelis as common threats to their security and survival. He focused on water demand due to population growth, urbanisation and food needs, as well as the changing supply due to the impact of climate change on precipitation, soil erosion, drought and desertification in the region. He suggested that these common challenges to human security should become an object of functional cooperation within the region, and that these efforts may contribute to a long-term environmental conflict avoidance.

This chapter builds on these conceptual and theoretical considerations and trend projections of demand factors (population growth, urbanisation, agriculture) as well as crucial natural supply factors of air (climate change), soil (degradation, erosion, desertification) and water (scarcity, degradation, salination) for the three countries with borders at the Gulf of Aqaba: Egypt, Israel and Jordan. These three countries have entered into bilateral peace treaties in 1979 between Egypt and Israel and in 1994 between Israel and Jordan. In this chapter Palestine or the presently Occupied Palestinian Territories in Gaza and in the West Bank are included.

Drawing lessons from 60 years of peace in Europe after World War II, this chapter develops a conceptual proposal for functional cooperation that is inspired by the ideas of Mitrany, Marshall and Monnet. They broke out of the perennial cycle of war and with their ideas of cooperation among former enemies they contributed to the emergence of functional and intergovernmental cooperation and supranational govern-

B. Morel and I. Linkov (eds.),
Environmental Security and Environmental Management: The Role of Risk Assessment, 25–48.

ance. This implied a voluntary transfer of national sovereignty to a new institution (EU) that is neither a state nor a federation of states.

This chapter suggests that scholars and technical experts in Egypt, Israel, Jordan, and Palestine should recognise and jointly study the longer-term environmental challenges as common threats, challenges, vulnerabilities and risks for "human, societal and environmental security" for all people in the region and they should start developing trans-boundary regional and non-military solutions that address the scarcity and degradation of water and soil for agriculture, resulting in major increases in food import needs due to both growing demand (population growth) and declining supply (regional impact of climate change on precipitation) for the people in the Near East. This chapter will address two initial questions:

1. Is there an awareness and understanding of the urgency of these non-military environmental challenges among scientific experts, policy analysts and policy makers in this region?
2. Is there a willingness for functional cooperation among the people of Israel and Palestine to address these common challenges jointly by functional cooperation?

Due to the nearly exclusive focus on the unresolved Middle East conflict, in both Israel (Kam 2003) and in neighbouring Arab countries (Selim 2003) but also in Turkey (Aydin 2003) security has been conceptualised narrowly as primarily national military threats while human security perspectives on economic, societal and environmental threats, challenges, vulnerabilities and risks have hardly been discussed (Brauch 2003; Brauch/Selim/Liotta 2003).

But irrespective of the lack of public awareness and understanding of the common future non-military challenges that cannot be addressed and solved by military means in the framework of zero sum strategies, there seems to be a willingness by the populations of Israel and Palestine for functional cooperation on major present human and environmental security issues.

2. Strategy for Recognizing Long-term Environmental Challenges

According to a poll of Israelis and Palestinians taken in June 2003 at the request of the World Economic Forum by Gallup International, a majority of ordinary Israelis and Palestinians were supportive of joint economic cooperation.[1] Among the key findings were: "66% of Israelis and 57% of Palestinians agree that mutual cooperation and joint projects between all nations of the region, including Israelis and Palestinians, in areas such as water, health, environment, tourism, etc., should start as soon as possible, even now before final peace agreements are reached. Both sides agree that it is important to

[1] See at: World Economic Forum: "New Survey Shows that Ordinary Israelis and Palestinians Are Keen to Move Forward with the Roadmap and Are in Favour of Starting Economic Cooperation", at: < http://www.weforum.org/site/homepublic.nsf/Content/New+Survey+ Shows+that+Ordinary +Israelis+and+Palestinians+Are+Keen+to+Move+Forward+with+the+Roadmap+and+Are+in+Favour+of+Starting+Economic+Cooperation >.

move forward on other areas of possible cooperation, such as water desalination and restoration, environmental issues, regional health projects and agriculture issues." This question was posed to both communities:

Q: The following are possible areas of cooperation between Israel and the Palestinian Authority. Thinking about each one, please tell us whether you are in favour or against cooperation between the two entities in each area:

	% in Favour	
	Israelis	Palestinians
Water desalination and restoration	**78**	**77**
Regional health projects	79	73
Environmental issues	**82**	**68**
Regional tourism	76	55
Agricultural issues	**79**	**72**
Regional trade relations	76	64
Communication, technology and hi-tech	67	68
Cultural and sports relations	72	51
Free movement of workers between the two sides	56	77

If these figures are representative for the public opinion in both communities, indicating a willingness for functional cooperation among both people, what have been the obstacles that little functional cooperation has survived the resumption of the conflict in September 2000? Has the narrow conceptualisation of security in both Israel and Palestine prevented the perception of the common environmental challenges that will affect the young generation on either side during their lifetime?

3. Security Perceptions: From Military to Human and Environmental Security

According to Arnold Wolfers (1962): "Security, in an *objective sense*, measures the absence of threats to acquired values, in a *subjective sense*, the absence of fear that such values will be attacked." The perception of security threats depends on the traditions and mind-set of policy-makers. The English school has distinguished three basic traditions that of a:

- *Hobbesian pessimist* (realism) where *power* is the key category (narrow concept);
- *Kantian optimist* (idealism) where *international law* and *human rights* are crucial; and
- *Grotian pragmatist* where *cooperation* is vital (wide security concept).

With the end of the Cold War, many authors (Buzan/Waever/de Wilde 1998) have observed a widening and a deepening of the security concept in many OECD countries, while a narrow security concept still prevails in the US (Brauch 2001) and in the Middle East (Selim 2003; Kam 2003) that has been influenced by the perception of a "security dilemma" (Herz 1959).

Brauch (2000, 2003, 2005) distinguishes between six factors contributing to global environmental change (GEC) that interact in linear or chaotic ways and may contribute to environmental scarcity of soil, water and food that intensify environmental degradation and my result, taking the specific national and international context into account, in environmental stress. In figure 1 six factors contributing to GEC are illustrated in a "survival hexagon" (Brauch 2005) consisting of: a) three *demand-side factors*: 1) population, 2) urban systems (industries, services, pollution, health), 3) rural systems (securing food and fibre); and b) three *supply-side factors*: 4) air (climate change), 5) water (degradation, scarcity) and 6) land (soil and ecosystem degradation).

Security dimension ⇒ Level of interaction (reference point) ⇓	Military	Political	Economic	Social	Environmental ↓ (longer-term environ-. mental challenges)
Human →					Cause and victim
Societal/Community					↓↑
National (short-term threats)	Middle East discourses on "security dilemma"				↓↑ „survival dilemma"
International/Regional					↓↑
Global/Planetary →					GEC

Table 1: Vertical Levels and Horizontal Dimensions of Security (Brauch 2003: 55)

Figure 1: Interactions in the Survival Hexagon (Brauch 2005: 15)

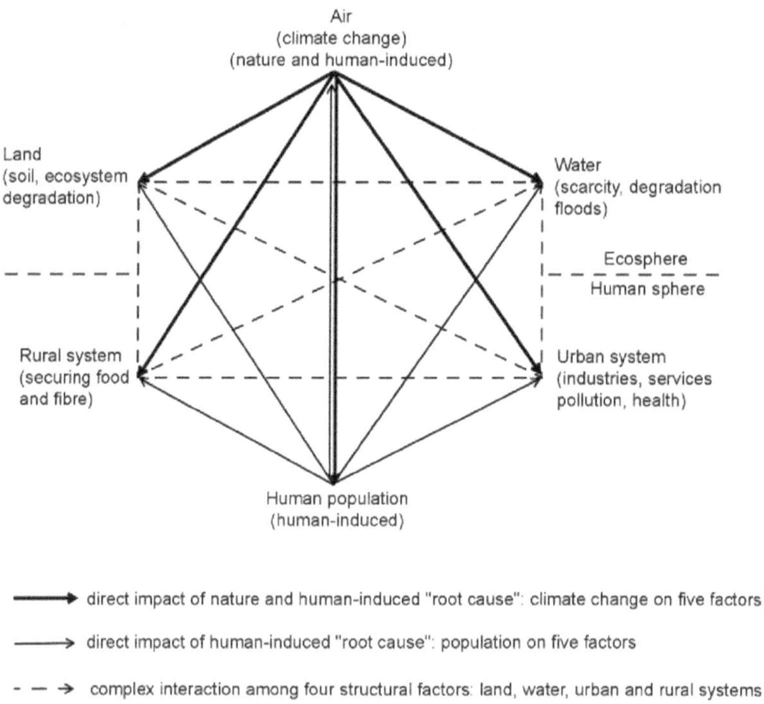

Depending on the system of rule and the level of economic development, the interaction between state, economy and society differ, as will the role of knowledge to enhance the national coping capacities for adaptation and mitigation. Climate change increases the probability and intensity of extreme weather events (floods, drought) and thus increases internal displacements and migration. Again both factors may contribute or cause domestic crises that may escalate to different forms of low-level violence (figure 2).

Figure 2: Model combining GEC, Environmental Stress and Extreme Outcomes

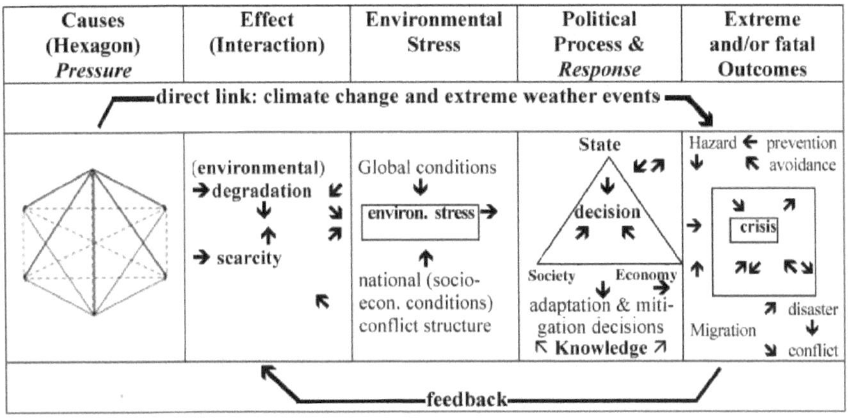

By shifting the attention of both Arabs and Israelis from the narrow national and military security concept that focuses exclusively on the unresolved Middle East conflict to a wider human security concept and its present and longer-term economic, societal and environmental challenges, the common security challenges that will affect the young and future generations may be perceived. However, this requires common and future-oriented research by climate, soil and water specialists as well as population, urbanization and food specialists on the complex linear or chaotic interactions of these factors and on their probable implications for the security of the countries in the region in 2050, 2100 and beyond.

While the proposals of Marshall and Monnet dealt with the problems of the past and started with functional cooperation on coal and steal, two indispensable components of the old war economy, the author suggests that experts from all countries in the region should address the foreseeable non-military and non-political environmental challenges, vulnerabilities and risks. A mutual recognition of these common dangers may be a precondition for anticipatory learning in all societies and for establishing better capacities for coping with these challenges by improved adaptation and mitigation measures. Such functional cooperation should address:

- the demographic changes and their implications for urbanisation, housing and transport;
- the regional impact of global climate change (temperature increase, sea-level rise) for precipitation and agricultural yield changes, as well as for human health (heat waves);
- the common problems of water degradation (salination) and scarcity;
- soil erosion, degradation and desertification and agriculture;
- food demand, crops for arid regions, improved irrigation processes;
- and potential of renewable energies, i.e. for desalination of water and for alternative pollution-free transport systems.

As a first step, international donors should assist the setting up and financing of centres of advanced learning and excellence in the region – for example in the Gulf of Aqaba in a transborder area in Egypt, Israel and Jordan – where the best experts in the region should be given excellent working conditions to address jointly the common environmental challenges and to look for common solutions to enhance the coping capacities and the resilience of the people in the region.

In a second step, as in the case of the Marshall plan, external donors should conditionalise their financial support for the build-up of cross-border infrastructures with components in different countries that would only function based on cooperation. Any attack on components of these life-support systems in the own or other country would discredit and hurt the attacker.

4. Unconventional solutions to cope with water scarcity

Three unconventional solutions to cope with water scarcity are:

a) to reduce the share of "green water" for irrigation by importing food (virtual water); or
b) to desalinate brackish and seawater for agricultural purposes;
c) to import drinking water from abroad by ship or water pipelines.

The first option has existed in the region since the 1960s due to the declining self-sufficiency rate in food (especially cereals) and the food import needs have been projected by FAO (Bruinsma 2003) to increase dramatically until 2030. To cut back irrigation for agriculture seems to be economically feasible for Israel but hardly for its Arab neighbours where a high percentage of the population is still employed in the agricultural sector. The second option is costly and requires much energy, either fossil or renewable. The first two options (virtual water, desalination) are inevitable. The third option to import drinking water from abroad, has been realised between Turkey and Israel, and it has been suggested for the Sinai and possibly also for Gaza by transporting water from the Nile.

While desalination using oil as an energy source is used in the Arab Gulf and in Libya, it is more costly for countries with limited (Egypt, Israel) or without oil and natural gas reserves, such as Palestine and Jordan. Hotels in the Sinai (Egypt)[2] and in the Gulf of Aqaba already operate small desalination units. Israel has several small desalination plants[3] and has been building in 2004 its first major desalination plant in

---

[2] See Mousa Abu Arabi, MEDRC R&D Project Manager: Desalination Growth in the MENA Region, in: Watermark, Issue 20, June 2003, at: <http://www.ewatermark.org/watermark20 /mousa.html>. About 50% of the global desalination capacity exists in the MENA region, with Saudi Arabia, USA, Libya and UAE leading.

[3] See: Israel, Ministry of Infrastructure, at: <http://www.mni.gov.il/english/units/Water/NonconvetionalWater ResourcesandConservation.shtml>. Israel has many small and medium desalination plants for processing brackish and seawater for water supply in the Arava Valley and the Gulf of Eilat. The largest produces 44,000 cum/day from brackish groundwater and seawater, meeting drinking water requirements of Eilat. The first sea water desalination plant, with a capacity of 10,000 cum/day was commissioned in 1997.

Askelon using natural gas[4], and both Jordan[5] and Palestine[6] have included desalination in their water plans as real options.

However, there are huge unused renewable solar energy potentials in the deserts of Sinai, Negev and Jordan for solar thermal electricity generation that could be used for desalination. This author proposes to develop existing technologies further, to optimise them, to create an economy of scale and to use the huge regional solar potential for coping with the rising water scarcity. To realise this goal, this technological option is linked with the political experience of Europe after 1945. But instead of focusing on coal and steel (Schuman Plan 1950), the functional cooperation in the Near East should focus on four commodities that are indispensable for human survival in the 21st century: water, soil erosion, agriculture, and energy.

4.1 REGIONAL TECHNICAL AND ECONOMIC POTENTIALS FOR SOLAR THERMAL AND WIND POWER ENERGY

The region has a huge renewable solar (thermal, photovoltaic) energy potential in the Sinai, Negev and the Jordanian desert, of wind power (Red Sea), city and agricultural waste that may be used for electricity generation, desalination and cooling. So far both technologies of solar thermal electricity generation and of desalination as well as pilot plants for solar desalination exist (Saudi Arabia). Solar thermal electricity generation has been developed first by Luz Industries (now Solel, Israel) in the Mohave desert (California) in the 1980s.[7] But until 2004 no large installation was built in Israel[8], Egypt or Jordan.

---

[4] See on the Ashkelon Plant at: <http://www.water-technology.net/projects/israel/>. The construction of the seawater reverse osmosis (SWRO) plant began in April 2003. In 2000, Israel launched a Desalination Master Plan that envisaged several seawater plants producing an annual total of 400 million m³ of desalinated water by 2005, rising to 750 million m³ by 2020. The Ashkelon facility is the first of a series of large-scale seawater desalination units with others planned for Ashdod, Palmahim, Kishon and Caesarea. The design uses advanced SWRO technology and state-of-the-art energy recovery systems to reduce operating costs and water prices ($0.527/m³). The total project costs are approximately $212 million.

[5] See the proposal by Murakami and Musiake for hydro-powered reverse osmosis for cogeneration in the Jordan valley, at: <http://www.unesco.org.uy/phi/libros/efficient_water/ wmurakami.html>. In 1999, there was hardly any desalination plant in Jordan. There are two potentials: a) brackish water in several areas in Jordan, and b) the seawater in the Gulf of Aqaba, see at: < http://www.ewatermark.org/Watermark8/hashmite.html >.

[6] See: Nahed Ghbn, assistant to the PWA chairman: "The Palestinian Water Authority and Its Desalination Activities", at: <http://www.ewatermark.org/watermark19/NoToparticle2.html>. The PWA has worked on two small seawater desalination plants in Gaza that are financed through grants from the French and Austrian governments. With support of USAID an integration aquifer management plan has been developed. A larger desalination plant was proposed with a capacity to reach 150,000 m³ per day by 2020.

[7] See on the website of the producer of the first major plant: <http://solel.com/energy/article1/>.

[8] However on 24 July 2004, the construction of a 1 bn solar thermal power plant in the Negev was announced. See: Dalia Tal: "Solel Solar Systems to build $1b solar power station in Negev".

On the Mediterranean coast a solar thermal plant could be built in the Sinai and the desalination plants in Gaza and on the Mediterranean Coast of Israel (figure 3A). In the Gulf of Aqaba large solar thermal plants could be built both in the Sinai, desalination plants on the Jordanian coast and a major agricultural training centre in Eilat in Israel. Gaza, Israel and the West Bank could be supplied with solar thermal electricity from the Sinai and with desalinated water from the Mediterranean coast in Sinai, Gaza and Israel while Jordan, Israel and the West Bank could also be supplied with desalinated water from the Gulf of Aqaba (figure 3B).

Figure 3A: Jordan basin

Figure 3B: Gulf of Aqaba

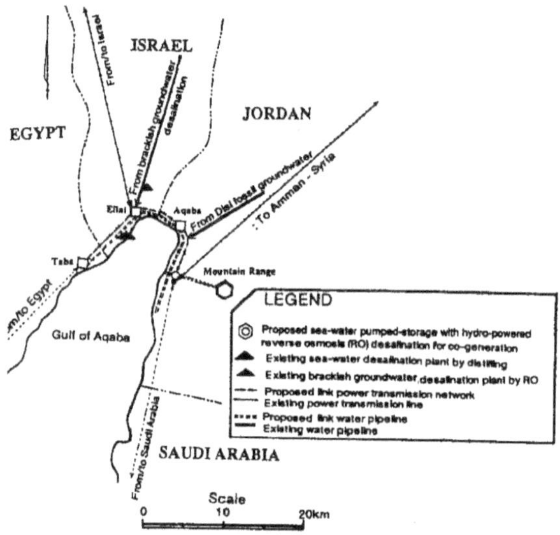

Only cross-border multinational infrastructures should be supported by grants and credits. Israel, Palestine and Jordan depend on fossil energy imports for electricity generation, transportation and desalination. Due to the projected global increase in demand for oil and gas, the prices for both will rise during the 21st century. Thus, electricity and in the future hydrogen produced by solar thermal energy may become economically competitive within ten to twenty years once an economy of scale exists. All components could be developed, and produced locally, creating new industries and jobs for Israelis, Palestinians, Egyptians and Jordanians. The donors should insist on the build-up of a multinational cross-boundary infrastructure that should be managed jointly involving all parties as was done with the Marshall Plan where the US insisted on German participation in the distribution and management of the assistance. Two geographical pilot projects may be considered in the border between Egypt, Gaza and Israel, and in the Gulf of Aqaba linking Taba, Eilat and Aqaba.

---

at: <http://solel.com/press/wp/ globes7-04/>. Upon completion the plant will produce some 500 MW electricity for Negev communities.

Cooperation should start with a joint multinational research centre addressing future challenges of climate change, soil erosion, water scarcity and degradation and modern sustainable agriculture for arid and semiarid regions involving Egyptians, Israelis, Palestinians and Jordanians that could form the basis of a Technical University in the Gulf of Aqaba financed by external donors and the four governments with the goal to develop joint concepts, methods and technologies to enhance the coping capacities for adaptation and mitigation to deal with the regional implications of global environmental change. The Joint Israeli-Jordanian Bridging the Rift Desert Science Center that is being planned under the academic guidance of Stanford and Cornell University could serve as a model for the following proposal.[9]

4.2. SOLAR THERMAL POWER TECHNOLOGIES

Solar thermal power technologies "use solar radiation to achieve high temperatures and to generate steam or air with high energy density, which can then be used for electricity generation and other purposes" (Trieb et. al. 2002). Four alternatives exist (figure 4): a) Fresnel concentrators, b) parabolic trough (400-600 °C), c) the solar tower concept with surrounding heliostat field (1200 °C, up to 50 MW), and d) solar dish (for small applications up to 50 kW).

According to some proponents the economic lifetime of such systems is at least 25 years, and the energy payback time of a solar plant ca. 0.5 years (Trieb et. al. 2002). Luz (now Solel) built 9 solar thermal stations in the Mojave desert in 1984 with about 354 MW (30 to 80 MW each), that produced electricity at a price of 12 c/kWh. The new technology may decrease the price to 10-5C/kWh. During the 1990s, Spain (CIEMAT) and Germany (DLR) cooperated in Almería (at the Platafroma Solar Almería PSA) to develop this technology further.

---

[9] In early 2005 no joint Israeli-Arab educational institutions exists. In March 2004, Israel and Jordan announced to launch a joint academic institution, see BBC at: <http://news.bbc.co.uk/2/hi/middle_east/3545877.stm>. This Bridging the Rift Desert Science Center - on the border between Israel and Jordan between the Dead Sea and the Gulf of Aqaba - is planned to open in 2009. This science centre is the idea of Israeli, Jordanian and American academics and business people and it will be funded by private donors, backed by Stanford and Cornell universities. One of the sponsors, Mati Kochavi, a New York-based Israeli businessman, said science could provide a common language between Israeli and Jordanian students. "Until now there has been nothing to bring Israelis and Arabs together beside the battlefield", he told the Associated Press news agency. "Besides science, there is something which is mutual to the region - how to make something out of the desert." After four years of planning a "free education zone" was set up at the border. This bi-national campus requires that Israelis would be covered by Israeli law, while Jordanians would follow their laws. In October 2004 an Israeli-Jordanian committee met to map out a site for the "Bridging the Rift" bi-national university, a joint project of Stanford and Cornell universities. Professor Marcus W. Feldman is overseeing the project at Stanford. See at: <http://www.jewishsf.com/content/2-0-module/displaystory/story_id/23996/format/html/displaystory.html>. See for additional information: <http://www.news.cornell.edu/features/BTR/BTR_cover.html>; see also Islam online: <http://www.islamonline.com/cgi-bin/news_service /middle_east_full_story.asp?service_id=652>.

Figure 4: Technological alternatives for Solar Thermal Electricity Generation

Due to the low performance in implementing its EU obligations under the Kyoto Protocol, Spain adopted a national energy law that requires 25% of renewables in electricity generation by the year 2012 (Brauch 2000a). The new Spanish electricity law requires utilities to feed-in the electricity of independent producers and to guarantee them higher prices. These Spanish laws have acted as a magnet to attract foreign capital to launch joint ventures with Spanish companies. In the framework of the Kyoto Protocol that entered into force on 16 February 2005, the Joint Implementation (JI) mechanism will attract foreign companies with high CO_2 emissions to invest in Spain in these technologies. Solel (Israel) was among the first companies to explore the Spanish solar energy market.[10] In non-Annex I or II countries under the UNFCCC that have ratified the Kyoto Protocol (both Israel and Jordan) the Clean Development Mechanism (CDM) offers a similar framework for attracting outside investments or joint ventures both to deduct the saved emissions in their home country while simutaneously contributing to a knowledge and technology transfer.

[10] See at < http://solel.com/news/pasc/>.

4.3. APPLICATION OF SOLAR THERMAL POWER TECHNOLOGIES IN ISRAEL AND IN EGYPT

Prof. Faiman of the Ben-Gurion National Solar Energy Institute described the status of solar research in Israel: “Solar-thermal power … is under investigation at Ben-Gurion University (parabolic troughs and a parabolic dish) and at the Weizmann Institute (solar furnace and central receiver tower), the latter with the active participation of industry. … The Weizmann Institute Central Receiver Tower… consists of a field of 64 so-called ‘heliostat’ mirrors, each of approximate area 50 sq.m. that re-direct the sun's rays to a boiler, or some other suitable receiver, mounted on a tower some 50 m. in height.”[11]

In November 2001, the Israel Ministry of National Infrastructures decided “to introduce to the Israel electricity market until 2005 the CSP as a strategic ingredient, with a minimal power unit of 100 MWe. There is an option to increase the CSP contribution up to 500 MWe at a later stage, after the successful operation of the first unit.” According to Solarpaces “The investment in the first unit is expected to be 200 million US$, with an estimated cost of 9 c/kWh for the electricity of the first unit and an expected reduction to 7 c/kWh when the 500 MWe unit is completed. The construction and operation of the first unit will create around 1000 jobs during the construction and 120 permanent jobs for the operation and maintenance of the plant.... In February 2002, the IEC management allowed the construction of a 100 MWe solar power plant at a 250 million US$ investment.”[12]

In July 2004, after the decision of the Israeli Public Utilities Authority (Electricity) to approve premiums for private electricity producers, “Solel Solar Systems has decided to build a solar power station in the Ashalim area in the Negev. The power station will initially produce 150 Megawatt of power for 50,000 homes. Upon completion, it will produce 500 MW of power for Negev communities. Under the plan, Solel will establish a consortium to finance and build the power station, which it estimates will cost $1 billion. … The premium is intended to attract more companies to the market. The cost of the first stage of the project is $250 million.” Solel’s excecutive Mandelberg said “that while Israel has plenty of sunlight available for exploitation, it has little land, and most of the Negev is occupied by the IDF and consequently is unavailable for use. This fact hinders the construction of solar power stations”.[13]

In 1995, the Ministry of Electricity & Energy (MOEE) in Egypt completed an assessment study for the Solar Thermal Electricity Generation (STEG) potentials. In 1996, the Egyptian Cabinet of Ministers approved the MOEE plans for the first ISCCS [Integrated Solar Combined Cycle System] power plant. In 1997, two pre feasibility studies were performed for the first ISCCS plant at Kuraymat with a capacity of about 150 MW through the European Community funded INTERSUDMED project for “Renewable Energy Electricity Generation in The Southern Mediterranean”. Egypt offi-

---

[11] See: David Faiman: “Solar Energy in Israel”, at: <http://www.jewishvirtuallibrary.org/jsource /Environment/Solar.html >.

[12] See at: <http://www.solarpaces.org/ ISRAEL_SEGS.htm>, last updated 15/11/03.

[13] See at: Solel’s News Page the press release of 29 July 2004 < http://solel.com/news/globes/>.

cially requested Global Environment Facility (GEF) to support the project. GEF may provide a grant of 50 million US $ to cover a substantial part of the incremental cost in comparison to the least cost conventional alternative producing the same annual electric energy. In June 2000 the final feasibility study report for the first Egyptian Solar Thermal Power Project (ISCCS) was submitted that concluded: a) the capacity of the first ISCCS is 127 MW; b) the capacity of the solar component is 31 MW; c) the produced electric energy is about 900GWh/year; d) the solar contribution is about 9% of the annual generated electricity, and d) the total investment is about 120 Million US$. In January 2000, the World Bank confirmed the decision of the GEF counsel to support this project with a US$ 49 million grant. The chairman of EEA [Egyptian Electricity Authority] confirmed that this plant was a firm element of Egypt's power extension planning, and that EEA/NREA [New and Renewable Energy Authority of Egypt] wants to lay the path for subsequent 300MW plants, which shall be connected to the grid in 2005 and 2007. In September 2003 Fichtner Solar GmbH was chosen as a consultant to assist in negotiations with the best offer for the project. The plant is expected to be operative in 2006. The ISCCS power plant (127 MW) is considered as a first of a series of hybrid solar fossil fuel power plants targeting to install about 750 MW capacity in an overlapped time frame by the year 2010.[14] In October 2002, the Egyptian Energy Minister, Dr. Hassan Younis, pointed to the promising potential of renewables for electricity generation:

> The Suez and Red Sea areas have rather high wind speeds all the year round. Today we have 68 MW capacity already in operation as wind farms in Zafarana. ... Wind farms of total installed capacity of 450 MW up to the year 2010 are planned to be constructed at Zafarana. Egypt ... receives more than 300 days of sunshine per year. The solar thermal gas electricity generation plan up to the year 2010 includes the construction of three power plants. The first plant of 127 MW solar thermal capacity is expected to be commissioned by 2006. The development of the renewable energy resources will result in optimal exploitation of natural energies available in Egypt.[15]

Thus, Egypt has been among the first five countries considered by the GEF for a solar thermal installation. Thus, the technology that has been successfully demonstrated for two decades in the US, will finally be applied in the Near East - both in Egypt and in Israel – and thus demonstrate its huge energy potential in the region.

4.4. COMBINED SOLAR POWER AND DESALINATION PLANTS

Combined solar thermal power and desalination plants with proven technology would combine: a) a steam turbine co-generation system and b) a thermal seawater desalination. Trieb, Nitsch, Kronshage, Schillings et. al. (2002) estimated that "a 200-MW plant of this type with 7.500 full load operating hours/yr under conditions of Dubai would deliver approximately 1.5 bn. kWh/yr of electricity and 60 million m^3 of fresh-

---

[14] See at: <http://www.solarpaces.org/ISCCS_EGYPT.htm>; and at: <http://www.eere.energy.gov/troughnet/de-velopment.html#egypt>.

[15] See at: <http://www.worldenergy.org/wec-geis/wec_info/structure_organisation/ea/cairo/spc021024Younes. Asp>.

water at approximately 4.3 €-cents/kWh and 1.30 €/m^3 of water, water for 50.000 and electrictiy for 250.000 people, with estimated costs of 800 million €.[16]

In the Executive Council of the *The Middle East Desalination Research Centre*, Muscat, Oman that has conducted desalination R& D in many crucial technological Omani, Israeli, Palestinian, Jordanian, American, Japanese and other experts closely areas, cooperate on technical solutions for the problems of water scarcity in the region, including on projects dealing with solar thermal desalination.[17]

Figure 5: Solel's Combined Solar Power and Desalination System

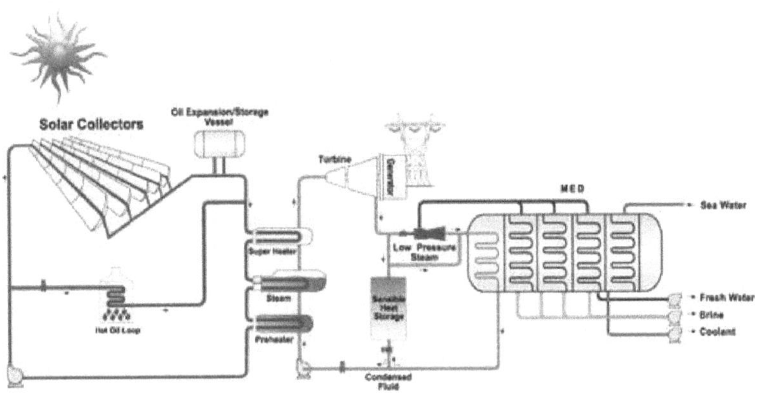

The system consists of: a) a solar field, b) a steam generator, c) a power block, d) a backup storage system, e) a heat storage unit (optional) and f) a MED Multi Effect Distillation plant.

Solel, the Israeli company that pioneered of solar thermal electricity in the 1980s has also designed solar powered desalination plants (figure 5). On their website they claim that "Solar thermal desalination is an effective, economical and proven method of providing both power and fresh water. It is most effective when direct radiation is available and when the cost of electrical power is more than 7 cents kWh. It is particularly suitable for resort, recreation sites and remote locations that require

[16] See at: Trieb/Nitsch/Kronshage/Schillings/Brischke/Knies/Czisch, 2002, at: <http://www2.dlr.de/TT/system/publications/2002_EUROMED-SPaD.pdf>; see also at: DESWARE, *Encyclopedia of Desalination and Water Resources*, on: "Renewable Energy Systems and Desalination" at: < http://www.desware.net/des19.aspx > and "Economics of Water and Energy Technology", at: < http://www.desware.net/desa9.aspx >.

[17] See at: < http://www.medrc.org.om/>. The Centre has developed a Middle East and North Africa (MENA) University and Research Institution Outreach Program in which leading universities from Egypt (Al-Azhar University, Hydraulic Research Institute) Israel (Ben-Gurion University of the Negev, Technion-Israel Institute of Technology); Jordan (Jordan University of Science and Technology; Hashemite University; Royal Scientific Society; Water and Environment Research and Study Center (WERSC); Saudi Arabia (King Abdulaziz City for Science and Technology; King Fahd University of Petroleum and Minerals); Kuwait (Kuwait University; Kuwait Institute for Scientific Research); Morocco (University IBN Tofail); Oman (Sultan Qaboos University); Qatar (The University of Qatar), Tunisia (University of Sfax).

autonomous power and desalination".[18] As a hybrid plant, a solar thermal desalination plant can rely on sunshine during the day and would need a backup, e.g. natural gas or waste heat (or at a future date possibly even solar hydrogen) for the night. Solel pointed to these components of the combined Solar Power and Desalination System

4.5. DESALINATION PLANTS IN THE GULF OF AQABA

Small-scale desalination plants have existed for many years in Egypt between Sham el Sheik and Taba especially for large tourist hotels and to supply drinking water for Eilat in Israel. However, besides rapid urbanisation and coastal development, power and desalination plants, recreation and tourism may pose threats to the marine environment of the Gulf of Aqaba if the environmental constraints are not fully taken into account during the planning and construction process.

Since the peace treaty with Israel in 1994 desalination plants have been discussed in Jordan for the Aqaba region. One proposed plant is to solve the water supply in the Aqaba zone.The Jordanian Water Ministry and USAID developed Aqaba's wastewater treatment plant ($30-35 million) to irrigate parks and for industrial purposes. A further project to convey water from the Disi aquifer at a 4 million m^3/y will provide Aqaba with drinking water, and meet the water demand of the industrial and tourism sectors in the Aqaba Special Economic Zone (ASEZ) for the next five years. The total investment cost of proposed hydropowered seawater reverse osmosis desalination plant in Aqaba was estimated at US$ 389.4 million. In addition to the three countries who signed peace treaties: Egypt, Israel and Jordan; Palestine should also be involved from the outset and possibly also Saudi Arabia who has techical expertise with a pilot plant on solar desalination (figure 6).

Figure 6: Pilot Solar Thermal Desalination Plant in Saudi Arabia

[18] See: „Solar Powered Desalination" at: <http://solel.com/products/dpower/desalination/>.

4.6. FINANCIAL INSTRUMENTS: GEF AND CLEAN DEVELOPMENT MECHANISM

While both technologies already exist in the region, their realisation has so far only been discussed in the framework of national energy and desalination plans and not as an element of an emerging functional peace infrastructure supported by outside donors. This is where the available technologies, the European political experience and political strategies for building a lasting infrastructure for addressing the long-term environmental challenges for the region should merge. In the framework of the Clean Development Mechanism of the UNFCCC and of the Kyoto Protocol a financial instrument exists. So far only Israel and Jordan have ratified the Kyoto Protocol and would thus qualify for CDM projects. Egypt has signed the Kyoto Protocol on 15 March 1999. Once Egypt has ratified the Protocol the Clean Development Mechanism will become a tool for attracting foreign investments in the framework of the UNFCCC. Solar thermal energy plants and solar desalination plants in the Sinai and in the Jordanian desert close to the Gulf of Aqaba could offer important CDM projects.

4.7. FEASIBILITY STUDIES ON INITIAL STEPS TOWARDS SOLAR THERMAL ENERGY AND DESALINATION

As part of a functional cooperation to address the environmental challenges of the future, the following three initial steps are proposed:

Step 1: With support of international donors, a project may be launched with a study on bilateral cooperation between Egypt and the PNA, as well as between Egypt and Jordan on their water needs, on technological options and on the economic costs of desalination with fossil and renewable energy sources. Such a feasibility study should assess the energy potential and costs of a research and development centre in Sinai for an integrated solar thermal desalination infrastructure for both Sinai and Gaza, as well as for the Gulf of Aqaba.

One goal of such a feasibility study could be to assess the costs for a Euro-Mediterranean R&D Facility for hybrid desalination with gas and solar thermal energy. The goal should be to assess the costs for a major desalination plant in Gaza that would use the solar electricity to be produced in Egypt in the Sinai. Thus, as a first step a purely intra-Arab infrastructure could alley Palestinian fears that Israel would control the tap. Such a first step could significantly improve the water and health security in the Gaza strip.

Step 2: In cooperation with the Middle East Desalination Research Center (MEDRC) where Arab and Israeli institutes have cooperated on desalination technologies, a second feasibility study may be launched on the development of trilateral hybrid gas and solar thermal electricity generation and desalination plants for the joint water needs of Sinai, Gaza and the Negev. Potential sponsors for such a multinational water and energy infrastructure could be GEF, the World Bank, the IMF, EIB, EU, USAID and Japan as well as the Arab Development Fund and Arab Gulf countries.

Such a step could reduce the reliance of Israel on water from Lake Tiberias or Kinneret for greening the Negev. The longer-term goal could be the development of a trilateral functional community for developing a joint integrated infrastructure for water and energy, with vital components in Sinai, Gaza and in the Negev (to the extent the IDF permit) to enhance water and food security. This would require a level of trust that presently does not exist. Thus, a strategy should be considered that gradually moves from national self-reliant strategies for water and energy to more integrated or interdependent infrastructures.

Step 3: Later on, the global environmental challenges affecting all countries should be addressed by focusing on the water needs and technological potentials. A third economic feasibility study may focus on the desalination infrastructure in Jordan for the West Bank in the Gulf of Aqaba and for water pipelines on Jordanian territory to supply the West Bank with desalinated water. In the framework of the Euro-Mediterranean Partnership, a small pilot project may be conceived and the potential of the Clean Development Mechanism as an additional funding instrument may be assessed to attract foreign investments in the framework of the Kyoto Protocol. Several project developers have the technical and financial expertise to assess the viability of different options.

So far different projects have been suggested for desalination plants in the Aqaba Region, e.g. by G. Fishelson (1994, 1994a) among others in the framework of the suggested a) Red Sea-Dead Sea Canal and of b) the Mediterranean–Dead Sea (North Route). Murakami (1995: 167) analysed a hydro-powered reverse-osmosis desalination in water resource planning in Jordan (Aqaba-Disi), as well as a solar-hydro power and pumped-storage co-generation in hydro-powered reverse osmosis desalination as an interstate development project in the Jordan River basin (Murakami 1995: 202). According to the Jordan Times (17.4.2002), the Jordanian Ministry of Water and Irrigation was studying a plan for a first sea-water desalination plant in Aqaba for the Aqaba Special Economic Zone (ASEZ).

5. Perspective for Functional Cooperation of Water, Energy, Desalination and Food Experts

This cooperation could start building on existing foundations of cooperation among water, energy, desalination, agricultural and food specialists within the region with the goal to create regional interdependence that requires daily cooperation. This cooperation may focus at several components to be discussed below

5.1. SETTING UP OF A RESEARCH CENTRE ON REGIONAL IMPACTS OF GLOBAL CHANGE

So far scholars from many countries in the Middle East have not actively participated in the research projects on global environmental change (e.g. in the four programmes IGBP, IHDP, WCRP, DIVERSITAS). In Arab countries and in Israel the public awareness on the regional impact of global environmental change (GEC) has been rather low (Newman 2007; Dajani 2007; Twite 2003, 2007). As GEC will have severe impacts on all countries in the region, research on the regional implications of global environmental change is needed to create the public awareness and to initiate the coping capacities. This could become a task of a multi-national *Research Institute on*

Global Change in Taba, Eilat and Aqaba (to be funded by private and international donors).

The planned *Bridging the Rift Desert Science Center* at the border of Israel and Jordan between the Dead Sea and to Gulf of Aqaba to be operated by Stanford and Cornell universities at the border of Israel and Jordan could serve as a model. Such a centre could become a key project of the new EU-sponsored *Anna Lindh Foundation for Euro-Mediterranean Cooperation* that has been established in Alexandria in 2005.

5.2. TOWARDS A JOINT GRADUATE TECHNICAL UNIVERSITY OF THE GULF OF AQABA

Around such a high-level regional research centre, several technical institutions of higher learning with scholars from all four countries including outside scholars should be considered. At a later stage even a joint Technical University of the Gulf of Aqaba with graduate schools in Egypt, Israel and Jordan may be conceived addressing those technologies most in need to enhance the coping capacities of all four countries to adapt to and to mitigate against the projected regional impacts of global environmental change. Such integrated advanced research centres could address the following areas for functional scientific cooperation:

1. Centre of excellence for water management (e.g. to be funded and operated by USAID or UNESCO 2003);
2. Developing most advanced renewable energy technologies relevant for the region (e.g. to be funded by the European Commission and to be operated by a group of European Universities);
3. Schemes for desalination (e.g. to be funded and operated by Japan);
4. Sustainable food production (e.g. to be funded with donations by the private food industry);
5. Sustainable tourism (e.g. to be funded by a consortium of the tourism industry);
6. New urban environments for jobs and living (to be funded by governments and to be operated by UNEP and UN-HABITAT).

5.3. STEPS FOR FUNCTIONAL COOPERATION IN ADDRESSING COMMON ENVIRONMENTAL CHALLENGES

The sponsors of such a joint research infrastructure for recognising the common environmental challenges and for developing the adaptive and mitigation technologies could be the EU countries, the USA and Japan, with support of major international financial institutions (World Bank, IMF, EIB). To realise this functional scientific cooperation the following ten steps could be considered:

1st Step: Problem Recognition and Creation of Awareness: Establishment of an international Research Centre on Regional Impact of Global Environmental Change to Mitigate Environmental and Human Security Risks;

2nd Step: Creating the Knowledge Basis for Mitigation: Establishment of an International Technical University of the Gulf of Aqaba with international departments and faculty in Taba, Eilat, Aqaba.

3rd Step: Setting up a tri-national integrated infrastructure, e.g. in **Taba**: a Centre and Laboratory on Renewable Energy: solar and wind funded by the EU; in **Eilat**: a Centre on Agriculture in Arid Regions in close cooperation with the Desert Research Center in Egypt and the Blaustein Institute on Desert Research (Israel) with US financial support, and in **Aqaba** a Centre for Hydrology and Desalination to be supported by Japan.

4th Step: Supplying Fossil and Renewable Energy. Initially such a pilot desalination plant could be operated with solar thermal energy and natural gas from Egypt or oil from Saudi Arabia as a backup. This fossil backup could gradually be reduced with energy generated from the burning of city waste and from wind energy along the Red Sea, as well as from solar hydrogen once this should become economically feasible within 15 or 25 years.

5th Step: A longer-term goal could be the creation of a joint infrastructure for a local solar hydrogen economy across the Gulf of Aqaba as is presently being planned by Daimler Chrysler, Norskhydro and the Icelandic Electricity Company for Iceland based on hydrogen based on hydrogen from hydro and geo-thermal power.

6th Step: Cooperative Mitigation of Water Scarcity by a joint training institution for water experts on water efficiency, and building joint water desalination plants to serve all four countries; and finally

7th Step: Creating new jobs and supplying food by joint research and training institutions for agriculture, irrigation, and desertification specialists for arid regions (ICARDA), and Centres for IT, computer, software industry.

8th Step: Develop Joint Advanced Medical Research Centres where integrated teams of specialists and nurses would serve patients from all four countries. Besides desalination, the realisation of regional health projects had the highest support among Israelis and Palestinians.

9th Step: Build New Sustainable Cities by developing sustainable tourist centres based on renewable desalination, and developing sustainable cities with a low emission transport system, solar cooling and energy generation, as well as waste based electricity generation.

10th Step: Gradually create a pride in joint achievements and create a culture of cooperation and tolerance.

These ten small steps of functional cooperation among scientists and experts should avoid any national political preconditions. As in the case of the Marshall Plan, the donors should develop a united strategy and avoid to become victim of national politics in the region. These scientific institutions should be international institutions, and they could be under the management of UN agencies (e.g. UNESCO, UNEP, UNU) or of high-level professional organisations, or private academic institutions (consortium of universities) that would guarantee with the support of the major donors both high academic standards and political independence.

6. Recognising Common Challenges and Potential for Functional Cooperation Responding to New Common Threats

This functional strategy of creating awareness and joint regional coping capacities will not realise the "big dreams" of either side that cannot be mutually realised. These steps rely on the "small hope" (Dajani 2004) on what is scientifically possible, politically acceptable and economically feasibility with outside assistance. Multilateral frameworks may also assist in the post-conflict environmental reconstruction, especially of the OPT. Based on the functionalist credo that *form follows function*: the process should start with modest functional cooperation in areas the people already support such as water, environment, health, and food.

This requires also a gradual shift in the mind-set of policymakers on either side to gradually shift from narrow military to wider human security concepts. A precondition is a recognition of the mutual challenges to survival (*Awareness creation*), Thus, collaborative research should address these joint challenges, establish joint scientific and technological capacities in the region, use the energy potential of deserts for its greening and for the protection of the climate. The development of scientific, environmental and economic partnership building measures may contribute to a potential spill-over from functional cooperation to conflict resolution, and thus create preconditions for the development of confidence-building measures for the political and military realm.

Whether the Middle East conflict is a permanent conflict depends on the worldview and mind-set of the observer or policy-maker and his or her preferred means in dealing with this conflict. The continued asymmetric cycle of violence will not produce peace but only continued hatred. It may be worthwhile to study the cases of successful peace-building and of overcoming centuries of conflicts, e.g. between Germany and France. It may also be worthwhile to study the relevance of innovative political concepts of Mitrany, Marshall and Monnet in shifting European policy after 1945 and of Gorbachev after 1990 from conflict towards West European and Pan-European cooperation.

This may require to overcome the traditional Hobbesian worldview and mindset by maintaining, creating, building on and developing regional functional networks of water managers in Israel and Palestine on joint groundwater aquifers, of energy and food specialists as well as sustainable urbanisation experts. The building of common institutions requires a certain degree of trust among the partners. The Gulf of Aqaba could become a laboratory for a joint regional development. Such a process of cooperation could start with the technical education and expand to the economic sector, and finally hopefully contributing to a political spill-over.

The search for common strategies for "human survival" that create joint coping capacities to adapt to and to jointly mitigate against the regional impact of global environmental change requires to overcome state-centred security concepts based on power (military means). Civilian society may contribute to a gradual awareness for the common environmental security challenges with the goal of a stable "human security". However, such a problem solution may require high political courage, President Sadat showed in 1979 by going to Jerusalem, and entering into a peace treaty with Israel. A "new thinking" with a new generation of leaders, as with Gorbachev may gradually evolve, destroying some of the big dreams of either side that can never be realised but that will only create more misery and deeper wounds on either side.

Let me return to my basic hypothesis that ideas mattered to overcome centuries of wars and conflicts between France and Germany, and to transform the Cold War peacefully. If "wars start in the minds of men" (UNESCO), then ideas mattered in Europe and they will matter in the Middle East as well, once there is a willingness to break out of the confinements of Hobbesian strategic thinking. The proponents of the *Geneva Initiative*, of the *Ayalon-Nusseibeh Plan: Vote for Peace* and of the many other peace plans have challenged the mind-sets of the past. Many small steps and new conceptual ideas are needed to break the ice.

Unilateral steps may contribute to a gradual reduction in tensions (GRIT) if they are part of a multilateral strategy. In the medium or longer-term there will be no alternative but to return to a multilateral peace process whatever its structure may be. One of the preconditions of success of the Marshall Plan was that the donor, in this case the US government, used its conditionalised aid wisely. A strong and unified strategy of all donors and equal treatment of all recipients may be a necessary prerequisite.

Grants and credits should be conditional on the development of multilateral regional functional infrastructures with a premium for cooperation and sanctions for violation that would hurt the violator with the suspension of assistance. In the past, conflicts and crises have been the time for learning and conceptual innovation. Thus, the present crisis may produce the conditions for a new "small hope" for a step-by-step implementation.

The conceptual ideas for multilateral functional projects should be developed by joint functional teams of scientists from Egypt, Israel, Jordan and Palestine. A multinational NGO consultation and planning process could be supported by the EU in the framework of the Euro-Mediterranean partnership or its Anna Lindh Foundation. Other functional projects may also be developed with the support of private foundations in Europe, North America (e.g. of the Carnegie, Ford or UN Foundation), and in Japan (by the Sasakawa or Nippon Foundation).

7. Summary and Conclusions

Let me summarise the key arguments and hypotheses of this paper in ten points:

1. The key goal of the paper has been to develop conceptual ideas to gradually overcome the cycle of violence in the Middle East by increasingly recognising the common regional impacts of global environmental change by addressing them jointly through a network of coordinated functional cooperation of water, soil, food, energy and health specialists of Egypt, Israel, Jordan and Palestine.
2. Periods of crises and conflicts have often been periods of conceptual innovation. During the World War II, during the early and late Cold War and during the Korean War new conceptual ideas were developed by Mitrany, Marshall, Monnet and during the stagnation in the Soviet Union by Gorbachev that fundamentally changed the political thinking and strategy in Europe.

3. New ideas mattered in Europe and fundamentally changed the political context. Such a "new thinking" and conceptual ideas are needed in the Middle East conflict overcoming the "big dreams" and developing the "small hope" by functional cooperation of experts to visibly improve the quality of life of both Arabs and Israelis in the region.
4. There seems to be a will among Israelis and Palestinians in support of functional cooperation on desalination, health, environmental, agricultural, tourism and on other issues.
5. Future environmental challenges in the region may become as severe for many young Arabs and Israelis during the next few decades that may only be addressed jointly.
6. There is a need for a fundamental shift in perception of subjective and objective security threats, challenges, vulnerabilities and risks in the security discourses within all countries in the region: from a narrow military threat-based national security view to a wider security perspective that includes the economic, societal and environmental security dimensions and the human being as well as humankind as referent objects.
7. The European experience of functional cooperation on coal and steel matters that started in 1951/1952 during the Korean War and gradually expanded to economic and nuclear cooperation in the Rome treaties of 1957 and later to other policy areas. However, cooperation should address not the most sensitive "war industries" but the technological opportunities to cope jointly with the environmental challenges of the future.
8. Such a functional cooperation will not produce miracles and will not immediately break the cycle of violence. It must be robust and guided by conditions that abide to all countries in an equal and equitable manner and must be directed by a consortium of donors of the international financial and political institutions that must be detached from domestic politics both in the donor and recipient countries.
9. If wars and conflicts start in the "minds of men" then these minds of the citizens and policy makers must be gradually changed to recognise the new common environmental challenges, to initiate a process of anticipatory learning by fostering the gradual emergence of joint coping capacities to address and to avoid those challenges to become threats to the security and survival of all people in the region.
10. While the donors should attach clear conditions on their support of cross-country functional cooperation in the region the recipients should be persuaded to accept the support without political links to the prior realisation of their respective "big dreams" for a prior peace settlement that only serves their own "national" or community interest.

Due to developments on either side a new window of opportunity may open in the Middle East. These conceptual ideas will only be of relevance if they reflect some of the interests of the people in the region. However, to translate abstract conceptual ideas into political proposals for this region with one of the longest unresolved conflicts, remains a difficult task.

8. Proposal for a Network to Develop Feasible Functional Concepts

Given the ongoing conflict and the daily mourning over the dead and the wounded this functional perspective may appear unrealistic to many colleagues in the region due to experience and lack of trust. But in 1947, the ideas of Marshall and in 1950 those of Monnet or in 1986 those of Gorbachev were perceived by some - who were victims of their own mind-set - as dreams and by others as propaganda.

To move from abstract academic conceptual ideas to policy relevant proposals may require a low-key consultation process. My initial operational proposal is very modest:

1. A group of functional (water, soil, food, energy) experts from Egypt, Israel, Jordan and Palestine may be formed and a few experts from Europe, Japan and North America should be added. This group should look for funding for several meetings outside the region.
2. These experts should be given a clear task: to explore areas where functional cooperation among experts in the region already exists, where it appears to be possible and where it may be needed to address the challenges of the future.
3. These experts should be asked to develop a priority list of concrete proposals for functional cooperative projects that appear to be feasible at present.
4. These experts should ask private foundations for seed money to develop concept or pre-feasibility studies on the most promising proposals for projects on functional cooperation.
5. These experts should present these feasibility studies at a conference to representatives of the Middle East Quartet, to international donors and private foundations.

Whether new ideas of functional cooperation matter in the Middle East to break out of the perennial cycle of violence depends on the political context. If such ideas should be attractive for experts in the region and if such a consultation process should be achievable then mutually acceptable policy relevant proposals should be ready once the next window of opportunity opens to resume the peace process in the Middle East. The *Bridging the Rift Desert Science Center* and the *Anna Lindh Foundation for Euro-Mediterranean cultural cooperation* could ease the way for moving ahead with scientific confidence and partnership building projects.

9. References:

Aydin, Mustafa, 2005: "Security Conceptualisation in Turkey", in: Brauch, Hans Günter; Liotta, P.H.; Marquina, Antonio; Rogers, Paul; Selim, Mohammed El-Sayed (Eds.): *Security and Environment in the Mediterranean. Conceptualising Security and Environmental Conflicts* (Berlin – Heidelberg: Springer 2003): 345-356.

Bogardi, Janos; Brauch, Hans Günter, 2005: "Global Environmental Change: A Challenge for Human Security – Defining and conceptualising the environmental dimension of human security", in: Rechkemmer, Andreas (Ed.): *UNEO – Towards an International Environment Organization – Approaches to a sustainable reform of global environmental governance* (Baden-Baden: Nomos): 85-109.

Brauch, Hans Günter, 2000: "Partnership Building Measures to Deal with Long-term Non-military Challenges Affecting North-South Security Relations", in: Brauch, Hans Günter; Marquina, Antonio; Biad, Abdelwahab (Eds.): *Euro-Mediterranean Partnership for the 21st Century* (London: Macmillan – New York: St. Martin's Press): 281-318.

Brauch, Hans Günter, 2000a: "Sustainable EU Energy Policy Between Market Liberalisation and Environmental Obligations. Implications for the Euro-Mediterranean Dialogue", in: Marquina, Antonio; Brauch, Hans Günter (Eds.): *Political Stability and Energy Cooperation in the Mediterranean.* Collection Strademed 10 (Madrid: UNISCI - Mosbach: AFES-PRESS, 2000): 71-150.

Brauch, Hans Günter, 2001: "The Mediterranean 'Space' Beyond *Geopolitique* and *Globalization.* Common Space – Divided Region". in: Marquina, Antonio; Brauch, Hans Günter (Eds.): *The Mediterranean Space ant its Borders. Geography, Politics, Economics and Environment.* Collection Strademed 14 (Madrid: UNISCI – Mosbach: AFES-PRESS): 109-144.

Brauch, Hans Günter, 2003: "Security and Environment Linkages in the Mediterranean: Three Phases of Research on Human and Environmental Security and Peace", in: Brauch, Hans Günter; Liotta, P.H.; Marquina, Antonio; Rogers, Paul; Selim, Mohammed El-Sayed (Eds.): *Security and Environment in the Mediterranean. Conceptualising Security and Environmental Conflicts* (Berlin-Heidelberg: Springer 2003): 35-143.

Brauch, Hans Günter, 2005: "Freedom from Hazard Impact: Developing the Environmental Dimension of Human Security", in: *Intersections* 2/2005, UNU-EHS.

Brauch, Hans Günter, 2005a: "Impact of Global Warming and Non-conventional Water Resources: A Conceptual Contribution to Conflict Avoidance", in: *Proceedings 2nd Israel-Palestinian International Conference of Water for Life in the Middle East- IPCRI* (Jerusalem: IPCRI).

Brauch, Hans Günter; Selim, Mohammed El-Sayed; Liotta, P.H., 2003: "Security Concepts for Cooperation and Partnership in the Mediterranean: Conclusions and Outlook for the 21st Century", in: Brauch, Hans Günter; Liotta, P.H.; Marquina, Antonio; Rogers, Paul; Selim, Mohammed El-Sayed (Eds.): *Security and Environment in the Mediterranean. Conceptualising Security and Environmental Conflicts* (Berlin-Heidelberg: Springer): 955-990.

Bruinsma, Jelle (Ed.), 2003: *World Agriculture. Towards 2015/2030. An FAO perspective* (London: Earthscan).

Buzan, Barry; Wæver, Ole; Wilde, Jaap de, 1998: *Security. A New Framework for Analysis* (Boulder-London: Lynne Rienner).

Dajani, Mohammad, 2004: "Big Dreams and Small Hope", oral presentation in the last session during the Third AFES-PRESS GMOSS Workshop on: Reconceptualising Security at the: 5th Pan-European Conference, The Hague, 9-11 September, see at: <http://www.bigdreamsmallhope.com>.

Dajani, Mohammad, 2007: "Conceptualisation and Debate on Environmental and Human Security in Palestine", in: Brauch, Hans Günter; Grin, John; Mesjasz, Czeslaw; Behera, Navnita; Chourou, Bechir; Oswald Spring, Ursula; Liotta, P.H. ; Kameri-Mbote. Patriacia (Eds.): *Facing Global Environmental Change: Environmental, Human, Energy, Food, Health and Water Security Concepts* (Berlin – Heidelberg – New York – Hong Kong – London – Milan – Paris – Tokyo: Springer-Verlag), in preparation.

Fishelson, Gideon., 1994: "The allocation of marginal value product of water in Israeli agriculture", in. Isaac, J.; Shuval, H. (Ed.), 1994, *Water and peace in the Middle East: Proceedings of the First Israeli-Palestinian International Academic Conference on Water, Zürich, Switzerland, 10-13 December 1992* (Amsterdam: Elsevier Science BV): 427-440.

Fishelson, Gideon, 1994a: "The water market in Israel: an example for increasing the supply", in: *Resource and energy economics*, 16(4) November: 321-334.

Herz, John H., 1959: *International Politics in the Atomic Age* (New York: Columbia University Press).

Kam, Ephraim, 2003: "Conceptualising Security in Israel", in: Brauch, Hans Günter; Liotta, P.H.; Marquina, Antonio; Rogers, Paul; Selim, Mohammed El-Sayed (Eds.): *Security and Environment in the Mediterranean. Conceptualising Security and Environmental Conflicts* (Berlin – Heidelberg: Springer 2003): 357-366.

Murakami. Masahiro, 1995: *Managing Water for Peace in the Middle East. Alternative Strategies* (Tokyo – New York – Paris: United Nations University Press).

Newman, David, 2007: "'In the Name of Security: In the Name of Pweace'. Environmental Schizophrenia and the Scurity Discourse in Israel and Palestine", in: Brauch, Hans Günter; Grin, John; Mesjasz, Czeslaw; Behera, Navnita; Chourou, Bechir; Oswald Spring, Ursula; Liotta, P.H. ; Kameri-Mbote. Patriacia (Eds.): *Facing Global Environmental Change: Environmental, Human, Energy, Food, Health and Water Security Concepts* (Berlin – Heidelberg – New York – Hong Kong – London – Milan – Paris – Tokyo: Springer-Verlag), in preparation.

Selim, Mohamed El-Sayed, 2003: "Conceptualising Security by Arab Maghreb Countries", in: Brauch, Hans Günter; Liotta, P.H.; Marquina, Antonio; Rogers, Paul; Selim, Mohammed El-Sayed (Eds.): *Security and Environment in the Mediterranean. Conceptualising Security and Environmental Conflicts* (Berlin-Heidelberg: Springer 2003): 333-344.

Trieb, Franz; Nitsch, Joachim; Kronshage, Stefan; Schillings, Christoph; Brischke, Lars-Arvid; Knies, Gerhard; Czisch, G., 2002: „Combined solar power and desalination plants for the Mediterranean region – sustainable energy supply using large-scale solar thermal power plants", in: *Desalination,* 153: 39-46.

Twite, Robin, 2003: "A Question of Priorities – Adverse Effects of the Israeli-Palestinian Conflict on the Environment of the Region over the Last Decade", in: Brauch, Hans Günter; Liotta, P.H.; Marquina, Antonio; Rogers, Paul; Selim, Mohammed El-Sayed (Eds.): *Security and Environment in the Mediterranean. Conceptualising Security and Environmental Conflicts* (Berlin-Heidelberg: Springer 2003): 563-572.

Twite, Robin, 2007: "Security and Environment – the Israel-Palestine conflict", in: Brauch, Hans Günter; Grin, John; Mesjasz, Czeslaw; Behera, Navnita; Chourou, Bechir; Oswald Spring, Ursula; Liotta, P.H.; Kameri-Mbote. Patriacia (Eds.): *Facing Global Environmental Change: Environmental, Human, Energy, Food, Health and Water Security Concepts* (Berlin – Heidelberg – New York – Hong Kong – London – Milan – Paris – Tokyo: Springer-Verlag), in preparation.

UNESCO, 2003: *Water for People – Water for Life. The United Nations World Water Development Report* (Paris: UNESCO; New York: Berghahn Books).

UNEP, 2003: *Desk Study on the Environment in the Occupied Palestinian Territories* (Nairobi: UNEP).

Wolf, Aaron T., 2000: "'Hydrostrategic' Territory in the Jordan Basin: Water, War, and Arab-Israeli Peace Negotiations", in: Amery, Hussein A.; Wolf, Aaron T. (Eds.): *Water in the Middle East. A Geography of Peace* (Austin: The University of Texas Press): 63-120.

Wolfers, Arnold, 1962: "National Security as an Ambiguous Symbol", in: Wolfers: *Discord and Collaboration. Essays on International Politics* (Baltimore: John Hopkins University Press): 147-165.

WATER RESOURCES MANAGEMENT AND ENVIRONMENTAL SECURITY IN MEDITERRANEAN TRANSBOUNDARY RIVER BASINS

J. GANOULIS
UNESCO Chair for Water,
Laboratory of Hydraulics, Department of Civil Engineering,
Aristotle University of Thessaloniki, 54124 Thessaloniki, GREECE

Abstract

Water has been recognized as a key natural resource for environmental security, socio-economic development and human well-being. In the Mediterranean area, sustainable water resources management is a major issue, given the semi-arid climate, the variability of hydrological characteristics and the fragile socio-economic conditions. The majority of the population around the Mediterranean lives in transboundary river basins. Sharing water and securing social and political stability in these regions present several technical and cooperative challenges.

In this paper, Multi-Criteria Decision Analysis (MCDA), based on integrated risk assessment, is proposed as a tool for conflict resolution in internationally shared water resources management. The case of the Mesta/Nestos River in South Eastern Europe (SEE) illustrates the methodology.

1. Introduction

Nowadays the concept of human security on a global scale may be extended from its traditional meaning of worldwide political and military security to also embrace the idea that every citizen should be able to benefit from sustainable socio-economic development From amongst different natural resources, water has been recognized as the key environmental resource for social security, economic growth and prosperity. Human security can therefore be seen to be related to environmental preservation (water, ecosystems and biodiversity) and to socio-economic stability and sustainable development. The concept of sustainable management of water resources was first mentioned in Stockholm in 1972, during the United Nations World Conference, and then at the Rio summit in 1992 with Agenda 21.

Historically speaking internationally shared water resources in transboundary river catchments have always been of importance. Rivers and lakes have often been used to determine frontiers between countries (e.g. the Rhine between France and Germany, the Rio Grande between the USA and Mexico and the Evros/Meric between Greece and Turkey). There have been numerous conflicts but also cases of cooperation over transboundary water resources. In many cases, one or several countries may

B. Morel and I. Linkov (eds.),
Environmental Security and Environmental Management: The Role of Risk Assessment, 49–58.

occupy parts of the upstream or downstream area of the river catchment. This makes the issue of water sharing even more complicated (e.g. the Nile between Egypt and the Sudan, the Middle East conflict over the Jordan River and the Danube between many European countries).

On a global scale, the importance of transboundary water resources is far from negligible: according to reports submitted to the UN, about 50% of the world's landmass (excluding Antarctica) is located in internationally shared water catchments. About 40% of the world's population lives in internationally shared water catchments, extending over more than 200 international river basins.

In the Mediterranean, transboundary water resources are extremely important. In SEE, 90% of the area lies in international basins. The Nile basin is shared by 10 countries from deepest Africa to Egypt. In North Africa and in the Middle East, transboundary aquifers are very important.

The aim of this paper is to show how traditional engineering planning and design methods for reducing risks in water supply and management can be extended to consider environmental and social risks. Furthermore, a multi-objective decision-making methodology is suggested, in order to help resolve water related conflicts.

2. Main Issues in Transboundary Water Resources Management (TWRM)

TWRM involves addressing not only physical and technical issues but should also take into consideration social actors, institutions and administrative procedures. According to LeMarquant [1], five foreign-policy factors influence international water situations:

1) international posture of each country,
2) international law,
3) linkage between water and other issues,
4) mutual commitment (reciprocity), and
5) national sovereignty.

The main objective of effective TWRM is to satisfy the demands of all riparian countries, given the possibilities and limitations of water supply. This balance between supply and demand should take into consideration both water quantity and quality aspects and the protection of the environment. Water quantity and quality problems are very much inter-related and should be studied in an integrated framework. According to Frey [2], in order to understand the origin of serious conflicts over international water systems, three main factors should be considered:

1) the importance of water (both in quantity and quality),
2) the relative power of the actors, and
3) the respective riparian position of the countries.

2.1 THE ENGINEERING APPROACH

This approach has been developed mainly by engineers and management experts. Depending on the number of objectives and decision-makers and their combination, models may be formulated as *optimisation, multi-objective trade-off* computerized codes or on the basis of the *team and game theories*. Most of these models are based on the fundamental notion of Pareto optimality and are predictive in the sense that they suggest a quantitative "optimal" situation, which should be to terminate a conflict by finding an equitable resolution between the countries involved.

Recent advances and related theoretical developments in this area can be found in the literature, including the application of the fuzzy set theory. However, the success in practice of this kind of *engineering* or *rational modelling* is mainly dependent on the interested actors' and countries' acceptance of the model assumptions, which rely on a set of prescribed objectives, and the relative weights or preferences between conflicting goals. In the real world this is not usually the case, and therefore, there is a need to develop better, easier-to-use, interactive and reliable predictive models for TWRM.

2.2 THE INSTITUTIONAL APPROACH

This approach is used mainly by law experts and political analysts, who focus on describing the anatomy of a given situation of conflict or cooperation. They determine the function of different parameters and factors influencing the behaviour of each country, such as the political perception of the importance of water, the international image and status of the country and also social and institutional issues. Such models, including the behaviour of institutional structures, international negotiation strategies, alternative dispute resolutions and political models are very useful. They are mainly prescriptive and not predictive. They do not necessarily give a quantitative output (such as costs and benefits), but are extremely important for understanding the processes and analysing the origin and evolution of conflicts or cooperation.

Many alternative negotiation strategies are available to modify a complex framework of TWRM issues. Decision makers and those who may negotiate on their behalf have a choice of six universal negotiation strategies:

1) "Win-Win" solutions or Positive sum benefits
2) "Lose-Lose" solutions or Negative sum benefits
3) "Win-Lose" negotiations or Zero-sum benefits
4) Unilateral creation of new facts
5) Conflict and threats of violence
6) No action, causing opportunity costs from neglect and/or delayed decisions.

The choice of a particular negotiation technique is always subject to political considerations and controversy. Preferences depend on the balance of power among transboundary stakeholders and the cost of concessions. The more powerful and wealthy stakeholders can resort to the creation of facts with minimal risk of counteraction by weaker and impoverished neighbours. They also can afford to make gestures of friendship through "Win-Lose" agreements in the interest of enhancing regional stability [3].

It may be mathematically proven that "Win-Win" agreements result in positive benefits for both parties and consist of the best trade-off between alternative solutions. The so-called "Prisoner's Dilemma", well known in the literature, gives insight to the fact that failure to reach an agreement between interested parties may increase benefits to each individual party but will decrease the total benefits. This is because when each party acts independently it will tend to over-use the resource. Cooperation schemes may provide better net benefits to both parties.

However, "Win-Win" solutions may not always be sufficient when considering cases where natural water resources are limited. In these cases regional networks of water stakeholders can play a very important role.

By combining the expertise and state-of-the-art knowledge of different scientific communities and disciplines, such as engineering, economics and environmental and social sciences, regional partnerships may contribute to the development of new methods and models in order to more efficiently resolve conflicts and controversial issues in TWRM.

3. Environmental Risk Assessment and Management

In a typical problem of technical failure under conditions of uncertainty, there are three main questions, which may be addressed in three successive steps.

1. When should the system fail?
2. How often is failure expected?
3. What are the likely consequences?

The first two steps are part of the uncertainty analysis of the system. The answer to question 1 is given by the formulation of a critical condition, producing the failure of the system. To find an adequate answer to question 2 it is necessary to consider the frequency or the likelihood of failure. This can be done by use of the probability calculus. Consequences of failure (question 3) may be calculated in terms of economic losses or profits.

It has been largely accepted that the ***probability of failure*** may be considered as one simple definition of the engineering risk. As explained in [4] we should define as *load* ℓ a variable reflecting the behaviour of the system under certain external conditions of stress or loading. There is a characteristic variable describing the capacity of the system to overcome this external load. We should call this system variable *resistance* r. A *failure* or an *incident* occurs when the load exceeds this resistance, i.e.

FAILURE or INCIDENT : $\ell > r$

SAFETY or RELIABILITY : $\ell \leq r$

In a probabilistic framework, ℓ and r are taken as random or stochastic variables. In probabilistic terms, the chance of failure occurring is generally defined as risk. In this case we have

$$\text{RISK} = \text{probability of failure} = P(\ell > r)$$

Uncertainties and risks may be quantified by using probabilities or fuzzy sets, and can be used as a tool for helping decision-making processes [4], [5]. The Integrated-Risk Analysis-Method [5] is one MCDA technique that can be used in TWRM for managing different conflicts. The steps to be undertaken for applying this methodology are the following [4]:

1. Define a set of ***alternative actions*** or ***strategies***, which includes structural and non-structural alternative options.

2. Evaluate the outcome risks or ***risk matrix,*** which estimate the risks corresponding to each particular objective (technical, environmental, economic and social)
3. Find by use of an averaging algorithm the ***composite risk index*** for technical and ecological risks (eco-technical composite risk index) and social and economic risks (socio-economic composite risk index).
4. ***Rank the alternative actions***, using as a criterion the distance of any option from the ideal point (zero risks).

4. Application of MCDA techniques in TWRM

MCDA techniques are gaining importance as potential tools for solving complex real world problems, because of their inherent ability to consider different alternative scenarios, the best of which may then be analysed in depth before being finally implemented. [6], [7], [8], [9], [10], [11].

In order to apply MCDA techniques, it is important to specify the following:

- **The objectives**, which indicate the directions of state change of the system under examination and which need to be maximized, minimized or maintained in the same position.
- **The attributes**, which refer to the characteristics, factors and indices of the alternative management scenarios. An attribute should provide the means for evaluating the attainment level of an objective.
- **The constraints,** which are restrictions on attributes and decision variables that can or cannot be expressed mathematically.
- **The criteria,** which can be expressed either as attributes or objectives.

As shown in Figure 1, the three pillars of sustainability, i.e. the economic, social and environmental criteria, can be defined hierarchically, starting from some basic indicators, which are then aggregated into second and three level indicators.

Figure 1: Social, economic and environmental attributes, objectives and goals.

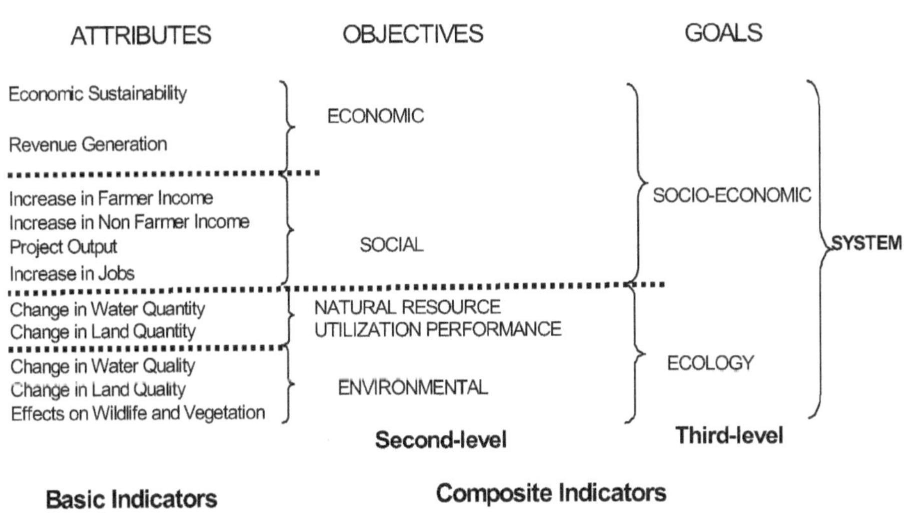

The methodology we propose addresses two fundamental issues in TWRM, which are conflict situations at two levels [12]:

(a) conflicts among attributes, in particular, economic, technical, environmental and social first-level indicators
(b) conflicts among different countries' strategic goals

This MCDA approach may be applied in three steps:

(1) in the *first step,* each country proceeds separately and evaluates alternatives according to its own attributes, objectives and goals (Figure 2).
(2) in the *second step,* the different attributes used by the different countries are first traded-off and then alternatives are ranked according to the composite goals (Figure 3)
(3) the *third step* is based on the aggregation of the countries' different goals in order to obtain a consensus between them (Figure 4).

Figure 2: Ranking alternative options by each country separately.

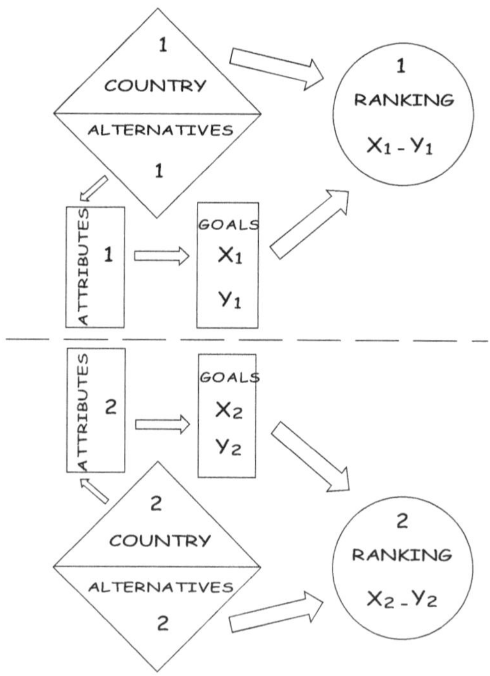

Figure 3: Compromising countries' attributes for conflict resolution.

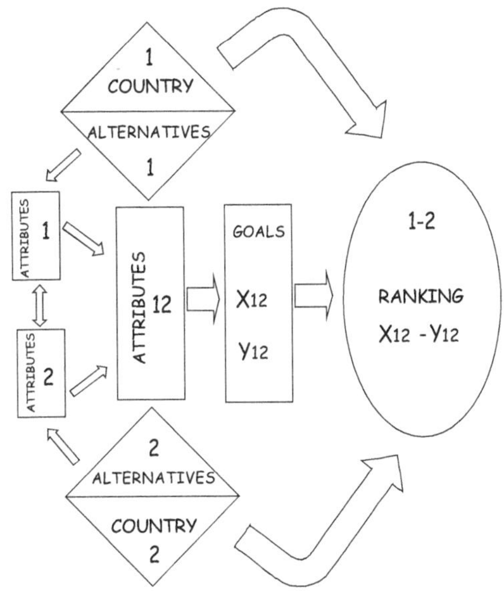

Figure 4: Compromising countries' goals for conflict resolution.

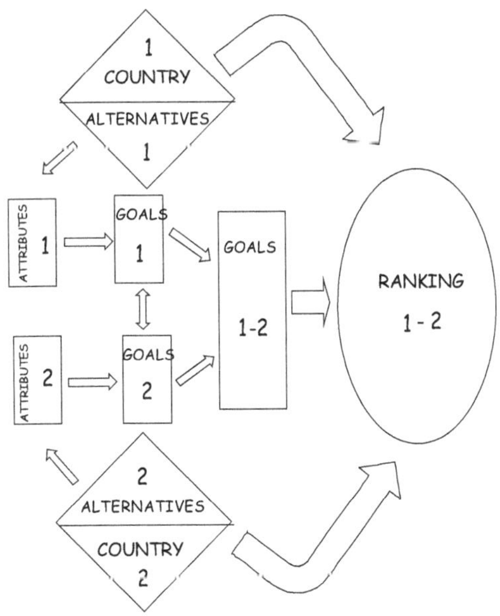

It may be expected that steps 2 and 3 will produce similar results, but these will be different to the rankings produced by each country separately in step 1.

To illustrate the methodology in practice, results from the case of the international Nestos / Mesta River, which flows between Greece and Bulgaria, are briefly presented below. In Figs. 5 and 6, points 1, 2, 3 and 4 represent alternative projects proposed by one country and 5, 6, 7 and 8 those proposed by the other. For both approaches, i.e. trading-off countries' attributes (Fig. 5) or countries' goals (Fig.) the same alternatives 3, 6, 7, 8 are located close to the ideal solution (maximum reliability), with ranking order 36-8-7.

As an extension of the present methodology, two different types of uncertainties may be taken into consideration:

(a) uncertainties in attribute and goal indicator values
(b) uncertainties due to different preference functions (weights) of the decision makers or interest groups.

Figure 5: Conflict resolution by trading off countries' assessment of attributes.

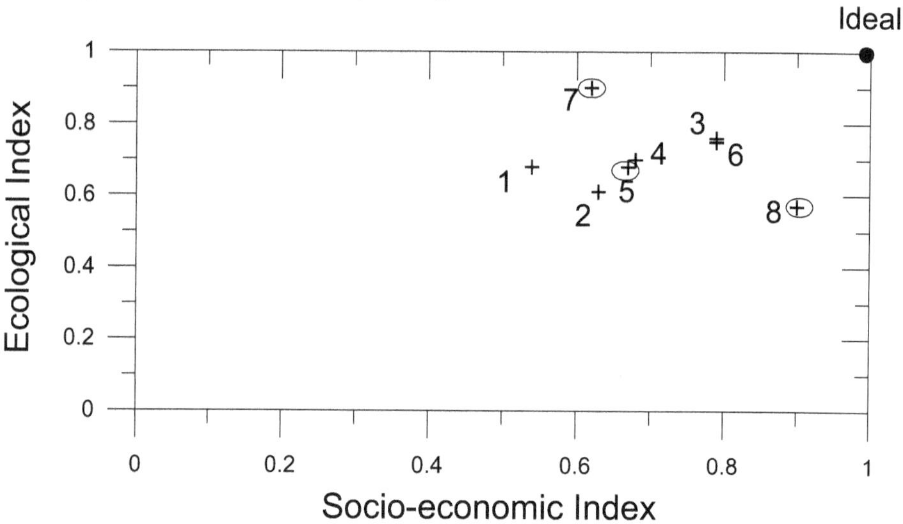

Figure 6: Conflict resolution by trading off different assessment of countries' goals.

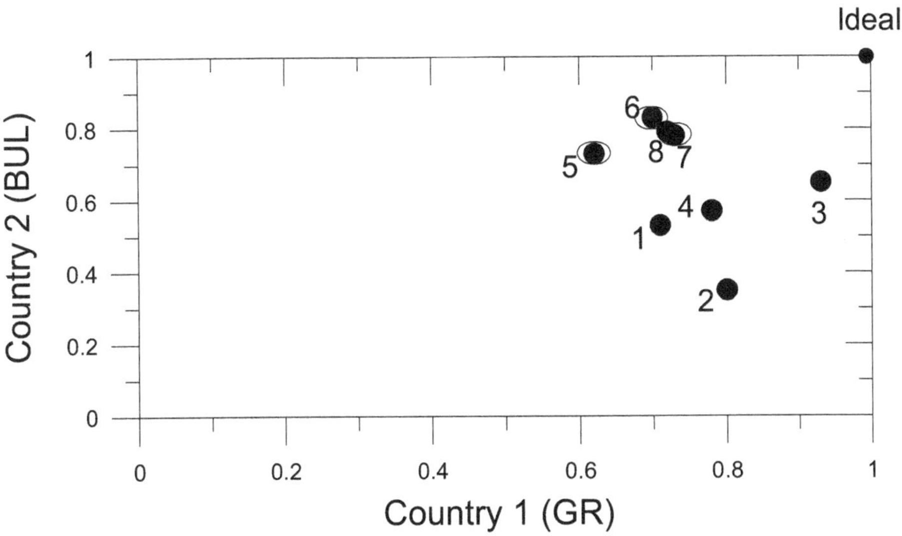

4. Conclusions

Conflicts in sharing transboundary water resources are usually due to the fact that countries use either different attributes or different goals to evaluate impacts from alternative strategies

The methodology we propose is based on a combination of Integrated Risk Analysis and MCDA techniques adapted to conflict resolution. Trade-offs are made either at the level of countries' different appreciation of individual attributes, or at the level of countries' different goals.

The methodology is easy to use and the results obtained are fair, transparent and simple to communicate to decision makers.

5. References

1. Le Marquant, D. G. (1990) International Development of the Senegal River. *Water International*, **15** (4): 223-230
2. Frey, F. W. (1984) Middle East water: the potential for conflict or cooperation. In Naff T. and Matson, R. C. (eds.) *Water in the Middle East,* pp. 180-198, Westview Press, Boulder, Co
3. Eaton J. W. and D. J. Eaton (1996) Negotiation Strategies in Transboundary Water Disputes In: *Transboundary Water Resources Management: Institutional and Engineering Approaches*, pp. 37-45, NATO ASI Series, Springer-Verlag, Heidelberg, Germany
4. Ganoulis, J. (1994) *Engineering Risk Analysis of Water Pollution: Probabilities and Fuzzy Sets.* WILEY-VCH, Weinheim, Oxford, NY, 306 pp.

5. Ganoulis, J. (2004) Integrated Risk Analysis for Sustainable Water Resources Management In: *Comparative Risk Assessment and Environmental Decision Making,* 275-285, NATO Science Series, Kluwer Academic, The Netherlands
6. Goicoechea, A., D. R. Hansen and L. Duckstein (1982) *Multiobjective Decision Analysis with Engineering and Business Applications*, J. Wiley, New York, 519 pp.
7. Bogardi, I. and H. P. Nachtnebel (1994) *Multicriteria Decision Analysis in Water Resources Management*, IHP, UNESCO, Paris, 469 pp.
8. Duckstein, L. and E. Plate (eds.) (1987) *Engineering Reliability and Risk in Water Resources,*_E.M. Nijhoff, Dordrecht, The Netherlands, 565 pp.
9. Fang, L., K. W. Hipel and D. M. Kilgour (1993) *Interactive Decision-Making - The Graph Model for Conflict Resolution*, J. Wiley and Sons, N.Y.
10. Fraser, N.M., and K. W. Hipel, (1984) *Conflict Analysis: Models and Resolutions*, North Holland, N.Y., 377 pp.
11. Vincke, P., (1989) *L'aide multicritere a la decision*, Editions de l'Université de Bruxelles.
12. Ganoulis, J., L. Duckstein, P. Literathy and I. Bogardi (eds.) (1996) *Transboundary Water Resources Management: Institutional and Engineering Approaches.* NATO ASI SERIES, Partnership Sub-Series 2. Environment, Vol.7, Springer-Verlag, Heidelberg, Germany, 478 pp.

THE ROLE OF COMPARATIVE RISK ASSESSMENT IN DECISION ANALYSIS MARSHLANDS OF MESOPOTAMIA AND AFFECTED RIPARIAN COUNTRIES

C. F. MAXWELL
Senior Environmental and Infrastructure Management Officer, Basrah, IRAQ

Abstract

The Marshlands of Mesopotamia are caught in the middle of regional water scarcity. The marshlands in Iraq are at the bottom of the flow of water from Syria, Turkey, and Iran before it drains into the Arabian (Persian) Gulf. As a result the marshlands reflect the impacts of dams, population growth, and over extraction of water from neighboring countries which result in reduced flows in the Tigris and Euphrates rivers and associated wetlands. The proposed dike to bisect the marshlands on the boundary line between Iran and Iraq creates a critical transboundary issue. This dike would impede flow of water into the marshland on the Iraqi side, one of the remaining pristine marshes in Iraq, a wetland unique to the world. These impacts to the marshlands cause security issues, unemployment, loss of biodiversity, economic hardship, and loss of wildlife habitat. In this paper I illustrate the cause and effect relationship of policy and practice of the countries and resulting impacts. The marshlands of Iraq are at the bottom end, as a result provide a benchmark of the environmental universe and serve as a spring board for debate and potential for positive change to Middle Eastern stability and security. The causes for the demise of the marshlands provide historical information of current effects. Through a comparative risk assessment of 8 recommendations developed under the UNEP[1] Early Warning and Assessment Technical Report (2001); The Mesopotamian Marshlands: *Demise of an Ecosystem* results are identified and positive change for the marshlands of Iraq and the affected riparian countries is possible for those recommendations that will minimize the major and/or catastrophic consequences of impacts to the water resources, the public, ecology, and biodiversity of the Mesopotamian Marshlands and the affected riparian countries are selected through comparative risk assessment.

1. The Garden Of Eden – Past And Present

About 30,000 BCE, the Great Ice Age still held most of Eurasia in its grip, and it caused the sea levels to fall by 400 feet so that what is now the Persian Gulf was dry land, all the way to the Strait of Hormuz. This was also the transition from Neanderthal to modern Man. Eden wasn't born then. From 15,000 BCE, rains diminished drastically in the Gulf region and then returned from 6,000 BCE to 5,000 BCE. About 6,000 to

B. Morel and I. Linkov (eds.),
Environmental Security and Environmental Management: The Role of Risk Assessment, 59–78.

5,000 BCE, rains returned to the Gulf region. The land was again a paradise on Earth. Agriculture had been invented-in today's Iran and Iraq as groups of hunter-gatherers evolved in-to agriculturists. The Mesopotamian region produced the world's first written language. In Sumerian the word "Eden" meant simply "fertile plain." The word "Adam" also existed in cuneiform, meaning something like "settlement on the plain." Mesopotamia is the name ancient Greeks gave to present-day Iraq. It means literally 'between the rivers', referring to the Tigris and the Euphrates, which rise in the mountains of Anatolia to flow down to the Arabian Gulf. Its natural borders are formed by the mountain ridges of Anatolia to the north, the Zagros range to the east, the Arabian Desert to the south-west and the Arabian Gulf. The first large-scale communities began to develop after man learned how to control his water environment by means of canals and dikes. People began to profit from a system beyond subsistence to produce a surplus, diversify their cultural activities and live in increasingly large numbers in a new form of collective community, the city. The invention of cities may well be the most enduring legacy of Mesopotamia. Physical Eden vanished under the waters of the Gulf. Man had lived happily there. But then, about 5,000 to 4,000 BCE, a worldwide phenomenon called the Flandrian Transgression, caused a sudden rise in sea level. The Gulf began to fill with water and actually reached its modern-day level about 4,000 BCE, having swallowed Eden and all the settlements along the coastline of the Gulf. [4] Today the Marshlands of Iraq are caught in the middle of regional water scarcity. The marshlands are at the bottom of the flow of water from Syria, Turkey, and Iran before water from these riparian countries drains into the Arabian Gulf through the Tigris and Euphrates rivers. The marshlands reflect the consequences of impacts, good or bad, caused by flooding, wars, dams and over extraction of water, resulting in reduced flows into the wetlands.

2. Cause and Effect Relationships

The impacts of policies and perceived security threats have cause-effect ramifications.

2.1 IRAN

For example, in the case of Iran the first level of impact on the marshlands is through regulation of the flow of the Karkah River by dam construction. This would supply the Hawr al Azim, which is the area of marshland on the Iranian side, continuous with the Hawizeh marshes in Iraq. Maintenance of the wetlands will be dependent on the flow release strategy for the dam and this is as yet unknown. However the continued expansion of agriculture and the need from irrigation water on the Iran's side will undoubtedly reduce the volumes available to the marshes. The precise reasons for construction of a dike along the border by the Iranians are unclear but there are several possibilities of this cause and effect relationship: (a) the response to Iraq drying out their marshes on there side of the border and the need for Iran to retain the precious and diminishing water resource which would otherwise be 'lost' to Iraq. (b) The need to have greater security control in an area where smuggling and lawlessness is rife. (c) The need for improved transport and communication in an area of potentially rapid oil field development. This dike, once completed by Iran, will impede flow of water into one of the remaining pristine and unique wetlands of the world. The need for a porous border to maintain the ecological integrity of the marshland complex is obvious but this will only be achieved through careful attention to the design criteria of the dike together with the maintenance of flow both sides of the border.

2.2 TURKEY

Without doubt the greatest reduction in flows occurs in Turkey, where the Ataturk dam has a reservoir capacity greater than Euphrates total annual flow which, as a result, has significantly transformed basin ecology.

2.3 IRAQ

There are also impacts within Iraq itself. The past regime in Iraq have caused death to tens of thousands of Iraqi Marsh Arabs, destruction of their homes and unique way of life, flight of refugees to neighboring countries, displaced persons to urban centers exacerbating unemployment, loss of biodiversity, economic hardship, loss of water buffalo, and destruction of wildlife habitat.

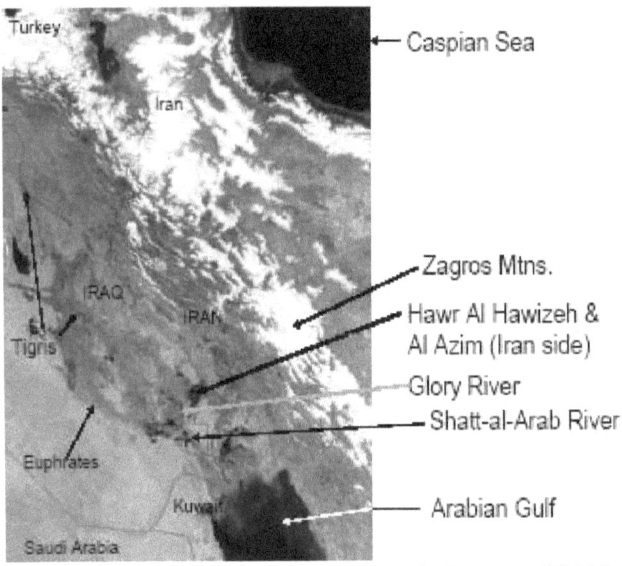

Recent Land sat image taken in 2004 of Marshlands, indicate winter rains and possibly melting snow from Iran's Zagros Mountains have begun to fill the marshes and shallow lakes of Southern and Eastern Iraq. Flooding in this region is an annual affair, though typically, the floods peak in April and May when spring snowmelt flows out of Iran. These images show that water levels had already begun to rise two months into the rainy season. In early January of 2004, re-diverted water combined with heavy rains caused intense floods near Nasiriyah, a little north of Lake Hammar on the shores of the Euphrates River. Over 8,000 people were evacuated, and at least 180 houses were destroyed.

The cause and effect relationship of climate change, of the vagaries of weather patterns, the policies and practices of countries, and the geopolitical uncertainties engendered, have and continue to have consequences to the social, economic, and environmental security for Iraq and its neighboring countries. What happened in the

Too much water

past? What is the situation now? What can be done about it? These are the questions that must be answered to find the solutions, ultimately linked to the marshlands of Iraq.

3. Impacts

Today, river flow into the Mesopotamian marshlands has been cut by 20-50 percent, and the spring floods that sustained the marshlands have been eliminated. The end result is what was once a lush wetland environment has been reduced by about 85 percent in area.

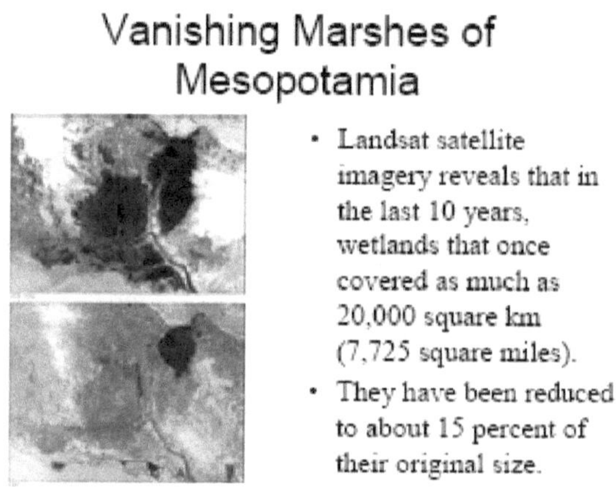

What was once a vast, interconnected mosaic of densely-vegetated marshlands and lakes, teeming with life, are now mostly lifeless desert and salt-encrusted lakebeds and riverbeds. Even for the 1,270 square km (490 square miles) of marshlands that still remain, quality of life has been adversely impacted by a decline in water quality. Human irrigation practices render the Tigris and Euphrates waters saltier than they originally were. And, with the boom in agriculture, not only in Iraq, but its neighboring countries, there have been dramatic increases in the levels of chemicals as well as a rise in urban and industrial effluents in the rivers. The impact of marshland desiccation and the discharge of polluted agricultural effluent via drainage works on the marine environment in the northwestern Arabian Gulf are also likely to be significant. Conversely, most of the load of natural sediments and silts the rivers used to carry now remains trapped behind multiple dams. Lower levels of silt decrease plankton and levels of organic carbon in the water which, in turn, adversely affect fish populations as well as soil fertility along the riverbanks and in the marshlands. All of these negative trends point to the inevitable demise of the Mesopotamian marshland ecosystem unless steps are taken soon to reverse the damage being done.

4. The Current Situation

A crucible of civilization, the marshlands have been home to ancient human communities for more than five millennia. The area's inhabitants are commonly known as the Ma'dan or "Marsh Arabs", whose population is estimated to range from 350,000 to 500,000. Heirs of the Sumerians and Babylonians, the Marsh Arabs act as a living link between the present inhabitants of Iraq and the peoples of ancient Mesopotamia.

The Marsh Arabs are Shiite Muslims, and their way of life is largely based on the traditions of the Arab Bedouin. They have evolved a unique subsistence lifestyle that is firmly rooted in their aquatic environment. Most of the Ma'dan is semi-nomadic, but some of them are settled in villages. Their settlements are located on the edges of the marshes, or stand on artificial floating islands that are regularly reinforced with reeds and mud. The marshes of southern Iraq were the largest wetlands in the Middle East. Since the dawn of human history, they have been home to many species of plants, animals, and fish and a vital wintering ground for migratory birds.

- For 5,000 years, the Madan or Marsh Arabs lived in these marshes, and developed a unique way of life that tied them intimately to their environment – fishing and hunting, planting rice,barley, and wheat, and building the Madan's unique reed houses.
- That way of life has been all but destroyed now. Since the 1950s, there have been 32 dams built in the Tigris-Euphrates river basin. Eight more are under construction and 13 more are planned.
- The destruction of the marshes – and the people who lived in them – accelerated rapidly after the first Gulf War. The former regime went to great lengths to build canals and drain the waters in order to punish the people of the marshlands for the Shiite uprising. Soldiers killed tens of thousands (perhaps

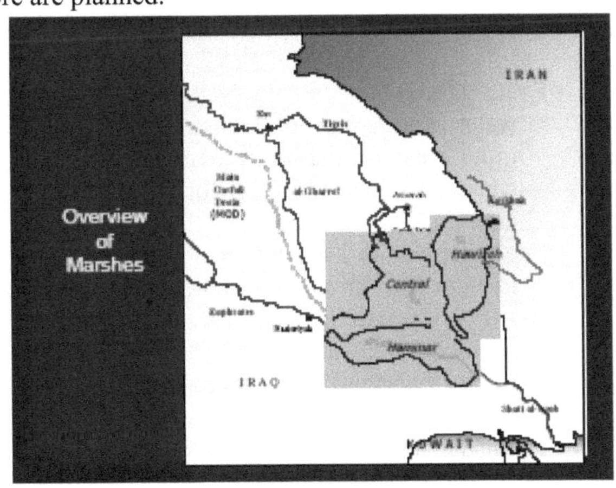

between 50,000 and 100,000), burned their communities to the ground, poisoned the water, destroyed their livestock, and placed unmarked land and water mines throughout the region.

- A population that had numbered more than a quarter of a million in 1990 was reduced to between 20,000 and 40,000. About 40,000 Marsh Arabs were refugees in Iran, while another 200,000 to 250,000 are internally displaced.
- Today, the marshlands that once covered 15,000 to 20,000 square kilometers now extend to less than 2,000 square kilometers. The Central Marsh has been reduced to three percent of its original size. The Al Hammar Marsh is down to six percent of its original size. Only the al Hawizeh Marsh, along the border with Iran, remains sizeable, and even that has been reduced by two-thirds. Hawr al Azim is the area of marshland on the Iran side, continuous with the Hawizeh marshes in Iraq.
- The decline of the marshes has meant a decline in water quality, too. Increased irrigation and the loss of spring flooding due to the dams add to the salinity of the water. Concentrations of fertilizer, pesticides, and human and chemical waste – all of which marshlands normally filter - - are steadily rising. With the decreased flow in the Tigris and Euphrates, the Shatt al-Arab estuary now carries sea water farther and farther upstream.

5. The difficulty in reversing this process

- First, we must be realistic. Neither the Tigris nor the Euphrates carries as much water as they once did. Upstream countries, Turkey, Syria, and Iran – in that order – control the headwaters and the water-flow from the dams and in each country demand for water for agriculture and development is sure to grow as their populations grow.
- Iraq's population has been growing rapidly, too, and the demands of those who live upstream from the marshes can be expected to grow accordingly.
- There is no international agreement or comprehensive river basin plan for managing the water from the Tigris and Euphrates, nor is there a mechanism yet for balancing Iraq's competing claims on these waters and the land that depends on them.
- The marshlands have been closed until now to outsiders since 1988. An accurate data base of actual conditions does not exist.

- There is anecdotal evidence that some Marsh Arabs are beginning to return. But what do they return to? How do you deal with issues of land and water rights, of culture and community when conditions have changed so dramatically?
- No one knows how the soil has been affected by the draining of the marshes or whether they can be effectively restored.
- The hydrology and ecology of wetlands are extraordinarily complex and exceedingly delicate. Even in the United States, there is considerable uncertainty whether the Everglades can be fully restored.

6. What might be done?

- Does this mean we give up hope? Certainly not. What it does mean, however, is that we must be cognizant of the many difficulties involved and approach them in a very careful, scientific way.

- Lessons learned from the United Nations Environment Program (UNEP) post-conflict assessments demonstrate that environmental contamination and degradation have critical humanitarian consequences requiring consideration at an early stage in relief and recovery operations.
- Furthermore, the assessments have revealed the critical need to build institutional capacities for environmental management immediately after the conflict in order to screen the potential environmental impacts of reconstruction and development projects, and to ensure their sustainability.
- Because of the conflict UNEP could not reach definitive conclusions on these issues. A presence on the ground would be needed before environmental priorities can be assessed fully.
- From this first start by UNEP, USAID is following through and has been working on a strategy for how post-conflict development activities might be carried out. In regard to the marshlands, it would comprise several tracks, including:
  - A process to resolve the social, political, and institutional issues related to resettlement, property rights, economic opportunities, and social safety nets;
  - A reliable database, established through a series of technical assessments as a basis for program planning;
  - Immediate interventions to manage and protect the marshlands that still exist;

- Restoring destroyed areas through targeted rehabilitation pilot projects;
- Partnerships with other donors, the private sector and non-governmental organizations;
- Cooperation with Iraqi scientists, officials from the new government, environmentalists, and, of course, the Madan people themselves, to develop their expertise in wetlands management; and prepare a comprehensive rehabilitation and monitoring plan.

7. Marshlands Restoration Program

7.1 BACKGROUND

Prior to drainage of the Iraqi marshes, the main elements of the local economy were based on their biological diversity: livestock, fishing, hunting, and mat-making. Agriculture was seldom done in the mashes. Yet, as the marshes were drained, agriculture became more important as a key source of income and nutrition for former marsh dwellers.

During and subsequent to the years of drainage, agricultural activities were disrupted and strictly controlled. Displaced people grew wheat almost exclusively as it appears that the regime forbade the marsh dwellers from cultivating anything else. With the demise of the former regime, there is opportunity for the former marsh dwellers to improve agriculture in the drained marshes.

7.2 DESCRIPTION OF FARMING ON DRAINED MARSHLANDS:

The international and domestic political circumstances of the last two decades in the southeastern and southern parts of Iraq have forced marsh dwellers to abandon their former lifestyle to become farmers. The transition from marsh to drained land has not been easy for the former marsh dwellers. Faced with challenges particular to their land and water conditions, the former marsh dwellers eke out a difficult existence. Yet, surprisingly, all former marsh dwellers interviewed for this program confessed that if given a choice between living in the marsh as they had done for millennia and living on drained marsh land cultivating crops, they would choose agriculture. They believe that agriculture provides the potential to improve their living conditions in ways and in magnitude that marsh living could not.

The farmers cultivating the dried marshes, like their counterparts living on the edge of the marsh face the same constraints as most farmers in Iraq. There is a lack of inputs, especially fertilizer, transportation is difficult, and they receive little or no extension advice. Yet, the circumstances resulting from drying of the marshes present other more difficult challenges to these people,challenges which are serious enough to warrant specially assistance programs to these population groups. Those distinctive challenges facing marsh dwellers turned farmers are:

- **Lack of experience in farming**: For millennia, the marsh dwellers lived off fish, livestock and the vegetation in the marshes. They had limited experience with farming grain crops. The events of the past two decades forced them to dispense with most economic activities they did as marsh dwellers in favor of cultivating crops as their maim source of income. But, they made this change without any training or help from the Ministry of Agriculture or other agencies.
- **Soil and water constraints**: There are significant and growing problems of salinization of soil and water in the dried marshes. Further, little is known about the physico-chemical changes in these saline/sulfidic soils when drained, such as when these soils desiccate, burn, and are subsequently sown to annual crops and pastures (wheat and barley). Lack of knowledge about these soils makes it difficult to provide good advice on crop selection, site selection or level and type of input.
- **Displacement**: Most marsh dwellers were forced to move from their homes more than once and often 10 or more times over the past 13 to 20 years. The psychological and emotional effects of this tragedy are compounded by the near total loss of assets and resources. Under any circumstances this would be hard to overcome, but in the Iraq of the past two decades, it has been virtually impossible for these people to recover their losses.
- **Marketing and agribusiness related problems:** Production and marketing systems of crops and livestock in Iraq have not kept pace with worldwide technological developments due to wars, the embargo and UN sanctions
- **Livestock:** Livestock and dairy have always been an important part of the marsh economy, but with the drainage of the area, the traditional means of production

were eroded and disrupted. In response to the drainage and to their general inaccessibility to water, people in the marshes made a dramatic shift to different livestock management patterns – from water buffalo to sheep. Livestock herding remains a critical endeavor in the marshes, both as a source of basic nutrition and income for families, but serious constraints exist.

- **Fishing:** The economic importance of fishing increased until the drainage of the marshes. Water Buffalo were the basis of household wealth in some tribes, but, where fish markets were accessible, fishing played an increasingly important role in the local economy. A combination of factors, including the Iran-Iraq War in the mid-1980s and the building of the drainage system in the 1990s, diminished the flow of nutrients to the marshlands and its fish and led to a precipitous fall in catches. The situation worsened with the massive displacement of the local population, and restrictions on access to those marshes still under water.
- **Primary Health Care:** The marshlands have always been one of the most remote areas in Iraq, historically outside the general control of the government and with a record of the fewest and least developed services provided. During the 1990s, medical services were virtually non-existent in much of the region, largely the consequence of a deliberate national policy to persecute the local population. Nowadays, throughout the marshes, the situation has not dramatically improved. Medical services are very limited, with only a few hospitals in the surrounding area and clinics offering no more than the most basic health care. Demanding cases have to be taken to the cities. Most clinic buildings are abandoned or in a state of bad disrepair. Only clinics in the marshes are still operational, but they provide only the most basic medicines and with few qualified personnel. They are likely to have a doctor's assistant and/or male nurse. As a result, women are likely to shun the clinics.
- **Sewage:** Towns and villages in the marshlands lack wastewater treatment facilities. In the larger towns, people have small outdoor toilets with pipes that directly go into cesspools or covered pits adjacent to the houses. During periods of rain or high use, the pits back up and sewage can enter the house. Cesspools offer few advantages since they are filled with raw sewage, have severe odor problems, and are a health hazard.
- **The available water can restore only a proportion of the marshlands:** The current amount of water delivered to northern Iraq via the Tigris is 40-44 BCM and via the Euphrates 13-14 BCM. The principal additional supply from Iran is the Karkhah, but this has now diminished as a result of dam construction, which stores about 8 BCM, and may be reduced further as a result of dike construction along the Iran-Iraq border. Restoring half of the original marsh would require roughly 25 BCM to meet the evaporation loss in southern Iraq, not accounting for the amount of water required to maintain through flow and permanent water bodies in the marshlands. That level of restoration would consume nearly half of the total available supply in the catchments and would represent an unrealistic allocation given other sector needs in the country. Clearly, there are decisions to be made by the

Government of Iraq regarding the priority locations and extent of areas to be restored.

7.3 THE IRAQ MARSHLANDS RESTORATION PROGRAM:

The program is a twelve-month effort which supports the restoration of the ecosystem through improved management and strategic reflooding and provides social and economic assistance to the local population. These two tracks – environmental and developmental – are necessarily integrated, reflecting the historic harmony of the marshlands and their indigenous population. The Iraq Marshlands Program operates at both a national and a marshlands level. At the national level, it works directly with and under its lead government entity, the Ministry of Water Resources. The following efforts are designed to support the key decision-making by the government about marshlands management and restoration, generally, and the operations of the ministry, specifically:

- *Strategic planning for CRIM* – assisting the Center for the Restoration of the Iraqi Marshlands as a newly-formed entity to determine its vision, mandate, responsibilities, programs, and organization within the context of the Ministry of Water Resources and other involved ministries
- *Hydrologic basin modeling* – enabling a partnership between the U.S. Corps of Engineers and the Ministry of Water Resources to develop the first hydrologic model of the Tigris-Euphrates basin
- *Hydraulic modeling* – focusing on one major marsh in the South to develop a profile for understanding the dynamics of reflooding and assisting in the designation of the marsh as a World Heritage site
- *Database development* – developing a database within CRIM to house water, soil, vegetation, social, and economic information about the marshlands for ongoing analysis and decision-making
- *Water and soil laboratory* – developing laboratory facilities and a capacities for soil and water analysis
- *International study tours and short courses* – building capacity in and commitment to wetland management and marsh restoration through study tours to the Everglades, Louisiana, and the Danube for officials from different ministries and from the research and university communities
- *Comprehensive donor strategy* – preparing a consensus-driven plan for long-term marsh restoration that reflects national and international views and commitments

In the marshlands, the program proposes to implement six discrete, but closely related, activities focusing on:

- *Integrated marsh management* – assessing the nature and distribution of reflooding and develop management options for future actions
- *Agricultural production and agribusiness* – diversifying agricultural production in the marshlands with higher value crops through demonstrations and nurseries

- *Livestock and dairy production* – addressing major sector constraints with veterinary services and the introduction of a new forage crop and supporting women and girls through education and incomegeneration initiatives
- *Capture fishing and fish farming* – increasing the population of high-valued fish in the marsh waters and encouraging aquaculture to increase incomes
- *Primary health care* – extending health care and related educational services to the marshlands populations, given that few clinics operate and that medical services are rarely available
- *Constructed wetlands* – building engineered wetlands to dramatically improve wastewater systems, public health, and overall water quality in the marshlands using low technology and natural processes

7.4 PROGRAM OBJECTIVES

The overall objectives of the program are the following:

- Improve the social and economic lives of marsh dwellers;
- Assess the success of wetland restoration and help guide government policies and decisions;
- Strengthen Iraqi commitment to marshland management; and
- Reach national and international consensus on long-term restoration strategies.

8. INTERNATIONAL AGREEMENTS

In 1980, a Joint Technical Committee on Regional Waters was created by Turkey and Iraq, on the basis of a former protocol (1946) concerning the control and management of the Euphrates and the Tigris. Syria joined the committee afterwards.

The Euphrates represents the most critical issue for Iraq's water strategy as more than 90% of its water comes from outside the country (against only 50% for the Tigris). According to an agreement between Syria and Iraq (1990), Iraq shares the Euphrates' waters with Syria on a 58% (Iraq) and 42% (Syria) basis, based on the flow received by Syria at its border with Turkey. Since Turkey has unilaterally promised to secure a minimum flow of 15.8 km^3/year at its border with Syria, this agreement would de facto represent 9 km^3/year for Iraq. Up to now, there has been no global agreement between the three countries concerning the Euphrates' waters.

As the marshlands are an integral part of an international river system, implementation of any future restoration initiative hinges not only on remedial actions within Iraq, but equally on the cooperation of all riparian states.

9. COMPARATIVE RISK ASSESSMENT (CRA)

Background: The United Nations Environment Programme or UNEP (2001); Partow. H. *The Mesopotamian Marshlands: Demise of an Ecosystem* recommended in the Early Warning and Assessment Technical Report, eight recommendations to mitigate the negative impacts to the Mesopotamian Marshlands and affected riparian countries.

The Iraq Marshlands Restoration Program funded by the U.S. Agency for International Development (USAID) address to varying degree the UNEP recommendations. This is a twelve-month program which supports (1) the restoration of the ecosystem through improved management and strategic reflooding and (2) provides social and economic assistance to the local population. These two tracks – environmental and developmental – are necessarily integrated, reflecting the historic harmony of the marshlands and their indigenous population.

Comparative Risk Assessment is used to select the most critical UNEP recommendations. Only those recommendations will be selected which minimize the severity of consequences of impacts to the water resources, the public, ecology, and biodiversity of the Mesopotamian Marshlands and the affected riparian countries.

Criterion: The following criterion is used to rank the comparative risks for inaction for each of the eight UNEP recommendations.

- Biological, chemical, physical or radiological assaults that has the potential to cause harm;
- An incident or situation that can lead to the presence of a biological, chemical, physical or radiological assault (what can happen and how); and
- The likelihood of identified hazards causing harm in exposed populations, ecology, and biodiversity in a specified timeframe, including the magnitude of that harm and/or the consequences.

9.1 RANKING SEVERITY AND LIKELIHOOD OF CONSEQUENCES:

Level of Severity: The severity of consequences for not addressing impacts to the riparian countries water resources are quantified as; Insignificant, Minor, Moderate, Major, or Catastrophic. There is little value in expending further effort considering very small risks and for that reason only those recommendations that reflect as Major or Catastrophic results are identified.

Level of Risk: A Comparative Risk Assessment is made of the following UNEP recommendations to determine the level of risk that would result in a significant impact at the Catastrophic and/or Major level only. As stated above, there is little value in expending further effort considering very small risks, if action is not taken to carry them out.

The likelihood and severity can be derived from technical knowledge and expertise, historical data and relevant guidelines. An example of descriptors that can be used to rate the likelihood and severity for calculation of the Risk Score is given in Table 2. A 'cut off' point must be established above which all hazards will require immediate attention. There is little value in expending further effort considering very small risks.

Table 1 – Risk Scoring Table [3]
(*The risk score for a particular hazardous events = Likelihood x Severity of Consequences.*)

	Severity of Consequences				
Likelihood	Insignificant	Minor	Moderate	Major	Catastrophic
Almost Certain	5	10	15	20	25
Likely	4	8	12	16	20
Moderate	3	6	9	12	15
Unlikely	2	4	6	8	10
Rare	1	2	3	4	5

Table 2 - likelihood and severity categories Item: Definition: Weighting:

Item:	Definition:	Weighting:
Almost certain	Once a day	5
Likely	Once per week	4
Moderate	Once per month	3
Unlikely	Once per year	2
Rare	Once every 5 years	1
Catastrophic	Potentially lethal to large population	5
Major	Potentially lethal to small population	4
Moderate	Potentially harmful to large population	3
Minor	Potentially harmful to small population	2
Insignificant	No impact or not detectable	1

IMPACTS OF INACTION:

- Reduced discharge, flow patterns, quality of water of the largest river systems draining into the Gulf negatively affects the inland fresh water and marine eco-systems.
- Potential harmful effects on regional fish resources.
- The social and economic livelihood of the Marsh Arabs collapses.
- A significant wetland ecosystem will collapse with catastrophic effect on wildlife and biodiversity
- An estimated 66 species of birds that occurred in the marshlands in internationally significant numbers are at risk
- Desiccation of the last Mesopotamian marshlands in Iraq and Iran is continuing and will be lost if nothing is done.
- Desiccation of over 9,000 km2 of wetlands and lakes would affect regional micro-climate (rising temperatures, toxic dust storms, water scarcity, and human health).

Table 3: Summary of UNEP Recommendations – Severity of Consequences for Inaction[1]

UNEP RECOMMENDATION	SEVERITY OF CONSEQUENCE[2]
1. International Agreement(s) on the Sharing of Tigris and Euphrates Waters Severity = 5, Likelihood = 4; 5 x 4 = 20	CATASTROPHIC
2. Mitigating the Impacts of Dams on Downstream Ecosystems Severity = 3, Likelihood = 3; 3 X 3 = 9	MODERATE
3. Re-establishing the Flood Regime Severity = 4, Likelihood = 4; 4 X 4 = 16	MAJOR
4. Protecting Water Quality Severity = 4, Likelihood = 5; 4 X 5 = 20	MAJOR
5. Re-evaluating the Role of River Engineering Works Severity = 3, Likelihood = 2; 3 X 2 = 6	MODERATE
6. Designation of Protected Areas Severity = 5, Likelihood = 5; 5 X 5 = 25	CATASTROPHIC
7. Assistance and Repatriation of Marsh Arab Refugees Severity = 2, Likelihood = 5; 2 X 5 = 10	MINOR
8. Data Collection, Monitoring and Long-Term Capacity Building Severity = 2, Likelihood = 3; 2 X 3 = 6	MINOR

10. Acknowledgements

The "Comparative Risk Assessment" has been developed based on baseline information from the Action Plan for the Restoration of the Marshlands of Iraq, under the Development Alternatives, Inc. contract funded by USAID (2004). The marshlands assessment team included: Peter Reiss, Social Anthropologist; Development alternatives, Inc (DAI).; Curtis J. Richardson, Professor and Director, Duke University Wetland Center; Professor Edward Maltby, Royal Holloway Institute for Environmental Research (UK), and Ali Farhan, Program Manager (DAI), Strategies for Assisting the Marsh Dwellers and Restoring the Marshes in Southern Iraq; Dr. Jane Gleason, Agriculture, DAI; Dr. Robert Fitzpatrick, Soils, Research Organization (Australia); Dr. Ghayeth Hameed Majeed, Livestock, University of Basrah; Dr. Jan Vymazal, Marsh Management, Duke Wetland Center (Czech Republic).

11. References:

1. UNEP (2001). Partow, H.: *The Mesopotamian Marshlands: Demise of an Ecosystem.* Early Warning and Assessment Technical Report, UNEP/DEWA/TR.01-3 Rev. 1: http://www.grid.unep
2. UNEP (2003). *Desk Study on the Environment in Iraq*. United Nations Environment Programme: http://www.unep.org
3. WHO. Guidelines for Drinking Water Quality, Third edition, 2003: http://www.who.
4. Dr. Zarins. *Eden Theory*: Written by Dora Jane Hamblin. www.idolphin.org/eden/

[1] See Appendix A For detail of recommendations
[2] 2 See Table 1 – Severity of Consequences

APPENDIX A

RECOMMENDATIONs	SEVERITY OF CONSEQUENCES	LIKELI-HOOD	WEIGHTED RISK FACTOR	REMARKS
1. International Agreement(s) on the Sharing of Tigris and Euphrates Waters Negotiations involving all the riparian countries and supported by international facilitation need to be reinforced and/or initiated with the aim of establishing an international agreement(s) on the sharing of Tigris and Euphrates waters.	5	4	20	This calls for setting a new threshold for long-term water sharing and integrated management of the Tigris-Euphrates drainage basin. Determination of water allocation quotas should not be solely a function of development needs, but requires an "ecosystem approach" aimed at maintaining river flows at levels that sustain all forms of life and the ecological viability of the river system, particularly that of the marshlands.
2. Mitigating the Impacts of Dams on Downstream Ecosystems Mitigate to the extent possible the cumulative impacts on ecosystems, biodiversity of the dams on the Tigris and Euphrates rivers.	3	3	9	The main option available is to use environmental flow requirements, which includes managed flood releases, to recreate a semblance of the natural distribution and period of stream flow. Another alternative for environmental restoration, which has been increasingly used by a number of countries, is to decommission large dams. For planned dams on the Tigris these include anticipating and avoiding impacts by carrying out environmental impact assessments (EIA).

RECOMMENDATIONs	SEVERITY OF CONSEQUENCES	LIKELI-HOOD	WEIGHTED RISK FACTOR	REMARKS
3. Re-establishing the Flood Regime Develop a basin-wide water management plan designed to reinstate significant large-scale flooding in the Mesopotamian delta	4	4	16	Specifically, this calls for the development of a hydrodynamic model of the Tigris-Euphrates river system and its inland delta, based on sound hydrological and ecological data. Such a model is necessary for the elaboration of re-inundation plans and for the identification of priority target areas for restoration. In the preliminary stages, pilot programs need to be undertaken in order to evaluate their ecological, economic, social feasibility, as well as avoid unforeseen negative consequences of proposed remedial measures.
4. Protecting Water Quality Improve and/or maintain the quality of waters entering the marshlands to support the flora and fauna of the marshlands.	4	5	20	Saline irrigation waters compounded with urban and industrial effluent would therefore have a major negative influence on any rehabilitation plans. Within Iraq, salt concentrations in the lower Tigris and Euphrates are considerably augmented by influxes from the Tharthar, Habbaniyah/Razaza reservoirs due to their calcareous soils and high evapo-transpiration rates. Remedial actions to control salinity levels and prevent polluted waters from entering the marshlands are therefore prerequisite steps for restoration. With large amounts of the sediment trapped behind multiple dams, the fertility of waters nourishing the marshlands has inevitability decreased. A rehabilitation program would therefore need to consider the significance of sediment retention by engineering structures and propose necessary remedial measures.

RECOMMENDATIONs	SEVERITY OF CONSEQUENCES	LIKELI-HOOD	WEIGHTED RISK FACTOR	REMARKS
5. Re-evaluating the Role of River Engineering WorksConduct a comprehensive Environmental Impact Assessment (EIA) of the drainage works, associated agricultural projects needs to be undertaken and balanced allocation of competing water demands made within Iraq.	3	2	6	Regulation of water flow through upstream dam construction has effectively eliminated the seasonal flooding cycle driving the ecological dynamics of the marshlands. In view of the fact that several of the engineering structures built in Iraq in the 1950s were originally intended for flood protection purposes, there is a logical need to review whether their role has been rendered redundant by the recent dams built in Turkey and northern Iraq. In particular, the utility of the massive off-river water storage reservoirs of the Tharthar, Habbaniyah/Razaza needs to be re-evaluated and due consideration given to the modification of their future roles.Inside the marshlands, diversion canals, embankments, sluices, locks, dikes, polders and other hydraulic works should be modified or removed as necessary to ensure adequate water release into the marshlands.

RECOMMENDATIONS	SEVERITY OF CONSEQUENCES	LIKELI-HOOD	WEIGHTED RISK FACTOR	REMARKS
6. Designation of Protected AreasInitiate immediate conservation measures by both Iran and Iraq to protect the transboundary Hawr Al Hawizeh/Al Azim.	5	5	25	As the last remaining vestige of the Mesopotamian marshlands and a very important sanctuary for endangered wildlife and migratory birds. Possible conservation options include establishment of:- A mechanism for integrated marshland management between Iran and Iraq such as a transboundary 'peace park'.- Designation as a Ramsar Convention or Word Heritage site, or as a UNESCO Man and Biosphere (MAB) or national reserve(s).
7. Assistance and Repatriation of Marsh Arab Refugees Develop feasibility studies for the repatriation of Marsh Arab refugees and reintegration within a restored marshland environment including humanitarian relief.	2	5	10	The drying out of the marshlands has displaced 350,000 to 500,000 Marsh Arabs; of whom at least 40,000 have sought refuge in neighboring Iran, while the rest are internally displaced within Iraq.
8. Data Collection, Monitoring and Long-Term Capacity Building Conduct detailed studies on the impacts of marshland desiccation on local, regional environmental conditions and wildlife.	2	3	6	A long-term monitoring program based on regular field data collection, remote sensing imagery, aerial photographic surveys should be developed and implemented. In addition, the capacities of environmental administrations in Iraq, Iran both at the national and local levels will need to be strengthened with targeted training on wetland management and restoration provided.

A FUTURE FOR THE DEAD SEA BASIN: OPTIONS FOR A MORE SUSTAINABLE WATER MANAGEMENT

Elements for Environmental Risk and Security

C. LIPCHIN
Arava Institute for Environmental Studies, Ketura, Israel
D.N. Hevel Eilot 88840, ISRAEL

Abstract

The Dead Sea basin plays a major role for regional economic development (industry, tourism and agriculture). This potential is threatened by the steady disappearance of the Dead Sea. Since around 1930 the water level of the Dead Sea has fallen by about 25 m, about half of this alone in the last 20 years. The Dead Sea is the terminal point of the Jordan River watershed. As such, it serves as a barometer for the health of the overall system. Its rapid decline reflects the present water management strategies of the riparian and upstream countries.

Elements pertaining to environmental security, whereby a sustainably managed environment provides for social, economic as well as environmental benefits are evident with regards the Dead Sea. The decline for example, undermines its potential as a tourist destination, despite the enormous investment in hotel and resort infrastructures in Israel and in Jordan. The decline also raises ethical issues about the exploitation of water resources by present generations at the expense of this natural heritage to future generations.

This paper provides a preliminary analysis of a European Union funded project whose aims are to synthesize and assess existing physical and socio-economic data and to assess options for a better future for the Dead Sea. It will identify the patterns of water supply and use in the region, and the factors that control these patterns. The underlying assumption is that solutions for a more sustainable development than today scenario will not come from simply providing "more water for more development", but from a new land and water management system that is sensitive to social, cultural and ecological resources thereby providing security and stability across sectors and nations.

As a first step, the project team has established a system model that combines the physical and social dimensions of water use. Data, information and knowledge between the human dimension (economy, sociology etc.) and the physical dimension (hydrology, ecology, agriculture, water planning) are linked under changing scenarios. The model is an attempt to reflect the complexity inherent in the system through the mapping of human and physical connections. An understanding of these connections can lead to a more secure environment for both.

B. Morel and I. Linkov (eds.),
Environmental Security and Environmental Management: The Role of Risk Assessment, 79–91.

1. Background

1.1 THE "DEAD SEA" PROJECT

The project partners, from Israel, Palestine, Jordan, Austria and the UK, have taken an interdisciplinary approach in assessing options for sustainable water management in the Dead Sea basin. This is being done by working in close collaboration with engineers, ecologists, social scientists and economists and including intense participation of the public.

The project encompasses the development of a GIS-based database that contains harmonized and comparable physical, economic and social data, including consistent sets of maps that document the spatial dimension of current and projected water supply and demand sectors, and of land-use patterns that drive water supply and demand. It also will establish realistic development scenarios until the year 2020 with social, economic, technical and ecological constraints. The project team will also try to establish criteria for essential water requirements for nature and ecosystems, and to propose socially, economically and environmentally sound alternatives for irrigated agriculture, the dominant water consumer in the region. Further details about the project can be found on the project website: http://www.deadseaproject.org.

2. The Dead Sea Basin

The Dead Sea Basin is particularly appropriate for such a study. It has a size of about 44,000 km² and its watershed is shared by Israel, Jordan and Palestine (
).

Figure 1. The Dead Sea watershed (from EXACT study: Assaf et al. 1998)

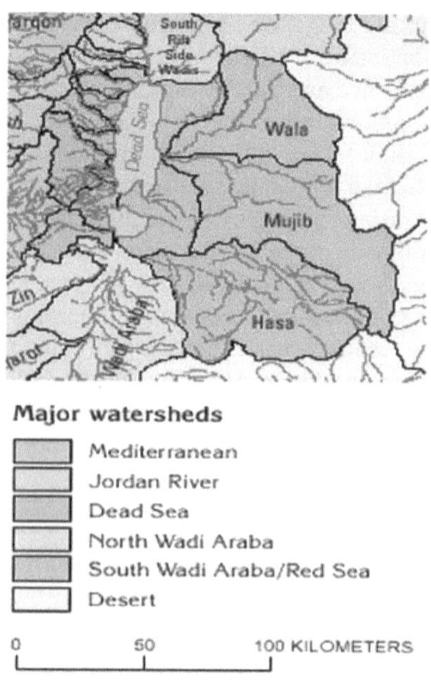

The basin plays a major role for regional economic development. Current economic activities in the basin are industrial (mineral extraction and water bottling), tourism and agriculture. The Dead Sea's mineral composition and the unique climate provide treatment for skin diseases, especially for psoriasis and atopic dermatitis (Schempp et al. 2000). The health and cultural features plus the unique landscape have made the area attractive for tourism.
Besides the regional relevance, the basin has a global

importance. Since 1998 there are efforts to promote the Dead Sea Basin as a UNESCO Man and Biosphere Reserve and a World Heritage site (Abu Faris et al. 1999) as it is a both a unique habitat for wildlife (particularly important around springs (e.g. Ein Fashkha and Ein Gedi) and wadis (e.g. Wadi Mujib) and a global cultural heritage site with some of the world's oldest continuous human settlements (e.g. the city of Jericho).

2.1 .PHYSICAL FEATURES

The Dead Sea is the terminal lake of the Jordan Rift Valley. Its surface is currently about 417 m below sea level which makes it the lowest point on earth. With a salinity of about 3,000 mg/l it is also the most saline water body in the world (Gertmann 1999). Rainfall is limited to winter months; it varies from about 500 mm/yr in the north-western highlands to less than 100 mm/yr in the valley floor (Isaac et al. 2000). Perennial storage in surface and underground water reservoirs is limited and vulnerable to pollution and depletion. Potential evapotranspiration in the valley floor is about 2,000 mm/yr, and actual evaporation from the Dead Sea surface is about 1,300-1,600 mm/yr (Stanhill 1984). The temperature is about 40°C in summer and 15°C in winter (Assaf et al. 1998). At the east and west there are steep escarpments, while in the north and south, the valley stretches gently upward along the Jordan River and along the Wadi Araba, respectively.

The historical Dead Sea consisted of two basins: the deep northern basin (which is now the only remaining Dead Sea proper), and the shallow southern basin from which the Dead Sea has retreated since 1978. The two basins are divided by the Lisan Peninsula.

2.2 PRESSURES ON ECOSYSTEMS

The land cover is mostly open with little vegetation. Sensitive areas include the Lisan peninsula, marshlands and wetlands at the northern and southern ends of the Dead Sea, the Wadi Mujib, the Ein Gedi oasis, and the Dead Sea itself (Fariz 2002). Lack of natural freshwater, expansion of human settlements, and inappropriate land use has affected these areas.

Waste waters from local domestic, agricultural, industrial and tourist activities flow directly into the Dead Sea. Raw sewage flows into the Dead Sea from Jerusalem-Bethlehem urban areas via the Wadi Nar (Kidron valley). Water shortage and land degradation are a problem all over the basin and these are likely to exacerbate with population growth (Rishmawi and Hrimat 1999).

2.3 THE DECLINING DEAD SEA

The most visible and most disturbing degradation is the decline of the Dead Sea water level and volume. Since around 1930 the water level of the Dead Sea has fallen by about 25 m, about half of this alone in the last 20 years (Anati and Shasha 1989; Assaf et al. 1998). In the past few years the rate of decline was 80-100 cm per year. The last available data from mid-2003 indicate a water level of -417 m (Figure 2). As a result of this decline, in the last 20 years the Dead Sea surface area has shrunk by about 30 %, and its north-south extent has shrunk from over 75 to 55 km (Anati and Shasha 1989). Since 1978, the Dead Sea has completely retreated from the southern basin, which presently consists only of artificial evaporation ponds.

Figure 2. Decline of Dead Sea water level 1976-2003 (Data from IL Hydrological Service)

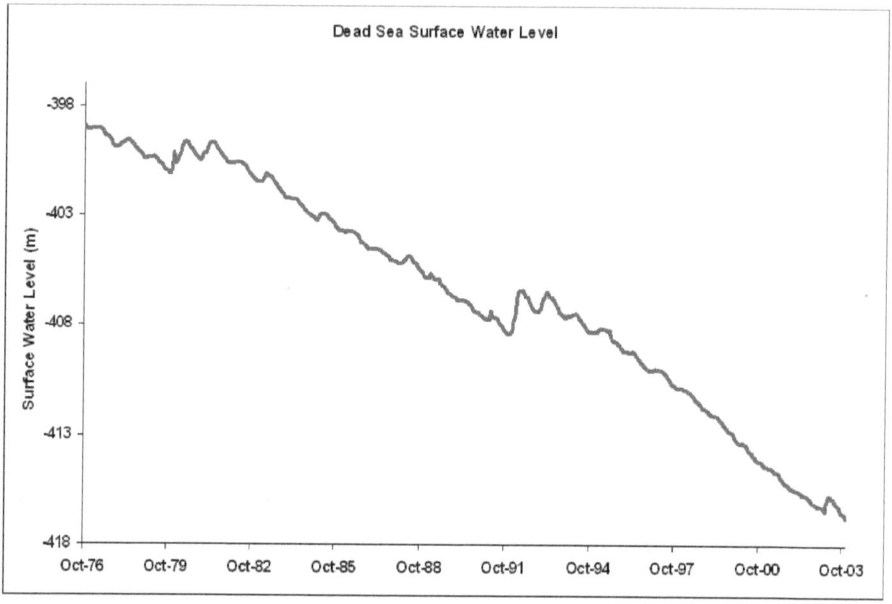

The reasons for this decline are well-known:

- First and foremost, the decline is a direct consequence of the declining freshwater input: this includes decreasing discharge from the River Jordan, increasing water use from natural springs and side wadis, and extensive use of aquifers that provide secondary water input (Klein 1985). Of all these factors, the River Jordan probably plays the biggest role. Insofar the Dead Sea's steady disappearance is a direct result of the water management strategies of the River Jordan riparians. While 100 years ago the River Jordan's discharge into the Dead Sea was about 1200-1300 million cubic meters per year (MCM/yr) of freshwater, it has been reduced to about 900 MCM/yr by the 1940's and now is not more than 100-200 MCM/yr of saline and polluted water (Hillel 1994; Rabi 1997; Al-Weshah 2000; Orthofer et al. 2001; Shavit et al. 2001). The main reason for this decline is that water from the Upper Jordan River as well as water from the Lower Jordan River tributaries (e.g. Yarmouk, Zarqa) has been blocked and diverted for urban and agricultural uses inside and outside the watershed.

- On top of the reduced freshwater input, more than 200 MCM/yr of water are pumped out of the Dead Sea into evaporation ponds in the shallow southern basin. It is estimated that the salt industries contribute 25 to 30 % of the present total evaporation rates (Wardam 2000).

It is not clear whether the Dead Sea water level has now come to equilibrium between the reduced surface and a reduced evaporation, or if it will continue to decline.

3.Elements of Environmental Security

3.1 CONSEQUENCES OF THE DECLINE

As a result of the lowering of the water level, the adjacent aquifers are seriously affected (Yechieli 1996). Sinkholes have opened up along the shoreline, caused by lowered water tables and groundwater over-exploitation (Bowman et al. 2000; Baer et al. 2002). These sinkholes are a serious threat to infrastructure development as they can cause, without warning, the collapse of roads, bridges and buildings. They also disrupt agriculture by making it unsafe to work with heavy machinery required for date harvesting. A number of date orchards have been shut down due to the threat of sinkholes.

Furthermore, the decline of the Dead Sea also affects the freshwater springs on its shores (e.g. Ein Fashkha, Ein Turiba) that support a unique biodiversity. The decline of the water level, in conjunction with the outbreak of the Intifada in September 2000, has also already had a serious effect on tourism. In Israel since 2000, international tourism to the Dead Sea has declined whereas domestic tourism in response to lower hotel and occupancy rates has flourished (Figure 3).

The current trend has a disastrous effect on the future situation. The growing population in all three countries will increase the pressure for the freshwater that currently remains unused. The possible re-settlement of returning Palestinian refugees will also increase water demand in Palestine. Palestinians demand as part of a regional water agreement that more water should be allowed for the Lower Jordan River and that this additional water should be usable for the Palestinian population. This, of course, means less for the Dead Sea.

In the next few years, there are plans for increasing tourism and industrial development of the area including the construction of over 50,000 new hotel rooms (Meunier 1999). In all three countries, development policies have disregarded impacts on the environment, indigenous people and small farmers. Essential water needs for nature were neglected; policies lacked incentives to promote local forms of environmental security and equitable access to natural goods and services. Water is increasingly allocated to the urban sector and to large-scale agriculture at the expense of the needs and rights of the rural and indigenous people. Consequently, the rural poor and indigenous communities are overexploiting land resources to sustain their livelihoods.

The declining Dead Sea undermines its potential as a tourist destination, despite the enormous investment in hotel and resort infrastructures in Israel and in

Jordan. For the fledgling Palestinian economy, the present state of the Dead Sea suggests that it may never have the opportunity to develop what should have been one of its more attractive tourist locations that could provide critical employment to a growing workforce.

Figure 3. International and domestic tourism occupancy rates at Dead Sea hotels in Israel (Data from the Israeli Hotel Association).

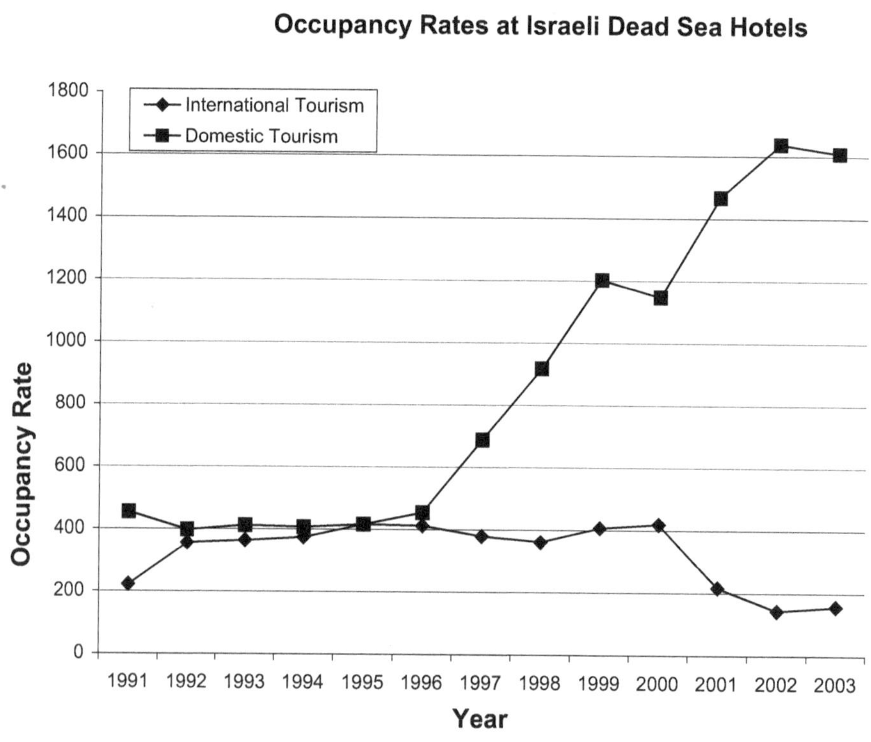

Furthermore, the decline of the Dead Sea raises ethical issues regarding the exploitation by present generations of water resources at the expense of the natural heritage for future generations. Many would argue that it represents an intolerable violation of the rights of future generations.

3.2 NEW WATER FROM THE PROPOSED RED-DEAD CANAL

There is concern in the region about the threat of a disappearing Dead Sea (EcoPeace 1998; Coussin 2001), but very little progress. Most options for solving the environmental and economical problems focus on the provision of "new water from outside", particularly through a canal that would connect the Red Sea and the Dead Sea ("Red-Dead Canal"). This 240 km conduit is expected to replenish the missing inflow, use the gravity pressure for desalination through reverse osmosis, and for production of electricity. Costs are estimated to be around 3 billion dollars (Pearce 1995). Among the questions which remain unclear are the environmental impacts of the canal, e.g. the chemical changes that will occur due to the mixing of the waters from the two seas.

This project is yet another example of engineering hegemony rife in the region. Yet, ultimately it is not a solution as it ignores the social and environmental impacts that such a project may cause and rather allows for the status quo to continue whereby water is diverted from the upper Jordan basin by both Israel and Jordan. If true environmental security is to be achieved then all elements need to be taken into consideration i.e.: social, environmental and economical concerns must be incorporated.

3.3 SCENARIO DEVELOPMENT AS A MEANS OF ASSESSING RISK AND SECURITY

The project will develop realistic development scenarios for the Dead Sea basin for the period 2000-2020. These scenarios will consider social, economic, technical and ecological uncertainties and constraints. It was particularly important to develop socio-economic indicators to describe the social dimension of development as this is often not taken into account by planners and policy makers. Activities have focused on establishing appropriate methods and tools for scenarios. First results have shown that it is necessary to define a limited number of development indicators.

These have been defined according to three driving forces impacting the region. These are: regional cooperation, investments, and the role of agriculture (Table 1). These three driving forces will be categorized into two levels each (high/low), so that in the end there would be eight scenario options. These scenarios options will be discussed with the regional development authorities and communities. One of the final scenarios will reflect "current/no major changes" situation, i.e. with low level of cooperation and investments, and with a high role of agricultural activities. At least two of the final scenarios will be adjusted to a situation that will reflect a "more sustainable than today" water management.

These driving forces can then be used to map out potential risks and securities for the Dead Sea basin. For example, if regional cooperation improves in the near future a likely result is an increase in development in the tourism and agricultural sectors. The former due to a more stable political climate and the latter due to a return of Palestinian refugees to Palestine. This bodes well for the security of the region but it may place an added burden on local water resources thus increasing the risk to the Dead Sea itself. From this simple example one can see the inherent complexity in such scenarios.

Table 1. Driving forces and parameters for scenarios (Source: Orthofer et. al., 2004).

Driving Forces	Parameters
A. Level of Cooperation	• Land administered by PS • Population - IL population & return of PS refugees • Volume of trade • GDP and HDI • Availability & price of desalinated water, treated waste-water, Water imported from outside • Water allocated to tourism sector, local labour force in tourism, number of tourists coming into the area
B. Level of Investment	• New dams • Rainwater harvesting • Efficiency of municipal & irrigation networks • Water allocated to the domestic & tourism sector • Volume/costs of wastewater & reused wastewater for agriculture, tourism (landscaping) and industry • Availability & costs of water from Red-Dead Canal
C. Role of Agriculture	• Subsidies for agriculture (water & general) • Income per unit of water used in agriculture • Investment in agricultural sector • Available labour force for agriculture • Protected area

Figure 4. Interrelation of system analysis components (Source: Orthofer et. al., 2004). . Note: The physical system determines the availability of water in different quality categories, the socio-economic system deals with factors that determine the water wishes for different uses, the water governance system analyzes why some social groups have their wishes fulfilled and others don't.

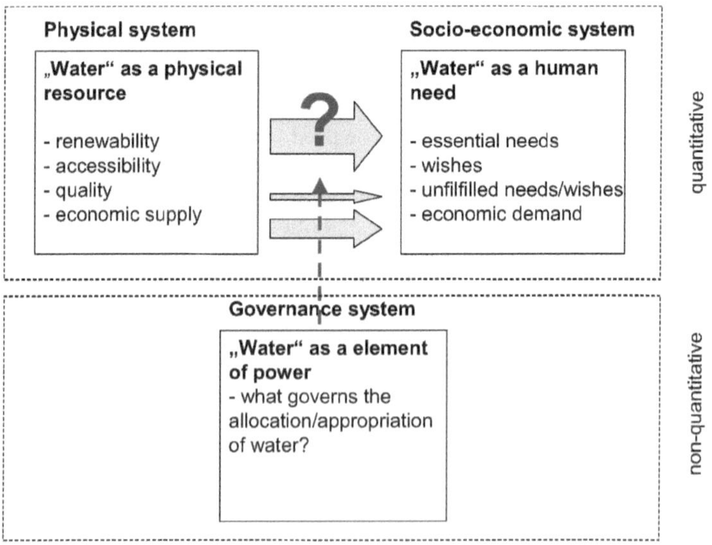

3.4 SYSTEM ANALYSIS AS A TOOL FOR ASSESSING RISK AND SECURITY

The purpose of the system analysis is to understand the complexity of interconnections of the water management system and its driving forces. This allows one to identify options for system changes. The system analysis has three elements (Figure 4):

- Physical system:
  This focuses on the physical dimension of water supply and use and the consequences on the environment (particularly nature, land/soil, and groundwater). The analysis also addresses the regional dimensions of the supply-use chain, and the exchange of water between the Dead Sea basin and the region outside the basin. In the end, the physical system analysis identifies the factors that determine the availability of four different water quality categories that are available. Water of low quality can be used for irrigation whereas water of higher quality is used for domestic purposes.
- Socio-economic system:
  This includes the analysis of social and economic issues of water wishes and water use. A wish for water is how much water a person/sector wants over and above what is necessary. Water for basic human services is defined as a need for water. Once this need is met, any further want for water is a wish for water. These wishes for water may be fulfilled or unfulfilled depending on the capacity of the person or institution to fulfil them. Water wishes can be satisfied ("fulfilled") or not ("unfulfilled water wishes") through the economic system ("demand") or through informal systems. Issues that influence wishes for water include gender powers, traditions, health, perceptions, rights, equity, and the role of communities, employment, benefits and income generation/distribution. The socio-economic system analysis thus identifies the factors that determine the water wishes for different uses (agricultural, municipal, domestic, tourism, industrial) including water for nature.
- Governance system:
  The analysis includes policies on national and regional levels, including institutional aspects of water governance and driving forces for policy changes. Other issues that are included are: traditional water rights, water policies in Israel, Jordan and Palestine, role and power of stakeholders, international dimension, and conflicts of interest. In the end, the water governance analysis will analyze the factors that determine why some social groups have their wishes fulfilled and others do not.

Figure 4 describes how these three systems are connected and how they influence one another. In order to minimize risk and maximize security one cannot look at these three elements separately. Consequently an integrated system perspective is crucial in understanding risk and security. The question mark in figure 4 indicates the inherent uncertainty that exists in any system. One can never completely remove uncertainty but a system analysis allows one to identify its source and work towards its reduction.

Within the "Systems Thinking" approach, a balanced system needs to have a negative feedback loop. This is the case with limited local water supply because in the long run more water availability will again result in more water wishes (Figure 5).

Figure 5. Aggregated causal-loop diagram for water use.
Note: In the local system (solid lines) the positive feedback-cycle (in black: more „Fulfilled water wishes" result in more „Economic benefit" and more „Water wishes") is counterbalanced through negative feedback loop (in gray: more "Fulfilled water wishes" result in less "Local water availability"). The availability of additional water from outside (dashed lines) bypasses the negative feedback and creates a new positive loop: In the long run, more water supply leads to more water wishes i.e.: the system is determined by pressure points from within and outside the system.

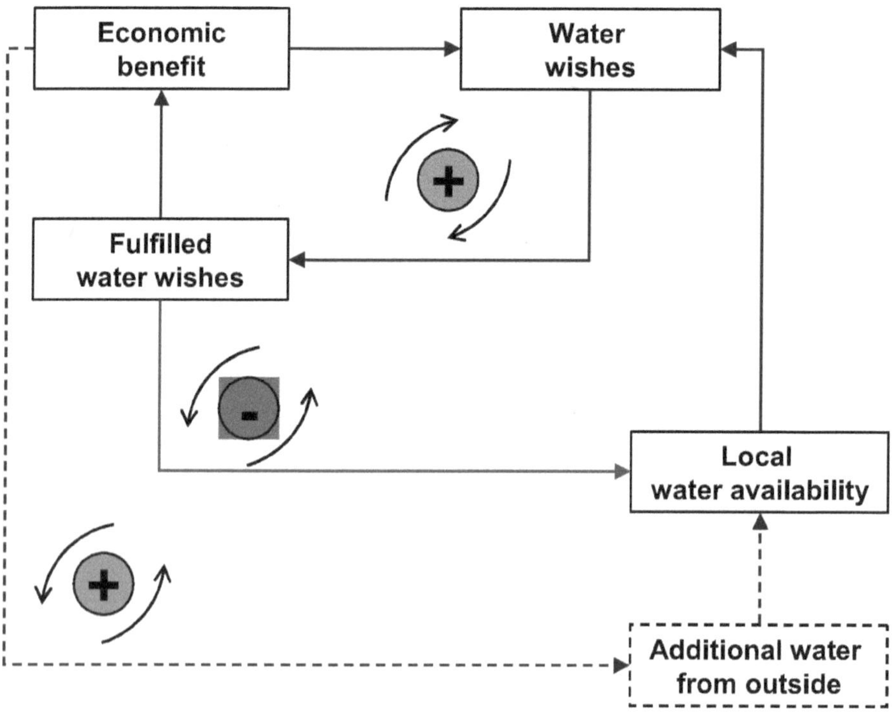

4.Conclusions

Future regional stability will depend on economic development, and the Dead Sea region will be able to make an important contribution. This will require cooperation among Israel, Jordan and Palestine. The region has a high potential for economic development, particularly for tourism. Yet water is one of the most important limiting factors in this regard (Nowitz 1980).

Given the fact that nature has made the Dead Sea the terminal sink of all freshwater sources in the Jordan River basin any withdrawal of water in the basin could be considered non-sustainable in terms of ecological sustainability.
A "more sustainable than today" water management should be possible. The "Dead Sea" project will contribute to provide an assessment of strategic options for such a more sustainable than today water management. An understanding of risk and security, and a means for quantifying them, are vital elements for the success of the project. Nonetheless, many questions remain unanswered:

- What is the carrying capacity of the system and its environmental resources?
- What is the impact of land use changes on the hydrological regime of the region? How can land use be optimized for a more sustainable water usage system?
- What are the competing interests and what are underlying factors for them?
- What are the essential water and land needs of nature that are required to preserve key processes?
- Are there possibilities to (partly) restore the natural inflow into the Dead Sea through a change in water management in Israel and Jordan?
- Do we have economic and technology development alternatives?
- How can the tourist sector, industrialization processes and modernization of agriculture be developed without threatening the quality of the environment of people's livelihoods and well-being?
- How can sustainable development plans provide incentives to promote local forms of environmental security and equitable access to goods and services?

In finding answers to these questions water management in the Dead Sea basin and elsewhere must be based on systemic solutions such as allocation priorities for different water qualities plus changes in water usage patterns. Solutions for sustainable development will not come from simply providing "more water for more development". Sustainable development will have to be sensitive to social, cultural and ecological resources as well.

5. Acknowledgements

The author wishes to thank Rudi Orthofer for his contributions to this paper. In addition, the author wishes to thank Khaldoun Rishmawi, Adi Afaneh, Raed Daoud, David Lehrer, Julie Trottier and Ernst Gebetsroither. Funding for the Dead Sea project is provided by the Fifth Framework Program of the European Commission.

6. References

1. Abu Faris, H., G. Bromberg, G. Fariz, S. Hoermann and M. Turner (1999). Moving Towards a Dead Sea Basin Biosphere Reserve and World Heritage Listings. Friends of the Earth Middle East, Amman, Jordan. Nov 1999. http://www.foeme.org/docs/dsbio.htm.

2. Al-Weshah, R. A. (2000). The Water Balance of the Dead Sea: An Integrated Approach. Hydrol. Process. **14:** 145-154.
3. Anati, D. A. and S. Shasha (1989). Dead Sea Surface-Level Changes. Israel Journal of Earth Sciences IJERAK **38**(1): 29-38.
4. Assaf, K. K., M. Ben-Zvi, J. S. Clarke, H. El-Naser, S. Kesslerm, M. N. Landers, M. F. Nuseibeh and C. J. Wipperfurth (1998). Overview of Middle East Water Resources. Water Resources of Palestinian, Jordanian, and Israeli Interest. Compiled by USGS for the EXACT Middle East Water Data Banks Project. US Geological Service, ISBN 0-607-91785-7. http://water.usgs.gov/exact/overview/index.htm.
5. Baer, G., U. Schattner, D. Wachs, D. Sandwell, S. Wdowinski and S. Frydman (2002). The lowest place on Earth is subsiding - An InSAR interferometric synthetic aperture radar) perspective. Geological Society of America Bulletin **114**(1): p. 12-23.
6. Bowman, D., D. Banet-Davidovich, H. J. Bruins and J. Van der Plicht (2000). Dead Sea shoreline facies with seismically induced soft-sediment deformation structures. Israel Journal of Earth Sciences **49**(4/2000): pp. 197-214.
7. Coussin, O. (2001). Wild and stunning and shrinking fast. The Dead Sea region is being undermined by industry and tourism. But is anybody paying attention? Ha'aretz English Edition. Tel Aviv, Israel. Thursday, 1 Feb 2001.
8. EcoPeace (1998). Final Symposium Report. Symposium on Promoting an Integrated SustEinable Regional Development Plan for the Dead Sea Basin, Amman, July 1998, EcoPeace - Middle East Environmental NGO Forum.
9. Fariz, G. H. (2002). Ecology of the Dead Sea Biosphere Reserve, Friends of the Earth Midddle East. **2002**.
10. Gebetsroither, E., A. Afaneh, C. Lipchin and D. Lehrer (2004). An analysis of the feedback structure driving water dynamics in the Dead Sea region. Presentation at MEDAQUA'04 INCO-MED Water Conference, 14-15 Jun 2004, Amman, Jordan.
11. Gertmann, I. (1999). The Hydrography of the Dead Sea 1999-2000. National Institute of Oceanography, Israel. Oceanographic and Limnological Research (IOLR), Haifa.
12. Hillel, D. (1994). Rivers of Eden. New York, NY - Oxford, UK, Oxford University Press: 355 pp.
13. Isaac, J., N. Hrimat, K. Rishmawi, S. Saad, M. Abu Kubea', J. Hilal, M. Owawi, G. Sababa, M. Awad, F. Ishaq and I. Zboun (2000). An Atlas of Palestine (The West Bank and Gaza). Applied Research Institute Jerusalem (ARIJ), Bethlehem.
14. Klein, C. (1985). Fluctuations of the level of the Dead Sea and climatic fluctuations in the country during historical times. Symposium: Scientific Basis for Water Resources Management, September, 1985,, Jerusalem, Israel, International Association of Hydrological Sciences. : http://water.usgs.gov/exact/overview/p4144.htm.
15. Meunier, J. (1999). Keeping the Dead Sea Alive: Toward a Regional Environmental Management Plan, International Development Research Center (IDRC) Project number 97-0217-01. September 24, 1999. http://www.idrc.ca/reports/read_article_english.cfm?article_num=519.
16. Nowitz, R. (1980). The Dead Sea as a Centre of Development. Kidma, Israel J. of Development **19**(5 (3)): 38-38.
17. Orthofer, R., R. Daoud, B. Fattal, M. Ghanayem, J. Isaac, H. Kupfersberger, A. Safar, E. Salameh, H. Shuval and S. Wollman (2001). Developing SustEinable Water Managemenent in the Jordan Valley. European Commission, DG Research, Final Project Report Project ERBIC18CT970161. August 2001.
18. Orthofer, R., C. Lipchin, J. Isaac and R., Daoud. (2004) The Dead Sea Project. Presentation at MEDAQUA'04 INCO-MED Water Conference, 14-15 Jun 2004, Amman, Jordan.
19. Pearce, F. (1995). Raising the Dead Sea. New Scientist: p 33-37.
20. Rabi, A. (1997). Ecological resources of the Dead Sea basin and their sustEinable use: problems and cooperation between the countries of the basin. International Conference on Transboundary Protected Areas as a Vehicle for International Co-operation, Somerset West, South Africa, 16-18 Sep 1997, UNEP World Conservation Monitoring Centre. http://www.wcmc.org.uk/protected_areas/transboundary/somersetwest/somersetwest-57.html.
21. Rishmawi, K. and N. Hrimat (1999). The use of high resolution satellite images to monitor the changes in the rates and directions of desertification in the southeastern part of the West Bank. The Second Palestinian Geographic Conference, Jerusalem.
22. Rishmawi, K., J. Isaac, D. Lehrer and A. Afaneh (2004). Possible socio-economic and environmental futures for the Dead Sea Basin. Presentation at MEDAQUA'04 INCO-MED Water Conference, 14-15 Jun 2004, Amman, Jordan.

23. Schempp, C., H. Dittmar, D. Hummler, B. Simon-Haarhaus, J. Schulte-Montig, E. Schopf and J. Simon (2000). Magnesium ions inhibit the antigen presenting function of huma epidermal langerhans calls in vivo and in vitro. Involvement of ATPase, HLA-DR, B7 molecules, and cytokines. Journal of Investigative Dermatology **115**(4).
24. Shavit, U., R. Holtzmann, S. Segal, A. Venghosh, E. Faber, I. Gavrieli and T. Bullen (2001). Water Sources and Quality Along the Lower Jordan Valley, Regional Study. Water Resources Quality: Preserving the Quality of our Water Resources., H. Rubin et al. (eds.), Springer. Berlin Heidelberg New York: 127-148.
25. Stanhill, G. (1984). Evaporation From the Dead Sea. A summary of Research till Sept. 1984. Mediterranean Dead Sea Project.
26. Trottier, J. (2004). Realistic Water Management Policy Recommendations in the Dead Sea Basin. Presentation at MEDAQUA'04 INCO-MED Water Conference, 14-15 Jun 2004, Amman, Jordan.
27. Wardam, B. (2000). Lowest Point on Earth is Getting Lower. Jordan Times, 22 Mar 2000. Amman, Jordan. March 22, 2000.
28. Yechieli, Y. (1996). Influence of the Changes in Dead Sea Level on the Groundwaters Around. Israel Assoc. Water Resources - Dead Sea Valley and Arava Meeting.

ARE STANDARD RISK ACCEPTABILITY CRITERIA APPLICABLE TO CRITICAL INFRASTRUCTURE BASED ON ENVIRONMENTAL SECURITY NEEDS?[1]

D.A. BELLUCK
Senior Transportation Toxicologist, FHWA/USDOT
400 7th Street, S.W. Washington, D.C. 20590 USA

R.N. HULL
Cantox Environmental Inc., 1900 Minnesota Court, Suite 130
Mississauga, Ontario L5N 3C9 CANADA

S.L. BENJAMIN
National Wildlife Biologist, Farm Service Agency, U.S. Department of Agriculture, 1400 Independence Avenue, S.W., Washington, D.C. 20501 USA

J. ALCORN
SAIC, 11251 Roger Bacon Dr., Reston, VA 20190 USA

I. LINKOV
Cambridge Environmental Inc.,
58 Charles Street, Cambridge, MA 02141 USA

Abstract

Risk levels calculated using accepted human health or ecological risk assessment paradigms are often compared to numerical acceptable or unacceptable risk levels found in statutes, administrative rules, guidelines or policies. In practice, the numerical results of systematic, rigorous, and transparent risk analyses are used as inputs into a risk management process that does not have the same performance attributes of the risk assessment process. The risk management process often transforms the definition of acceptable or unacceptable risk in a non-transparent manner resulting in inefficient multi-criteria decision-making and public confusion as to what constitutes acceptable or unacceptable risk. This paper discusses an approach to make such decision-making explicit and suggests that an acceptable/unacceptable risk threshold might be considered for each class of infrastructure based on its critical nature. This paper initiates a policy level analysis and discussion of this issue by presenting arguments to

[1] The work presented by the authors was performed in their private capacities and does not reflect the policies or views of their parent institutions.

B. Morel and I. Linkov (eds.),
Environmental Security and Environmental Management: The Role of Risk Assessment, 93–102.

support or refute the contention that critical infrastructure should be held to an explicit and less stringent acceptable/unacceptable risk standard. Strong arguments exist for both sides of the issue. It is the responsibility of risk managers to determine whether arguments for or against flexibility in establishing acceptable or unacceptable risk levels are sufficiently compelling for use in their jurisdiction.

1. Introduction

Facility risk assessments typically evaluate potential human health or ecological risks or hazards resulting from chemical releases to air, water and soil and subsequent direct or indirect exposures by human, animal or plant receptors. Using U.S. EPA's or similar government risk assessment guidance or guideline documents, risk assessors identify chemical releases by media, evaluate their environmental fate and transport, receptor exposure, and potential hazards or risks. Calculated hazards or risks are compared to acceptable hazard or risk levels (single values or ranges). If calculated risk exceeds a numerical limit then risk managers must determine if this exceedence represents unacceptable risk. Risk managers use risk management principles to make such unacceptable risk determinations and, thereby incorporate non-risk factors into their decision-making.

It might be expected that acceptable risk levels would be more flexible for critical infrastructure based on national security, economic security, or other vital security needs. In contrast, facilities that are not vital, but produce discretionary goods or services, might be held to a more stringent standard. In practice, these important distinctions are not a part of the acceptable hazard or risk definitions used in human health or ecological risk assessments. In most cases, risk assessment-based acceptable risks and hazards are not separated into those for "critical infrastructure" versus those for "non-critical infrastructure." In addition, information about facility location, chemical and physical processes and activities, location-specific ecological and human population distributions and other national security information may not be made publicly available resulting in risk assessments that may not reasonably represent facility-related hazards or risks.

This paper discusses government evaluation of human health and ecological risks with relation to the determination of acceptable or unacceptable risk thresholds. It poses the following two questions:

1. Is it reasonable to establish or use acceptable/unacceptable risk thresholds based on the critical nature of a particular type of infrastructure? If yes, then
2. Is it reasonable to establish explicit acceptable/unacceptable risk thresholds based on the type of critical infrastructure prior to performing the risk assessment (i.e. explicit tools) rather than after completion of the risk assessment during the risk management decision-making process (implicit tools)?

The authors of this paper will take no position on this issue, as this is a conceptual treatise aimed at stimulating further discussion.

2. Lack of Explicit Use of Critical Nature of Infrastructure in Determining Acceptable Risk Levels

The authors of this paper have been unable to identify any existing risk assessment/management guidelines that explicitly provide a sliding scale of acceptable and unacceptable risk for non-critical facilities (e.g. production of widgets) or critical facilities (e.g. production of structures vital to the national economy) that are created as explicit risk management tools prior to starting a risk assessment. Rather, implicit risk management tools are used or created after completion of a risk assessment report as part of the risk management process.

Determining whether a non-critical facility should be judged differently from a critical facility is left to risk management, a decidedly values-based decision-making process which may lack the transparency and rigor of risk assessment and often does. Several important questions arise from this discussion, including:

1. Should different acceptable/unacceptable risk thresholds be established for non-critical and critical facilities, activities or processes?
2. Is it more equitable for a society to explicitly state its environmental security values in terms of acceptable risks for non-critical or critical infrastructure than to negotiate them as part of the site-specific risk management decision-making process following each risk assessment?
3. If it is determined to explicitly establish differential risk levels for non-critical and critical infrastructure, should acceptable/unacceptable risk thresholds be codified for non-critical and critical facilities, activities or processes?
4. If different acceptable/unacceptable risk thresholds for non-critical and critical facilities, activities or processes are not provided or codified in the risk assessment part of the process, should subsequent risk management decisions that implicitly use different risk thresholds for non-critical and critical facilities, activities or processes be formalized in a deliberative, systematic, and transparent manner?

While it is impossible for a short paper of this type to definitively respond to these questions, the basis for thinking about such answers will be provided to assist in future discussions of this emerging issue.

3. Defining Risk Levels for Critical Infrastructure: Pros and Cons

3.1 ARGUMENTS AGAINST DEVELOPING AN EXPLICIT DEFINITION OF ACCEPTABLE RISK LEVELS THAT ARE LESS STRINGENT FOR CRITICAL INFRASTRUCTURE

<u>Mitigation Options</u>

Currently established acceptable risk levels for all infrastructure development and operations are the result of decades of experience, statutory implementation, and litigation. There is no compelling legal reason to reduce human health and environmental standards based on a given location or technology selection decision. It has been repeatedly proven that there are numerous risk acceptable alternatives to any given infrastructure proposal. Thus, the need to locate and construct critical infrastructure in such a manner that human health or environmental criteria are compromised is a function of incomplete planning, poor facility location, and a failure to use or create engineering methods that would provide for the simultaneous construction or maintenance of a critical facility and achievement of human health and environmental standards or goals. For example, facilities can be constructed with: adequate pollution control equipment to reduce risk below established acceptable risk criteria; above-ground structures that do not damage sensitive habitat or block migration routes; and sufficient distance from residences to reduce unacceptable risks. The current risk management process allows for citizen input into a dynamic process and follows established statutes, laws, rules and regulations. Disturbance of this process could negate long-standing mechanisms for citizen participation and statutory objectives.

The current acceptable risk process allows for flexibility in the application of risk findings. Were a "bright line" approach using less stringent risk criteria to be adopted to accommodate critical infrastructure, it might not reflect risk uncertainty, population variation in susceptibility, community preferences and values and economic considerations that are part of the current non-bright line risk management process.

It must be recognized that options to mitigate potential risks and impacts come at a price. That price could be reflected in numerous was including money spent (e.g., building a road around a sensitive habitat) lost time (e.g., increased travel distance) and other factors where the accepted solution results in actions that are less than technically optimal. Trade-offs between technically optimal solutions and values-based options are part of the risk management determination process that may or may not be done in a systematic and transparent manner.

Serious Impacts and Sensitive Receptors

It could be argued that, where effects are likely to be serious and/or irreversible, a lesser level of proof might be sufficient to justify actions to remove or reduce risk drivers. For example, for risk assessments with limited amounts of field collected data for known human carcinogens, it might be argued that the potential serious or irreversible effects posed by a facility, activity or process should lead towards a very conservative risk management process, including the use of more stringent, rather than less stringent, acceptable risk measures. Thus, if moving in any direction, acceptable risk criteria should be made more stringent for certain classes of chemicals and their attendant infrastructure. In other words, a weak knowledge base must lead to more stringent standards on precautionary grounds (Keiding and Budtz-Jorgensen, 2004). Similarly, if a facility, activity or process severely impacts a large proportion of a habitat, or is predicted to adversely affect a significant proportion of a population or particularly sensitive species, more stringent risk levels may be warranted, particularly if these species/habitats were not able to adapt, persist, or recover on their own or without some assistance (Hobbs and Kristjanson, 2003).

It is considered reasonable to be more protective of sensitive habitats than of those that are less sensitive. Sensitive habitats intuitively need more protection. These critical habitats deserve to be protected at levels greater than those that are common habitats. U.S. EPA (1997) notes the following partial list of sensitive environments that could require protection or special consideration:

- Marine Sanctuaries
- National Parks
- National Preserves
- Designated Federal Wilderness Areas
- State-designated Natural Areas
- Federal land designated for protection of natural ecosystems
- Areas identified under the Coastal Zone Management Act
- Coastal Barriers
- Sensitive areas identified under the National Estuary Program or Near Coastal Waters Program
- Critical areas identified under the Clean Lakes Program
- National Monuments
- National Seashore Recreational Areas
- National Lakeshore Recreational Areas
- Habitat known to be used by Federal designated or proposed endangered or threatened species
- National or State Wildlife Refuges
- Spawning areas critical for the maintenance of fish/shellfish species within river, lake or coastal tidal waters
- Migratory pathways and feeding areas critical for maintenance of anadromous fish species within a river
- Reaches or areas in lakes or coastal tidal waters in which the fish spend extended periods of time
- State-designated areas for protection or maintenance of aquatic life
- Terrestrial areas utilized for breeding by large or dense aggregations of animals
- State land designated for wildlife or game management
- National river reaches designated as Recreational habitat known to be used by state designated endangered or threatened species
- Federally-designated or State-designated Scenic or Wild Rivers
- Particular areas, relatively small in size, important to maintenance of unique biotic communities
- Wetlands

Given the number and type of sensitive habitats and the need to protect them in an ethical manner, acceptable risk levels should be maintained at levels of high protection. This would be true for programs at various levels of government that protect unique habitats not covered by national programs.

3.2. ARGUMENTS FOR DEVELOPMENT OF EXPLICIT ACCEPTABLE RISK LEVELS THAT ARE LESS STRINGENT FOR CRITICAL INFRASTRUCTURE

Toxicological and Risk Uncertainty

It is intuitively obvious that not all calculated human health and environmental risks are equally probable. For example, there may be two different facilities, each releasing chemicals to the air with calculated inhalation risks greater than 10^{-4}. However, one facility releases only "probable" human carcinogens and the other "known" human carcinogens. Should the acceptable risk levels for these facilities be exactly the same?

One could argue that the acceptable risk levels for the probable human carcinogens should be much less stringent than those for known human carcinogens because the probable human carcinogens may not be human carcinogens and the known human carcinogens may only be carcinogenic at certain doses or via a particular exposure route-(e.g. for a chemical found to be carcinogenic in occupational settings at high levels, an argument could be made that there is no evidence that it will cause cancer in humans at relatively low level background exposure levels). Given the uncertainties associated with risk levels derived using the current risk assessment process, from a risk management policy standpoint, there is a critical need to develop explicit definitions of acceptable risk that account for uncertainty in methods, toxicology and results. These should take into account the need for critical infrastructure and recognize the uncertainties associated with risk modeling and the human and ecological risk assessment process and products. This is particularly important to consider because risk numbers in the eyes of many are "bright lines" that delineate unsafe from safe, not as a potential "red flag" for risk managers who must weigh these risk findings with other risk management inputs and options. Therefore, identifying alternate acceptable risk levels, according to the level of uncertainty associated with the predicted risk (e.g., a less restrictive acceptable risk level for a probable versus a known human carcinogen) would compensate for the common tendency to treat risk findings as real values predicting actual health outcomes. Another way to achieve this same outcome would be to have risk management flexibility linked to systematic, comprehensive and transparent risk management decision documents that explicitly link each risk management decision to a particular set of logical arguments and proofs to support each decision.

From a toxicological and risk assessment standpoint, toxicity factors for chemicals are not precise and are often different between political units. The risk levels they represent may vary between government entities. For example, toxicity criteria for non-carcinogens vary internationally by one to three orders-of-magnitude resulting in different risk assessment findings and risk management activities for multinational firms (Rudel et al., 1994). For carcinogens, the risks may be as high as those reported or as low as zero because ambient exposure levels are below concentrations known to cause effects in humans and the inherent uncertainty related to high dose to low dose extrapolation. Thus, providing less stringent acceptable risk values for critical infrastructure, according to how the toxicity factors were derived, would balance off these identified problems.

Essentiality

What if the critical infrastructure under risk review is essential to economic security? Economic security results in sufficient medical care, nutrition, psychological health, and other tangible and intangible benefits. Is it appropriate to eliminate economic security to achieve conformance with policy based acceptable risk levels where actual impacts or risks are not provable or expected in the near future??

What if the critical infrastructure is a road that spans a river, marshland, and mountains and creates or improves a main logistical route, allows for coordination of critical materials and manufacturing, and provides for the rapid movement and deployment of national security assets? What level of calculated human health or environmental risks is equivalent to the benefits provided by construction of this critical infrastructure? Can there be a balance between added costs associated with avoiding a sensitive habitat, increased time and emissions due to a longer, less-direct route, and the need to meet the critical purpose of developing the infrastructure? We will not answer these questions. However, these are typical questions that a risk manager must confront and answer when balancing calculated risks against non-health risk factors.

3.3 MATRIX APPROACH TO DEFINING ACCEPTABLE RISKS BASED ON THE CRITICAL NATURE OF INFRASTRUCTURE

Acceptable risk assessment values for cancer and non-cancer endpoints in human health risk assessment and in ecological risk assessment are based on law, regulations, policies, court orders or other means. Since safety and acceptable risk are intellectual constructs, they are amenable to alteration or modification based on reasoned judgment. In an effort to factor in these constructs, modern risk management decision analysis uses both scientific and values-laden inputs to resolve societal risk issues in a transparent and systematic manner (Linkov et al., 2004).

There are no doubt countless rigorous systematic and transparent methods that could assist risk managers in ordering the multiple criteria that are necessarily part of their decision analysis process. Tables 1 and 2 provide a simple example of one such approach. Table 1 presents a receptor sensitivity/critical infrastructure matrix to establish case specific acceptable risk levels. Table 2 presents a risk probability/severity matrix for critical infrastructure to establish case specific acceptable risk levels. Both tables use qualitative descriptors and require the identification of an Acceptable Risk Level (ARLn) that meets the stated conditions (e.g. what would be the ARL for critical infrastructure of low importance having highly sensitive species vs. critical infrastructure of high importance having species of low sensitivity). Such tables allow risk managers the opportunity to establish ARLs based on perceived species sensitivity coupled to the critical nature of the infrastructure or risk probability of occurrence. These are inherently a mix of attributes such as science, policy, law, and values judgments that are not amenable to mathematical or other strictly quantitative scientific analyses.

Use of such tables allows real world risk management concerns to be addressed and acceptable risk levels established based on a matrix approach rather than a uni- dimensional approach that only recognizes a bright line paradigm. Using this method, risk managers have a systematic approach to decision analysis, one where the various factors used to reach acceptable risk decisions are systematically and transparently arrived at, rather than an amorphous process that uses a single point acceptable risk value as the starting point for non-transparent and non-systematic deliberations.

Table 1. Example of receptor sensitivity/critical infrastructure matrix to establish case specific acceptable risk levels.

		LEVEL OF RECEPTOR* SENSITIVITY			
		Low	Medium	High	Highest
Critical Nature Of Infrastructure	Low	ARL_1	ARL_2	ARL_3	ARL_3
	Medium	ARL_1	ARL_2	ARL_3	ARL_3
	High	ARL_1	ARL_1	ARL_2	ARL_3
	Highest	ARL_1	ARL_1	ARL_2	ARL_3

ARL = Acceptable Risk Level (numerical values to be entered here by risk analysts)
*Human or ecological receptors (e.g. plants, birds, fish, etc.)

Table 2. Risk probability/severity matrix for critical infrastructure to establish case specific acceptable risk levels.

		LEVEL OF RECEPTOR SENSITIVITY*			
		Low	Medium	High	Highest
Risk Probability Of Occurrence	Low	ARL_1	ARL_2	ARL_3	ARL_3
	Medium	ARL_1	ARL_2	ARL_3	ARL_3
	High	ARL_1	ARL_1	ARL_2	ARL_3
	Highest	ARL_1	ARL_1	ARL_2	ARL_3

ARL = Acceptable Risk Level (numerical values to be entered here by risk analysts)
*Human or ecological receptors (e.g. plants, birds, fish, etc.)

Explicit risk management, the result of cost and environmental concerns, can form the basis for the acceptance of greater risk (Russell and O'Grady, 1996). The development of explicit risk management acceptable risk criteria for critical infrastructure would encourage the clear and detailed identification and assessment of uncertainty in the risk assessment process. It would facilitate the preparation of development plans and allow for more rapid evaluation of acceptability. Explicit acceptable risk criteria for critical infrastructure would formally recognize the need for trade-offs to ensure national security, economic security, and environmental security. In other words, the value of science, when appropriate, is to allow us to relax the standards (Keiding and Budtz-Jorgensen, 2004). For example, zoning for different purposes (e.g. heavy industry, light industry, residences, schools, hospitals, parkland, etc.) could assist in defining acceptable risk levels based on the nature of the use rather than a one-size-fits-all approach. It could be argued that at present, the current regulatory acceptable risk approach overprotects certain areas (e.g. transportation corridors) while under-protecting others (residential complexes). If this were not the case, why then would regulatory agencies allow background risk levels to exceed calculated allowable risk limits in densely populated areas while stringently controlling risks in areas where similar exposure conditions do not exist?

International trade treaties might have the effect of stimulating the need to develop explicit and uniform levels of acceptable risk and risk assessment methodologies. "Trade treaties have introduced the principle that similar risks should be treated similarly, and that countries must achieve internal consistency in the levels of protection they afford against certain hazards within their territories. The problem is that there is no agreement on when risks are "similar" or when levels of protection are similar" (Walker, 2001).

4. Conclusions

This paper presents an initial discussion of the pros and cons for modifying acceptable risk levels based on currently available risk assessment and management practices and the transparent use of decision-making tools that allow for modifying acceptable risk levels based on the critical nature of infrastructure. It offers no answers, but should stimulate debate about this emerging issue.

There is currently no uniform definition of environmental security or implementation of its basic tenants with relation to the evaluation of acceptable risk and critical infrastructure. As discussed in this paper, there are numerous compelling reasons to switch to an explicit acceptable risk management methodology that takes into account the critical nature of a given infrastructure type or specific project. Reasons include increased efficiency, less time and cost between plan inception and execution, potential for lower construction costs, systematic and transparent decision-making, and defined planning expectations. These beneficial results could accrue from the creation of explicit acceptable risk criteria to accommodate the need for critical infrastructure.

There are also many compelling reasons, to continue using the current implicit risk management system, that do not take into account the critical nature of an infrastructure type or allow the acceptable risk level to be explicitly adjusted by risk managers based on non-numerical non-risk parameters. These include a variety of acceptable/unacceptable risk criteria at different political levels or under specific statutes representing local control of risk decisions, existing law, experience with the system, and proven flexibility in the risk management decision-making phase. The implicit acceptable risk management process is an administratively and legally established process with known methods to determine whether acceptable risks have been exceeded. In such cases, the acceptable risk level is looked at as a point of departure for consideration and not a bright line. This system is in place and working.

Implementing a new system could lead to expensive and unforeseen downstream impacts.

Determining whether the goal of environmental security is best served by rigid or flexible explicit or implicit acceptable/unacceptable risk thresholds is not a scientific question. It is a legal and values driven question. Within this system, governments provide the risk assessment process, scientists provide the factual foundation and technical reports, citizens provide factual and other input (e.g. values), and risk managers provide the determination of whether a given calculated risk level is acceptable or unacceptable.

Perhaps the best summary of the problem of acceptable risk is expressed by Hunter and Fewtrell (2001): "The notion that there is some level of risk that everyone will find acceptable is a difficult idea to reconcile and yet, without such a baseline, how can it ever be possible to set guideline values and standards, given that life can never be risk free?"

6. References

1. Hobbs, R.J. and L.J. Kristjanson. 2003. Triage: How do we prioritize health care for landscapes? Ecological Management & Restoration 4(Suppl): S39-S45.
2. Hunter, P.R. and L. Fewtrell. 2001. Acceptable Risk. World Health Organization Water Quality Guidelines, Standards and Health. Chapter 10.
3. Keiding, N. and E. Budtz-Jorgensen. 2004. The Precautionary Principle and Statistical Approaches to Uncertainty. *Int J Occup Med Environ Health* 17(1):147-51.
4. Linkov, I., Varghese, A., Jamil, S., Seager, T.P., Kiker, G., Bridges, T. 2004. Multi-Criteria Decision Analysis: Framework for Applications in Remedial Planning For Contaminated Sites. in Linkov, I. And Ramadan, A. eds Comparative Risk Assessment and Environmental Decision Making, Kluwer, 2004.
5. Rudel, R.A., Kinkaid, T.M., and B.D. Beck. 1994. Variability in International Criteria for Chemical Toxicity: Comparison of Methods and Criteria Developed by the U.S. Environmental Protection Agency (U.S. EPA), the World Health Organization (WHO), and the National Institute of Public Health and Environmental Protection in the Netherlands. Abstract of Meeting Paper. Society for Risk Analysis 1994 Annual Meeting.
6. Russell, C.S. and K. O'Grady. March 1996. Applied Risk Communication Within the Corps of Engineers. U.S. Army Corps of Engineers. Water Resources Support Center, Institute for Water Resources. IWR Report 96-R-14.
7. U.S. EPA (United States Environmental Protection Agency). June 1997. Ecological Risk Assessment Guidance for Superfund: Process for Designing and Conducting Ecological Risk Assessments. EPA 540-R-97-006. OSWER 9285.7-25.
8. Walker, V.R. 2001. Consistent Levels of Protection in International Trade Disputes: Using Risk Perception Research to Justify Different Levels. *Environmental Law Reporter,* available at http://people.hofstra.edu/faculty/vern_r_walker/Walker2.pdf

FROM GLOBAL WARMING TO WATER SCARCITY: WHAT ARE THE MOST URGENT ENVIRONMENTAL PROBLEMS OF THE REGION.

Report of the Discussion Group on Environmental Security

A. MOREL
Energy Environmental Study Institute (EESI)
122 C Street, Suite 630, Washington DC, 20001, USA

B. MOREL
Department of engineering and public policy Carnegie Mellon University, Pittsburgh PA 15213, USA

Abstract

The Environmental Security discussion group focused on the issue of water scarcity in the Mediterranean region, namely Israel, Palestine, Jordan, and Egypt, for a medium time-scale (10-20 years in the future). No consensus was reached as to whether climate change would have a significant impact, and what kind of impact. On the other hand, the inevitable increase of population pressure in the region is a source of considerable concern and should be enough impetus to begin adaptation strategies as soon as possible. The group's discussion addressed potential hazards and solutions for a medium time scale. There was consensus that, considering the security situation between Israel and its neighbors, the environmental problems of the region, which are serious and getting more so with time, can be addressed adequately only with the active participation of the international community. Increasing water use efficiency would be the most cost-efficient short term solution. Longer term solution includes large scale endeavors such as the increase desalination capacity. The members of this discussion group were representing 5 countries [Spain, Israel, US, Turkey, Belarus] from the public, private, and academic sectors.

1. Introduction

The working definition of environmental security decided upon was "The process of peacefully reducing human vulnerability to human induced environmental degradation by addressing the root causes of environmental degradation and its possible impact on human insecurity." The environmental issues were considered on a *regional* scale. The region of relevance for the workshop involves four neighboring countries: Israel, Palestine, Jordan, and Egypt. The main concerns addressed were water quality and availability or scarcity, as well as air quality. The goal was to identify adaptive policies

B. Morel and I. Linkov (eds.),
Environmental Security and Environmental Management: The Role of Risk Assessment, 103–107.

to mitigate their societal impacts. For the purposes of this discussion, scarcity was defined as a lack of infrastructural capacity, also termed *adaptive capacity*.

Societal factors under consideration were divided into supply- and demand-side. Factors affecting the water *supply* were desertification, pollution, and allocation of resources. Factors that would influence demand on water are population, urbanization/industrialization, GDP per capita, agriculture (i.e. food security), energy, and the availability of virtual water (outside sources/imports).

A complicating factor in this region is its political and economic heterogeneity. The large differences in standards of living and lifestyle between neighboring nations, translate in huge differences in use of waters and in the willingness to pay for it. Today water consumption varies widely between communities within the region. Water allocation has been for a long time a contentious issue in that region and will stay so for the foreseeable future. If water allocation is a divisive issue, water quality has the opposite effect. All the parties have a common interest in seeing an improvement in the water quality. One side-effect of the worsening of the security situation in the region has been to let environmental concerns take a back seat. This environmental neglect is a major reason for the recent degradation of the water quality in the region.

Although all the countries of the region are affected by the problem, they have very different adaptive capacity to deal with the issue. Israel has a substantially greater adaptive capacity than Palestine. But Israel should be expected to use its resources to mitigate the problem in its territory, not to take on itself to develop a regional solution for water supply. The amount of resources required would be too large for one country like Israel. Some form of international involvement is needed.

The ultimate goal before this discussion group was to offer feasible policy initiatives for the governments of the region, providing viable solutions to avoid future conflicts. Ideally this cooperation would need to occur in the political, economic, and human sectors in order to have the broadest impacts with input from scientific, political, and community organizations. These policies would need to address a broad range of environmental risks that would include human health and welfare. The consensus was this cooperation could not occur without international intervention, though the members of the discussion group were optimistic that states within this region would be working together by 2015-2020.

2. Discussion

Currently there are very limited water resources available for the inhabitants of this semi-arid to arid region and what is available is not being equitably consumed. For instance, Israeli's represent 10% of the population in the West Bank and they are responsible for 30% of the water consumption in the area. This reflects a variety of factors, including the economic disparity between the communities sharing that resource. Economic development of the region, increased industrialization exacerbates the problems associated with water scarcity and food security. Needed improvement of the standard of living of all the communities living there, is translating also in increasing water demand. This is in a context where there is already a water deficit while the local population is growing at a fast rate, and not only is water availability limited, but strategic water reserves are polluted, surrounding land is degrading, and the overall ecology of the region is on the decline.

Following the laws of economics, reduction in the supply of water will increase its commodity price having substantially negative impacts on the poorest in this region, who already often pay ten times more than the rich for water due to their lack of water distribution infrastructure. Allowing market forces to increase the price of water and dictate water distribution would not be a viable solution. Instead, a minimum amount of water required for survival should be available to all inhabitants at a subsidized price while any consumption above that would be subject to a higher price as a function of GDP.

There is a huge discrepancy in the region with the GDP of Israel at $16-17,000, Palestine at $800-1,000, and Jordan at $1,300. One concern for this discussion group was the discrepancy in water use technology for the different countries in the region. For instance, Jordan and Palestine currently 'waste' a great deal of water through evaporation, and generally practice more 'primitive' land use.

In order to successfully minimize or avoid future conflicts over water distribution in the region, the group suggested a variety of strategies. They fell under the following categories: conservation through targeted demand management and increased efficiency of water use, adaptation (i.e. modernizing agricultural water use), and risk reduction accomplished through a number of strategies and implemented policies.

3. Adaptation/Conservation

The agricultural sector and therefore food security in this region could potentially be hit hardest by a sudden drop in water availability; therefore adaptive strategies should be adopted immediately in this sector. Currently, crops are being watered from treated wastewater that is minimizing agricultural competition with municipal water consumption. In future, farmers can switch to less water intensive agricultural products, introduce technologies and means in order to promote the enabling environment for efficiency improvements. Improving efficiency of water use through changes in economic and financial activities would be necessary to adopt adequate systems for cost recovery. To address socio-economic discrepancy in the area a system of progressive water pricing and water marketing would need to be developed, such that water would be allocated to affordably provide for survival.

Certainly an option would be to decrease demand on water by encouraging migration of the region's inhabitants. Richer share-holders would be able to invest in improved supply technologies, i.e. desalination. Desalination seems a realistic solution for Israel in the near term, as its economy can more easily accommodate this more expensive form of water, than is the case for its neighbors and also because of its geographical situation. 'Virtual water' like the water imported from Turkey) enters also in the equation.

Conservation is obviously the cheapest solution though historically unpopular and considering the low level of water consumption in the region, it cannot lead to a general solution of the water problem in the region. Still whenever conservation can be introduced, it should as it is sometimes one of the most effective as part of a variety of adaptive strategies. So there should be a system of incentives established to make this an attractive option.

4. Risk Reduction

The aforementioned adaptive strategies may prove ineffective if not coupled to some risk reduction strategies. The local population is predicted to double in the next 20 years. The World Bank predicts that population in Israel will increase to 12.8 million and from 3.2 to 7 million in Palestine. Therefore any efforts to mitigate future potential population growth should be adopted, most likely through education (i.e. family planning). Curbing environmental degradation was another important concern, with emphasis on halting contamination of watersheds. This requires the implementation and enforcement of an aggressive environmental mitigation policy. One unresolved challenge is to find an organism with adequate authority to do that. Cooperation between the local governments is problematic. On the other hand to be effective, such a mitigation policy should be designed at the regional level. The geographical distribution of sovereignty in that region is extremely complicated and does not follow the logic of the environmental problems. This renders a close collaboration between regional authorities a necessity, or if that is not an option, potentially an international organism could substitute.

The discussion group was unable to reach consensus as to the urgency of the danger represented by global climate change on the region. Yet it was agreed that industrial emissions would need to be reduced to limit greenhouse gas emissions and release of heavy metals.

5. Conclusions

As mentioned earlier the financial responsibility for these strategies should be a function of GDP. With Israel being the richest country in the region, it will be expected to pay the greatest amount for these mitigation strategies. But realistically the size of the problem and the cost of any effective regional solution are such that there is a need for international involvement at an unprecedented level, in the form of resources and also in the implementation of any policy. Due to the current lack of regional cooperation for a variety of political reasons, the international community will be expected to take many strategic initiatives in order to begin the process of adaptation and risk reduction efforts.

Since the second Intifada, i.e. since the fall of September 2000, the dialogue on water issues in that region has been severely reduced, and when it comes to the issues aforementioned completely absent. The size and complexity of what needs to be done is such that those dialogues should resume as fast as possible to help clarifying the issues, and provide the international community with adequate data and facts, as well as a sense of urgency.

6. Postscript

The findings of this discussion group reflected or amplified the observations of a panel discussion involving the participation of Shmuel Brenner (former Israeli chief-negotiator for water issues with the Palestinian in the conference of Madrid, 1992) and of Alain Jubier (NATO program manager).

Shmuel Brenner voiced his concern that "The environment is being deteriorated and it is non-reversible." He gave a rough estimate of the economics of the problem: "It would cost roughly $2-3 billion to address the waste water problem in the West Bank or Gaza. The only way any viable solution will be offered is to work with the World Bank to back a project there. This situation, frankly, seems like a lost cause and the only way out is to decide which cities are the most dangerous to the aquifer. We are dealing with a per capita income of $2-5,000. Yet in the Oslo agreement there was a provision for waste water treatment for Palestinians." Then Dr Brenner went on stating that: "This is not only a question a foreign entity offering money assistance but *administrative* help will be necessary as well. Take for instance the United States and their Clean Water Act. Seventy percent of their funding can go to treat waste water. Here in Israel our water policy is dependent on this U.S. legislation, in order to implement it effectively the Israeli administration would like to dig for additional water and does not have money to clean water. The only way to effectively address this very pressing problem is to disassociate these environmental issues from political issues—get agreement on principle—and then worry about the political/economic implications." About the form that the administrative help could form, Dr Brenner said: "I think there is certainly a need for a 'wake-up NGO' to clean up the World Bank. If the discussion is around the whole area there will not be a government in a position to implement any policy shifts. Instead the outside organization will have to take the administrative lead, acting without any political interests."

Alain Jubier suggested that NATO could be a facilitator in the future. NATO tends to be seen strictly as a military alliance. What is often overlooked is the strong science program it has always supported. After 1992 the program was more focused on east/west relations with a focus on 'technology for peace.' NATO's role in the Middle-East will have to be dictated by the United Nations and will not take any actions without U.N. consent. As for the conflict in Iraq and Afghanistan, there is disagreement within the organization as to the role NATO should take. Many believe that there will not be any peace in the region without NATO intervention. Since November of 2003, NATO decided to shift from support of general science programs to only 'security through science' programs, which will be focusing on issues centering on counter-terrorism and other threats to security (such as the environment and population). Hence, NATO would be able to continue to support future conferences such as this one on environmental security.

Part 2.

Environmental Management

THE INTERCONNECTION BETWEEN THE BUILT ENVIRONMENT ECOLOGY AND HEALTH

H. S. KOREN
Carolina Environmental Program,
University of North Carolina, Chapel Hill, NC 2759-1105, USA

C. D. BUTLER
National Centre for Epidemiology and Population Health, Australian University, Canberra, ACT, 0200 AUSTRALIA

Abstract:

The built environment (BE) affects ecosystems, ecosystem services and human health and well being. While, formally, the BE ranges from the smallest hut to the largest city, this chapter focuses upon the health effects of urban areas, which increasingly are the preferred human habitat. Urban areas have many attractive and beneficial influences to human well-being. But at the same time, many effects of urban areas are harmful to well-being, and many are not even recognized as such. Most publications about these topics have described the effects of the BE separately, on *either* ecosystems *or* on human health. The interconnectivity between these two effects relative to BE is rarely studied. This paper focuses on the mutual influence and interactions between three related aspects of the BE which can impact ecosystems *and* human health: transportation, land use, and life style. It also explores some of the links between the BE, human health, and human security.

Transportation, especially when based on systems of private cars burning fossil fuels, is often the most important cause of air pollution in both developed and developing countries. Air pollution has many adverse health effects, including asthma and cardiovascular disease. Transport systems based largely on the use of private cars are a major contributor to global warming and to ecosystem degradation. This occurs directly, as the operation of vehicles releases greenhouse gases causing global climate change which is associated with altered temperature and rainfall patterns and rising sea levels. Warmer ocean temperatures are projected to increase the frequency and intensity of extreme weather events. These effects can also harm cold water fisheries, and otherwise degrade aquatic habitats. The heat-retaining nature of road surfaces and many buildings, together with the loss of vegetation, contributes to "heat islands" sometimes exacerbated by particulate-dense air pollution. Heat islands may create a double burden of pollution as people respond reactively, for example with air conditioning, rather than within urban redesign, such as exchanging black for green surfaces, fewer roads and more rooftop gardens. Transportation is therefore having an impact on ecohealth as well as human health, and these are interconnected.

B. Morel and I. Linkov (eds.),
Environmental Security and Environmental Management: The Role of Risk Assessment, 111–125.

Land use is greatly affected by urban sprawl practices that are responsible for degrading habitats, for altering ecosystem function, and for reducing biodiversity. Sensitive and critical habitats are often fragmented or sacrificed for roads, suburbs and industrial estates. Urban conurbations also pollute reservoirs, ground water, and stream networks with chemicals and pathogens, with numerous adverse health effects. Habitat loss and fragmentation are two of the most direct impacts of development on previously undeveloped land. Habitat fragmentation and an increased proximity of forest, agricultural land and human populations can promote interaction among vectors, pathogens, and hosts, and in some cases lead to increased infectious disease, including Lyme disease. Deforestation continues to increase in many developing countries, in part to supply affluent urban populations. These illustrations demonstrate that not unlike the transportation factor the BE also affects landuse, incurring major impacts on ecosystems and human health.

Changes in life style are a direct consequence of the effects of transportation and land use associated with many BEs. High automobile dependency is often characterized by reduced physical activity, and by diminished personal relationships between individuals and groups, a quality known as "social capital". A lack of physical activity combined with excessive caloric consumption commonly leads to overweight or obesity, in turn increasing the risk of many diseases, including Type II diabetes, hypertension, asthma, and cancer. The life style of many BEs exerts a toll on quality of life, including by increased noise disturbance, decreased air clarity and reduced contact with varied and stimulating natural ecosystems.

Human security is a widely recognized component of human well-being (Millenium Ecosystem Assessment 2003). Though the World Health Organization (WHO) definition of health does not explicitly include security, the WHO conceptualizion of health is much broader than the absence of physical and mental disease (WHO 1948). It follows that if human health is adversely affected by the BE, then human security will also be reduced, though it is acknowledged that many other factors also influence security. Some of the factors which influence non-health aspects of security, such as the level of crime or interpersonal violence, are also likely to be influenced by the BE, including through the quality and level of social capital and psychological health. As well, the health or wellness of a person is likely to influence that person's resilience in the face of threat. In general, healthy people will feel more secure. Finally, the community level of health can influence security, by influencing one's perception of personal disease risk, including in some cases, vector borne diseases.

In conclusion, adverse effects of the BE, including reduced air and water quality, degraded ecosystems and biodiversity, and the spread and emergence of infectious disease, are relevant to security. The quality of human life and the integrity of ecosystems are affected not only by direct stressors created by the BE which can affect them separately, but also by stressors derived from one or the other, thereby demonstrating the close interconnectedness between the environment and human health. This paper, therefore, highlights the complexity and the interconnections between the BE, ecosystem and human health, and security.

1. Introduction

1.1 ECOLOGY, HEALTH AND SOCIETY

Many paradigms have been used to explain health and disease, particularly of epidemics-- sicknesses affecting large numbers of people at the same time. These causal paradigms have included evil spirits, misdeeds, and the spells of malevolent enemies (Ahmad, 1998; Rahman, 1998). More recently, epidemics were often attributed to "miasmas" - large bodies of toxic air, often found near swamps and flooded areas. In the late 19th century, the miasmatic paradigm fell from favor, replaced by reductionist explanations stimulated by the growing power of epidemiology and microbiology. An impressive sequence of discoveries distinguished and explained, for the first time, epidemic diseases from cholera (Davey-Smith, 2002) to malaria (Nye and Gibson, 1997), yellow fever, tuberculosis (Ryan, 1993), polio, lung cancer and Minimata disease (Watts, 2001) due to infectious and toxic agents such as tobacco smoke and mercury. Most importantly, these causal models enabled effective methods of control, such as reducing mosquitoes, microbes and smoking. (Susser and Susser, 1986).

Powerful as these advances were, some theorists argued that other causal models still retained validity. These workers argued, for example, that causation could be considered as having both "proximal" and "distal" explanations, and that the appropriate explanation is a matter of the "focal depth" sought by the investigator (McMichael, 1999). Others argued that causal models that focused on individual behaviour did not always lend themselves as readily to public health improvements as did explanations of behavior at population levels (Rose, 1990).

The explosion of knowledge in medical science has led to numerous epidemiological sub-specialisations, each focusing on branches of knowledge formed from a single trunk. These branches have names such as infectious, genetic, nutritional, environmental and social epidemiology. All are valid and yet none is complete. Ecological explanations for health and disease are, similarly, incomplete. Yet, this paper argues, ecological models of causation usefully complement these other, related, causal models.

The interconnection among the environment, ecosystems and human activity has been the subject of numerous publications and meetings (DiGulio and Monosson, 1996; DiGulio and Benson, 2002; Fisher, 2001; Aron and Patz, 2001, Koren and Crawford-Brown, 2004). It is becoming increasingly clear that numerous issues that were previously thought of as independent of the environment are intimately connected to it. Human health, the economy, social justice, ecological processes and national security all have important environmental aspects whose magnitude and interconnections are not generally reflected in public policy. A system of public policy with two defining characteristics has evolved. First, human and ecological health protection generally have been treated as separate domains of policy, with significant differences in both the analytic methods used to characterize risks and the policies developed for risk reduction. Second, individual human health risks (e.g., of malaria, schistosomiasis or cholera) have been analyzed in isolation. The objective of risk assessment is to support decision making by assessing risks of adverse effects on human health and the environment from chemicals, physical factors, and other environmental stresses. With increased recognition of the need to more effectively protect both humans and the environment, it is time to consider a move to a more integrated, "holistic" approach to risk assessment. The positioning of humans as a part of a broader ecology can be traced to ancient peoples, including the Greeks and Native Americans, but the modern relationship between ecosystems, humans, "wellness" and disease owes much to René Dubos, a microbiologist who discovered gramicidin, the world's first antibiotic, when Dubos was working in New York in 1939. Unfortunately, gramicidin proved toxic to humans. In his later life, Dubos became well-known for his work to protect the global environment, including at the world's first great summit about this, held in Stockholm in 1972 (Ward and Dubos, 1972).

Box 1. The concept of ecosystem "services" (Daily, 1997; Millennium Assessment, 2003) has been developed to complement the more fundamental argument that ecosystems should be preserved because of their own "existence value"(Fox, 1990). Supporters argue that the concept that ecosystems provide "services" to humans adds a utilitarian reason for their protection. They suggest that many of the links between natural and human systems were once widely understood, by different forms of "folk ecology" (Ramakrishnan et al, 1998; Atran et al, 1999; Berkes, 2003). But urbanization has created a human world with contact between people and nature that is less common and less intimate than the past, disrupting the understanding of these connections, and partially insulating many humans from adverse ecosystem change.

Ecosystem "services" include the more obvious "provisioning" benefits of food, fibre and fuel. They also include the less well-known "regulating" services. For example, the excessive clearing of a forest can contribute to both flooding during heavy rain and aridity during periods of dryness, while an intact forest will reduce both of these extremes by acting as a sponge that both absorbs and releases water. Other regulating services include erosion control, climate modification, and water and air purification. As well, ecosystems provide important spiritual, recreational, and cultural "services." Many people find psychological and spiritual refreshment through contact with special and even sacred aspects of nature, such as visual landscapes, wilderness, the seashore and special groves and springs. Finally, ecosystems provide "supporting" services that make all of the others possible. These include soil formation, nutrient recycling and pollination. For example, pollinators from intact forest patches have been shown to substantially increase the yield of coffee in adjacent plantations (Ricketts et al, 2004). There is a growing concern that the quality and quantity of key ecosystem services are declining in many regions and in the world as a whole, especially in relation to the still growing global population (Crutzen, 2002; Millennium Ecosystem Assessment, 2003). There are particular concerns over the long-term capacity of global food production, including deep sea and coastal fisheries, and for the productivity of dryland ecosystems. More broadly, there are concerns over adverse social changes in response to, and perhaps exacerbating, ecosystem service depletion. The Millennium Ecosystem Assessment is an attempt to measure, forecast and respond to these concerns.

Less well known are essays in which Dubos prophesized that urbanization would lead to a greater tolerance of "starless skies, treeless avenues, shapeless buildings, tasteless bread, and joyless celebrations," and that this urbanization would harm human health (Moberg and Cohn, 1991).

Many of the issues and concerns raised by Dubos and others have been amplified and endorsed by the global environmental movement. There are increasing signs that these concerns are being appreciated by health researchers (Daszak et al. 2000, Frumkin 2001, DiGulioand Benson 2002, Jackson LE 2003, Jackson RJ, 2003, Frumkin et al. 2004). In the last decade a growing number of studies have found links between health and macro-environmental issues, such as emerging infectious diseases, climate change, and ecosystem "services" (see box 1).

In parallel, there has been an increased appreciation that good health is more than the absence of physical and psychological symptoms, including subjective feelings of contentment and security. A new paradigm is emerging in which health is viewed as a central component of well-being, where health is seen as contextual as well as individual, and where the natural environment is appreciated as a source of sustenance to be maintained (Epstein et al, 2003) rather than an enemy to overcome, or a mine to exhaust.

This paradigm, sometimes called "ecohealth," is still little known (Rapport, 1997). As mentioned, it is related to other health paradigms and complements rather than substitutes these causal models. Supporters of ecohealth argue that this new field has much to offer at diagnostic, therapeutic and preventitive levels. For example, if good health – even partly – depends on regular exposure and contact with non-human species, then insights from ecohealth may lead to changes to urban design that are not only aesthetically pleasing, but beneficial to health and well-being.

Groundbreaking studies have shown that hospital patients cure faster if their window faces trees instead of unattractive walls of buildings (Ulrich, 1984). Prison inmates are healthier when their windows face the outside landscape (Moore 1981) and ordinary citizens viewing savanna-like settings report feelings of tranquility, peacefulness or relaxation (Hanna and Coussens, 2001). Contact with pets improves health for survivors of myocardial infarction (Friedman and Thomas, 1995), reduces the incidence of minor health problems when comparing pet owners to petless subjects (Serpell, 1991), and improves the health of isolated and aged individuals (Siegel 1990).

Other insights from ecohealth include the "spillover" of animal microorganisms into human populations because of close contact (Daszak et al, 2000), the adverse impact to human, animal and economic health because of large-scale land-clearing due to fires, climate change and altered ecosystems in many places around the globe including South East Asia, South America and Africa (Epstein, 1999) and indirect effects to human health because of economic decline subsequent to ecosystem degradation. The emergence of variant Creutzfeld Jacob Disease, initially in the UK, occurred because of animal husbandry practices that flouted basic ecological principles – most obviously by forcing normally vegetarian cows to become cannibals (Prusiner, 1997). Other examples of links between intensive farming practices and periodic emergences of epidemic diseases that can infect humans include influenza (Oxford et al, 2002) and SARS (Webster, 2004).

The clinician can attempt to cure or alleviate these diseases in individual patients, and the public health physician can use means such as vaccines, quarantine and masks to quell epidemics. Social epidemiologists can describe how the illness varies by neighborhood and social class. However, effective control of many diseases will also benefit from the insights from ecohealth, including urban as well as rural and jungle settings.

1.2 HEALTH, SECURITY, AND HUMAN WELL-BEING

Outside health circles, the broad definition of health as "a state of complete physical, mental and social well being and not merely the absence of disease or infirmity" (WHO, 1948) is little appreciated; instead, health is often seen as a medical condition requiring pills and surgery. However, as Dubos recognized, health, which lexicologically is related to "wholeness" is far more than the absence of disease. There are many overlaps between the WHO definition of health and the more recently developed concept of human well-being (Millenium Ecosystem Assessment, 2003), including human security, an element vital to both. Like the air we breathe every few seconds, security is most obvious by its absence. Security takes many forms, such as having reliable access to food, shelter, warmth, physical safety, confidence, companionship, friendship and love. Security underlies peace of mind and enables planning. It makes us more fully human. Security is also related to ecosystems and to other aspects of our environment, including the BE. For example, certain spaces can be perceived as alien and threatening, reducing security.

Excessively rapid change, including the loss of familiar landmarks, also reduces security. Many indigenous populations have insecure access to their local ecosystems, upon which every facet of their lives depends. Sometimes, their resultant despair seems to be transmitted through multiple generations. Now, as the world faces unprecedented anthropogenic environmental change, some sectors are sensing a new form of insecurity, taking the form of concerns over climate change, lost species, adequate water (including quality) and new diseases.

1. 3 HEALTH, WELL-BEING, AND THE BUILT ENVIRONMENT

The built environment (BE) has been defined to comprise urban design, land use, and the transportation system, encompassing patterns of human activity within the physical environment (Handy et al. 2002). Patz et al. (2004) have recently stated that "the 2000 Census shows that 80% of the U.S. population now lives in metropolitan areas, with 30% living in cities of 5 million or more. The environmental issues posed by such large population centers have profound impacts on public health beyond the city limits." But, as well, the BE has many impacts on public health *within* the city limits.

The fields of city and urban planning/design and landscape architecture have generated a wealth of literature on the *physical* consequences of urban land use (U.S. EPA 2001). The state of ecosystems has also been extensively studied (e.g., http://www.heinzctr.org/ecosystems/). In contrast, the literature on the *health* effects from the BE is relatively limited. This chapter attempts to fill some of this gap.

1.4 URBAN SPRAWL

Urban sprawl entails the encroachment and diffusion of the periphery of a city across more and more rural land (Jackson 2003). Although largely driven by increasing population, together with sufficient financial means ("white flight"), the process is also motivated at least partly by a desire for bucolic contentment. Such a statement is seemingly attainable by a life spent in the more affordable and less densely populated urban fringe, which sometimes offers a glimpse to surrounding farmland and forest. Most planners and government rely on roads to cope with the ever-increasing distances caused by sprawl, in turn increasing the commuter pressure on more central roads, planned and built by earlier generations. Generally, roads are cheaper to construct and operate than public transport systems, especially in cities that cover very large, but comparatively less densely populated areas, and where cultural constraints also limit the attraction and economic viability of public transport systems.

As well as land, sprawl consumes significant amounts of other natural and man-made resources, including systems for delivering water and power, and for removing wastes. Importantly, sprawl usually adds to the total cost of travel and infrastructure provision, thus offsetting some of the efficiencies made possible in more densely populated BEs.

1. 5 THE BUILT ENVIRONMENT AND ENVIRONMENTAL HEALTH

Formally, the built environment, itself a specialized form of ecosystem, includes all buildings, spaces and products that are created or modified by people (U.S. EPA 2001). It thus includes homes, schools, workplaces, parks and factories. As well, the reach of the BE extends, tentacle-like, in all directions, encompassing overhead electric transmission lines, underground waste disposal pipes and (sometimes) subway trains, and across the country as highways (Health Canada, 1997).

In its broadest sense, environmental health comprises those aspects of human health, disease, and injury that are determined or influenced by factors in the environment. Traditionally, this discipline has focused upon the study of the direct pathological effects of various chemical, physical, and biological agents. Increasingly, environmental health embraces the effects on health of the broad physical and social environment, including housing, urban development, land use and transportation, industry, and agriculture (U.S. Department of Health and Human Services, 2000).

2. Discussion

2.1 TRANSPORTATION: ENVIRONMENTAL AND HEALTH CONSEQUENCES

While vehicular travel has many benefits, such as mobility, privacy, convenience, and flexibility, it also has many unintended adverse environmental consequences, many of which only become obvious beyond a threshold number or density. One very important negative consequence is degraded air quality. The U.S. Environmental Protection Agency (EPA) has established National Ambient Air Quality Standards

(NAAQS) to protect public health, including the health of sensitive populations such as children and the elderly, from adverse effects of poor air quality (U.S. EPA 2001). Motor vehicles emit a variety of pollutants each of which contributing a large portion of CO and ozone precursors in particular. Vehicle travel also kicks up large quantities of particulate matter (PM) from roads (especially on unpaved roads in rural areas). Cars also emit hazardous air pollutants (HAP) or air toxics known or suspected to cause cancer and other serious health effects in humans and ecosystem damage. Persistent air toxics are of particular concern when aquatic ecosystems, as toxic levels can magnify up the food chain.

One of the most serious public and environmental health issues in Western countries is the worsening epidemic of asthma. In the United States the prevalence of asthma has doubled in the last twenty years (Mannino et al., 1998). More than 17 million people now report having the disease (Center for Disease Control and Prevention, 1998). Asthma has increased most rapidly in children less than 14 years old, who also account for the highest overall rates of asthma among the population at large. Trends toward increased prevalence, deaths, and costs of the disease have also been observed in many other countries (Kuehni et al, 2001).

Genetic susceptibility, viral and parasitic infections, and allergic status have long been known to be related to asthma incidence or severity (Koren, 1997). However, the dramatic increase in the frequency of asthma has also led investigators to consider other possible causal factors, including diet, lifestyle, air pollution and broad environmental factors (von Mutius, 2003). The role of some air pollutants in worsening symptoms in people with asthma is beyond any doubt. Increases in ambient levels of ozone (O_3), sulfur dioxide (SO_2), nitrogen dioxide (NO_2), and particulate matter $<10\mu$ microns aerodynamic diameter (PM10), as well as suspended sulfates, have been correlated with emergency department visits and hospital admissions for asthma. In a recent study, a team of scientists recruited a group of 3,535 school-age children in Southern California who did not have asthma, and followed them for five years (McConnell et al. 2002). The results of this and another study conducted in adults (McDonnell et al. 1999) found that chronic exposure to ozone not only worsens symptoms among people with asthma, but can also lead to the onset of the disease, among both children and adults. Evidence from the past 10 years also shows that sudden increases in ambient air pollution can cause as much cardiovascular as respiratory morbidity and mortality (Pope et al. 2004).

Another adverse health effect associated with the BE and with air pollution is the heat island effect. Urban areas have been shown to be warmer by 6-8° F than surrounding areas. This effect occurs because surfaces like dark roads and roof tops absorb heat from the sun and radiate heat back into the atmosphere. In addition urban areas are usually poor in vegetation that would provide shade and cool air. Excessive heat, especially in populations that are not inured to it (Frumkin 2002, 2004; Kalkstein 2002) is responsible for various health effects including fainting, hyperventilation and edema. Heat exhaustion is a more severe condition and presages fatal heat stroke. Heat may further contribute to the adverse effects of air pollution. Increased temperatures raise energy consumption, especially from air conditioners, thus triggering more emission of pollutants from fossil fuel burning power plants (U.S. EPA 2001).

There is growing agreement that the large volumes of "greenhouse gases" (especially carbon dioxide) released by human activity such as the burning of oil, gas and coal, are not only substantially altering the atmosphere, but causing global climate change. Predicted effects include altered temperature and rainfall patterns and rising sea levels. Warmer ocean temperatures are projected to increase the frequency and intensity of extreme weather events. Together, these effects are predicted to alter global patterns of land use, ecosystems and human activity. Transportation is a significant and growing source of greenhouse gas emissions. In the US, carbon emissions from transportation are projected to grow by 47.5 percent over the period 1996-2020 (U.S. EPA 2001).

Motor vehicle transport also has adverse effects upon water quality and ecosystems. Highway maintenance involves activities that can adversely affect the environment, such as road salting, use of solvents, and pesticides. Highway de-icing can adversely affect roadside vegetation, soil structure, drinking water supplies, and aquatic life. Paved roads and parking lots, both of which are necessary to support vehicular use, create the majority of impervious areas in the landscape. Impervious surfaces impede the recharge of groundwater from rainwater, funnel toxic chemicals into waterways, and can increase erosion, flooding and changed water temperatures (Jackson 2002). In turn, these effects can harm cold water fisheries, and otherwise degrade aquatic habitats. It is therefore evident that transportation can have far reaching effects on both ecosystem health as well as on human health which are closely interconnected.

2.2 LAND USE CONSEQUENCES FOR ECOSYSTEMS AND HEALTH

Development uses land space and modifies habitats and ecosystems. Land consumption rates in the United States, already high, are rapidly increasing. More land was developed during the five-year period from 1992–1997 than during the 10-year period that preceded it (U.S. EPA 2001). In addition to directly destroying areas of natural habitat, development usually fragments habitat and increases the invasion of introduced species, altering ecosystem function and changing biodiversity. However, development can sometimes take account of ecology, leading to the preservation of critical habitats such as wetlands, and creating greenways and buffer zones around sensitive habitat, thus preserving ecosystem integrity and create amenities for adjacent neighborhoods. Habitat loss and fragmentation are two of the most direct impacts of development on previously undeveloped land. Habitat fragmentation and an increased proximity of forest, agricultural land and human populations can promote interaction among vectors, pathogens, and hosts (Koren and Crawford-Brown, 2004), and in some cases lead to increased infectious disease, including Lyme disease in the northeastern part of the U.S. (LoGiudice et al. 2003). In other parts of the world changes in agricultural practices, made more intensive to feed large and growing urban populations, have contributed to the emergence of Nipah virus in Malaysia, cryptosporidiosis in Western countries, and a range of food-borne diseases in many parts of the world (Epstein et al. 2003, Patz et al. 2004). Deforestation continues to increase in many developing countries, in large part to provide materials to the BE.

In the U.S., approximately 75% of people live in coastal watersheds, with coastal urbanization and agricultural and industrial development increasing at rapid rates. As mentioned, urban development affects water quality through alterations to the natural flow of water within a watershed, particularly by increasing impervious surfaces and channeling stormwater runoff. The EPA estimates that 36 percent of the nation's lakes, rivers, and estuaries are impaired by pollution. Stormwater is often polluted by pathogens, pesticides, fertilizers from yards and farms, and a variety of other substances that have accumulated on impervious surfaces (Jackson 2003). According to research published in 2001 by Johns Hopkins University, more than 50 percent of waterborne disease outbreaks in the U.S. between 1948 and 1994 were preceded by extreme rainfall events (Curriero et al. 2001). Accelerating nutrient- and pathogen-enriched wastewater discharge accompanying coastal development is putting unprecedented pressure on estuaries that receive and process the bulk of land-based runoff. This has led to increased primary productivity or eutrophication, the symptoms of which pose a significant threat to coastal resources, ecological, and human health (Paerl 2002). Eutrophic conditions are evidenced by surface algal scums, reduced water clarity, odors, and dense algal growth on shallow water substrates. Algal blooms block the light needed by submerged aquatic vegetation, and also reduce oxygen availability. As a result, the habitat for juvenile fish and shellfish is reduced. Septic systems in low density suburban and rural residential development also reduce groundwater quality. Toxin producing dinoflagellates are found throughout the marine world and can cause a range of health effects from acute neurologic diseases (such as ciguatera and paralytic shellfish poisoning) to chronic dementia (such as amnesic shellfish poisoning from domoic acid). Disease is associated predominantly with the ingestion of contaminated fish and shellfish. However, disease can also occur through epidermal contact and inhalation (Baden et al., 1995). In the latter route, dinoflagellates are lysed by the surf and release toxins that may cause conjunctival irritation, rhinorrhea and nonproductive cough in persons who are close to the water (Morris et al., 1991). Cyanobacterial toxins have been associated with contact irritation after bathing in marine waters (Codd et al., 1989). This is yet another example of how the environment and human health are tightly linked. This linking is an important factor that needs to be considered in the context of public policies and risk assessment.

2.3 LIFE STYLE, QUALITY OF LIFE, AND HEALTH

A hallmark of urban sprawl is increased automobile travel (Frumkin 2002, 2004). Sprawl increases the distances between shopping, school, and places of employment and residence, and rarely if ever is provided with good public transport. Many parts of recently constructed BEs are centered on the automobile, in contrast to older cities and towns, which developed in an era when cars were either non-existent or rare, and where commercial and residential areas were mostly within walking distance.

For instance, parking lots are built as close as possible to their destinations in order to increase convenience and safety for motorists. Millions of Americans use a car to run almost every errand, and clearly appreciate the autonomy that car ownership offers. But the resultant automobile dependency has many adverse health effects. One of the strongest explanations for the declining health status of the U.S. population is its significant and continuing decline in physical activity. A sedentary life style is a well-established risk factor for cardiovascular disease and stroke, whereas physical activity prolongs life (Frumkin 2002, 2004). It is established that physical activity is protective against diabetes, hypertension, high cholesterol and some forms of cancer. In fact, Lagerros et al. (2004) have shown an association of moderate or vigorous recreational physical activity during adolescence and young adulthood with reduced breast cancer risk. Other studies have found that obesity increases the risk for cancer of the esophagus, colon, rectum, and breast (Key et al. 2004, Moore et al. 2004).

Lack of physical activity, combined with excessive caloric intake, is an important cause for being overweight, the prevalence of which has increased substantially in recent decades in many countries. From 1976 through 1994, the prevalence of overweight children and adolescents in the U.S. almost doubled (Jackson 2003). The death rate amongst overweight people is as much as 2.5 times higher than that of non-obese people. Obesity co-morbidities are a major problem as well, with an estimated $75 billion spent annually to treat these maladies. A major risk factor for obese people is Type II diabetes, and the current epidemic of Type II diabetes tracks closely with the increase in being overweight (Mokdad et al. 2004). It is most disturbing that the prevalence of overweight children and adolescents is increasing, probably at a greater rate than among adults. The increased prevalence of obese children is already reducing the average age of onset of Type II diabetes, and this is likely to cause an immense economic and health burden in the years to come, as the complications of Type II diabetes, including heart, kidney and eye disease, also become manifest at a younger age. This is likely even if the treatment of diabetes improves; of much greater benefit would be lifestyle changes that reduce obesity.

Automobile transport systems also generate substantial noise. Even moderate noise levels can cause serious psychological, social and other adverse effects including stress reactions which can lead to hormonal changes, increased blood pressure, increased risk of heart attacks and decreased well-being and quality of life (Riediker and Koren 2004). Reduced visibility consequent to air pollution decreases the recreational value of scenic sites. The clean air act of the USA appreciates these aesthetic and recreational aspects by including visibility as a factor for setting secondary national ambient air quality standards (U.S. EPA 2001, Riediker and Koren 2004).

Many aspects of the life style in the BE exert a toll on mental health. Stress-related problems accrue from long hours of commuting (Gee and Takeuchi 2004). Commuting stress may affect well-being and social relationships both on the roads and off the roads. As described above, there is also growing evidence that appreciation of the aesthetics of and connectivity to nature seem to positively influence human health (Frumkin 2001). In a recent article, Jackson (2003) has written: "Parks and gardens have long been noted for their restorative effects on both mental and physical health. E.O. Wilson has coined the term biophilia to express the apparently innate human attraction to nature, citing a widely-shared evolutionary explanation that relates pleasing, park-like settings to prehistoric cues for water and shelter."

Life style and quality of life are directly affected by degradation of landuse practices and transportation. Renewed appreciation of these issues may lead to future

BEs that encourage greater physical exercise, less automobile dependency and greater contact and interaction with both our fellow human beings and with nature, including more locally accessible and better protected sensitive habitats.

3. Conclusions

Understanding the relationships among population growth, development, security, the environment, human health and ecosystems is an important area of both scientific inquiry and environmental policy. The objective of risk assessment is to support decision-making by assessing risks of adverse effects on human health and the environment from biological, chemical, and physical factors, and other environmental stresses. For practical reasons, the methodologies for human health and ecological risk assessment developed independently. The potential applicability of understanding the complex interconnection between the environment and health and incorporating it into integrated human and ecological risk assessments makes this issue important and timely.

The BE presents society with a multitude of issues with consequences for the environment and human health. Most articles that discuss the BE focus on impacts on either the environment (e.g., http://www.heinzctr.org/ecosystems/) or human health and quality of life (e.g., Frumkin 2002, 2004) but seldom examine or recognize the interconnectivity that exists between the two entities. In this paper we have emphasized the effects of the BE on the environment considering the effects both on ecosystems and on human health and pointing out the interconnectivity between them. It is not further reasonable to separate ecohealth from human health. It is evident from the work reviewed in this paper that changes in the environment such as those due to the BE inevitably have consequences for ecosystems and human health.

The BE has many direct and indirect environmental effects. Stressors arising as a result of the BE directly affect habitat, ecosystems, endangered species, and water quality through land use changes, habitat fragmentation, and replacement of natural cover with impervious surfaces (Frumkin 2002, 2004; Jackson 2003). Development patterns and practices also indirectly affect environmental quality since the BE influences the travel decisions that people make. Many development patterns encourage increased use of motor vehicles, which is associated with decreased physical activity, leading to obesity and consequent increases in Type II diabetes, and cardiovascular disease, and increased cancer, as well as increased air pollution, including greenhouse gases that contribute to global climate change. Air pollution and climate change, in turn, can adversely affect respiratory illnesses, heat island effect, water quality and habitat.

This paper points out that the BE affects transportation, and land use, which consequently affect life style. These lead directly or indirectly to adverse health effects (Koren and Crawford-Brown 2004). Urban sprawl in particular has caused major changes in the life style of citizens in many countries. As well, sprawl appears to erode the sense of purpose that neighborhoods once delivered, damaging social capital and in turn reducing human wellness, security and quality of life (Riediker and Koren 2004).

In conclusion, there is significant evidence that compact, mixed-use development focused on mass transit can reduce vehicle travel and air pollution from motor vehicles. There is ample evidence that the BE matters to communities- not just for social and economic reasons, but also for environmental reasons of national concern. Issues related to our BE are growing in importance and, if left unaddressed, will make it difficult to meet national and global environmental goals. A richer and more detailed understanding of the BE and sprawl could yield development strategies that address the social, economic, fiscal, political, and environmental impacts of such growth.

4. Acknowledgment

The authors would like to thank Drs. Doug Crawford-Brown (UNC), and Laura Jackson (EPA) for their thoughtful and critical review.

5. Disclaimer

The research descried in this article has been reviewed by the National Health and environmental effects Research Laboratory, US Environmental Protection Agency, and approved for publication. Approval does not signify that the contents necessarily reflect the views and policies of the Agency, nor does mention of trade or commercial products constitute endorsement or recommendation for use.

6. References:

1. Ahmad K. (1998) WHO to improve epilepsy management in Africa. *The Lancet,* 355:1704.
2. Aron JL, Patz JA, eds. (2001) Ecosystem change and public health: a global perspective. Baltimore, MD, USA: Johns Hopkins University Press.
3. Atran S, Medin D, Ross N, et al. (1999) Folkecology and commons management in the Maya Lowlands. *Proceedings of the National Academy of Sciences,* 96(13):7598-7603).
4. Baden, DA, Fleming, LA and JA Bean. (1995) "Marine Toxins", *Handbook of Clinical Neurology.* 21:141-175
5. Berkes, F., 2003: *Sacred Ecology. Traditional Ecological Knowledge and Resource Management.* Taylor and Francis, Philadelphia, PA.
6. Center for Disease Control and Prevention, National Center for Health Statistics (1998)
7. Codd, GA, Bell, SG and WP Brooks (1989) Cyanobacterial toxins in water. *Water Sci Technol.,* 21:1-13.
8. Crutzen, P.J., 2002: Geology of mankind. *Nature*, 415, 23.
9. Curriero, F.C. Patz J.A., Rose, J.B., Lel, S. (2001). The association between extreme precipitation and waterborne disease outbreaks in the United States. *Am J Public Health,* 91(8): 1194-9
10. Daily GC, ed. (1997) Nature's Services: Societal Dependence on Natural Ecosystems. Washington, DC: Island Press.
11. Daszak P, Cunningham AA, Hyatt AD. (2000) Emerging infectious diseases of wildlife-- threats to biodiversity and human health. *Science,* 287:443-449.
12. Davey-Smith G. Commentary: Behind the Broad Street pump: aetiology, epidemiology and prevention of cholera in mid-19th century Britain. *International Journal of Epidemiology,* 2002;31:920-932.
13. DiGulio RT and WH Benson eds. (2002). Interconnections Between Human Health and Ecological Integrity. Society of Environmental Toxicology and Chemistry (SETAC), Pensacola. FL, USA.
14. DiGulio, R.T. and E. Monosson eds. (1996). Interconnections Between Human Health and Ecosystem Health. Chapman and Hall, London.
15. Epstein P. (1999) Climate and health. *Science,* 285:347-348.

16. Epstein, P.R., E. Chivian, and K. Frith, 2003: Emerging Diseases Threaten Conservation [editorial]. *Environmental Health Perspectives*, 111(10): A506-507.
17. Fisher, WS (guest editor) 2001. Human and Ecological Risk Assessment: Indicators in Health and Ecological risk Assessment (special Issue). Eds. BL Johnson, EL Calabrese, BG Callahan, PM Chapman, and RA Pastorok. HERA 7(5), October 2001. CRC Press.
18. Fox W. (1990) Toward a Transpersonal Ecology. Developing New Foundations for Environmentalism. Boston: Shambhala.
19. Friedmann E, Thomas SA. (1995) Pet ownership, social support, and one-year survival after acute myocardial infarction in the cardiac arrhythmia suppression trial (CAST). *American Journal of Cardiology,* 76:1213-1217.
20. Frumkin, H. (2001). Beyond toxicity: Human health and the natural environment. *Am J Prev Med,* 1:20(3), 234-244.
21. Frumkin H. (2002) Urban sprawl and public health. *Public Health Rep.*, 117(3): 201-17.
22. Frumkin H, L.Frank, and R. Jackson eds. (2004) Urban Sprawl and Public Health: Designing, Planning, and Building for Healthy Communities. Island Press, Washington.
23. Gee GC, Takeuchi DT. (2004) Traffic stress, vehicular burden and well-being: a multilevel analysis. *Soc Sci Med.* 59(2):405-14.
24. Handy SL, Boarnet MG, Ewing R, Killingsworth RE. (2002) How the built environment affects physical activity.
25. Hanna K, Coussens C. (2001) Rebuilding the Unity of Health and the Environment: A New Vision of Environmental Health for the 21st Century. Washington, D.C.: National Academy Press.
26. Health Canada. (1997) Health and the environment: Partners for life. Ottawa, Health Canada.
27. Jackson RJ. (2003) The impact of the built environment on health: an emerging field. *Am J Public Health*, 93(9):1382-4.
28. Jackson, L.E. (2003) The relationship of urban design to human health and condition. *Landscape and Urban Planning*, 64, 191-200.
29. Kalkstein LS. (2000) Saving lives in extreme weather in summer [editorial]. *BMJ*; 321(7262):650-651.
30. Key TJ, Schatzkin A, Willett WC, Allen NE, Spencer EA, Travis RC. (2004). Diet, nutrition and the prevention of cancer. *Public Health Nutr*, 7(1A):187-200.
31. Koren HS, Crawford-Brown D. (2004) A framework for the integration of ecosystem and human health in public policy: two case studies with infectious agents. *Environ Res.*, 95(1):92-105.
32. Koren, H.S. Environmental Risk Factors in Atopic Asthma. (1997) I. Genetic and Environmental Factors in Atopic Diseases. *Int. Arch Allergy Immunol*, 113: 65-68.
33. Kuehni CE, Davis A, Brooke AM, Silverman M. (2001) Are all wheezing disorders in very young (preschool) children increasing in prevalence? *The Lancet,* 357:1821-1825.
34. Lagerros YT, Hsieh SF, Hsieh CC. (2004) Physical activity in adolescence and young adulthood and breast cancer risk: a quantitative review. *Eur J Cancer Prev.*, 13(1):5-12.
35. LoGiudice K, Ostfeld RS, Schmidt KA, Keesing F. (2003) The ecology of infectious disease: effects of host diversity and community composition on Lyme disease risk. *Proc Natl Acad Sci U S A*, 100(2):567-71.
36. Mannino DM, Homa DM, Pertowski CA, Ashizawa A, Nixon LL, Johnson CA, Ball LB, Jack E, Kang DS. (1998) Surveillance for asthma--United States, 1960-1995. MMWR CDC Surveill Summ., 24;47(1):1-27.
37. McConnell R, Berhane K, Gilliland F, London SJ, Islam T, Gauderman WJ, Avol E, Margolis HG, Peters JM (2002) Asthma in exercising children exposed to ozone: a cohort study. *The Lancet*, 359(9304):386-91.
38. McDonnell WF, Abbey DE, Nishino N, Lebowitz MD (1999) Long-term ambient ozone concentration and the incidence of asthma in nonsmoking adults: the AHSMOG Study. *Environ Res.*, 80(2 Pt 1):110-21.
39. McMichael AJ. (1999) Prisoners of the proximate. Loosening the constraints on epidemiology in an age of change. *American Journal of Epidemiology,* 149(10): 887-897.
40. Millenium Ecosystem Assessment (2003) Ecosystems and human well-being. In: Ecosystems and Human Well-being. Washington: Island Press: 71-84
41. Moberg CL, Cohn ZA. (1991) René Jules Dubos. *Scientific American,* 264(5):32-38.
42. Mokdad AH, Ford ES, Bowman BA, Dietz WH, Vinicor F, Bales VS, Marks JS. (2003) Prevalence of obesity, diabetes, and obesity-related health risk factors, *JAMA*, 1289(1):76-79.

43. Moore LL, Bradlee ML, Singer MR, Splansky GL, Proctor MH, Ellison RC, Kreger BE. (2004) . BMI and waist circumference as predictors of lifetime colon cancer risk in Framingham Study adults. *Int J Obes Relat Metab Disord.*, 28(4):559-67
44. Morris, PD, Campbell, DS, Taylor, TJ and JI Freeman. (1991) Clinical and epidemiological features of neurotoxic shellfish poisoning in North Carolina. *Am J Pub Health,* 81:471-474.
45. Nye, E. and M. Gibson, (1997): *Ronald Ross: Malariologist and polymath.* MacMillan, London.
46. Oxford JS, Sefton A, Jackson R, Innes W, Daniels R, Johnson N. (2002) World War I may have allowed the emergence of "Spanish" influenza. *The Lancet Infectious Diseases*, 2(2):111-114.
47. Paerl HW. (2002) Connecting atmospheric nitrogen deposition to coastal eutrophication. *Environ Sci Technol.*, 36(15):323A-326A.
48. Patz JA, Daszak P, Tabor GM, Aguirre AA, Pearl M, Epstein J, Wolfe ND, Kilpatrick AM, Foufopoulos J, Molyneux D, Bradley DJ (2004) Working Group on Land Use Change and Disease Emergence. Unhealthy landscapes: Policy recommendations on land use change and infectious disease emergence. *Environ Health Perspect.*, 112(10):1092-8.
49. Pope CA 3rd, Burnett RT, Thurston GD, Thun MJ, Calle EE, Krewski D, Godleski JJ.(2004) Cardiovascular mortality and long-term exposure to particulate air pollution: epidemiological evidence of general pathophysiological pathways of disease. *Circulation*, 6:109(1):71-7.
50. Prusiner SB. (1997) Prion diseases and the BSE Crisis. *Science* 278:245-251.
51. Rahman F. (1998) Angels and spirits in the world of illness. *The Lancet* 352: 1218-1219.
52. Ramakrishnan, P., K. Saxena, and U. Chandrashekara, 1998: *Conserving the Sacred: For* Biodiversity Management. UNESCO, Oxford, IBH, New Delhi, 480 pp.
53. Rapport, D.J., (1997): The dependency of human health on ecosystem health *Ecosystem Health,* 3(4), 195-196
54. Ricketts TH, Daily GC, Ehrlich PR, Michener CD. (2004): Economic value of tropical forest to coffee production. *Proceedings of the National Academy of Science,* 101(34):12579-12582.
55. Riediker M, Koren HS. (2004) The importance of environmental exposures to physical, mental and social well-being. *Int J Hyg Environ Health*, 207(3):193-201.
56. Rose G (1990) Reflections on changing times. *BMJ,* 301:683-687.
57. Ryan F. (1993) The Forgotten Plague : How The Battle Against Tuberculosis Was Won--And Lost. 1st American ed. Boston: Little Brown.
58. Serpell J. (1991) Beneficial effects of pet ownership on some aspects of human health and behaviour. *Journal of the Royal Society of Medicine,* 84:717-20.
59. Siegel J. (1990) Stressful life events and use of physician services among the elderly: the moderating role of pet ownership. *Journal of Personality and Social Psychology,* 58:1081-1086.
60. Susser M, Susser E. (1996) Choosing a future for epidemiology: I Eras and paradigms. *American Journal of Public Health,* 86:668-673.
61. The State of the Nation's Ecosystems (2002). Cambridge University Press. http://www.heinzctr.org/ecosystems/
62. Ulrich RS. (1984) View through a window may influence recovery from surgery. *Science,* **224:**420–421.
63. U.S. Department of Health and Human Services. (2000) Healthy People 2010. Washington, DC, U.S. Department of Health and Human Services
64. U.S. Environmental Protection Agency Office of Policy, Economics, and Innovation,Our built and natural environment, January 2001. Publication No.: EPA 231-R-01-002. http://www.smartgrowth.org
65. von Mutius E (2003) A conundrum of modern times that's still unresolved. *Eur Respir J.*,22(5):719-20.
66. Ward B, Dubos RJ. (1972) Only One Earth; the Care and Maintenance of a Small Planet. [1st] ed. New York: Norton.
67. Watts J. (2001) Mercury poisoning victims could increase by 20 000. *The Lancet,* 358, 1349.
68. Webster RG. (2004) Wet markets—a continuing source of severe acute respiratory syndrome and influenza? *The Lancet,* 363:234-236.
69. WHO (1948). "Preamble to the Constitution of the World Health Organization as adopted by the International Health Conference, New York, 19-22 June, 1946; signed on 22 July 1946 by the representatives of 61 States (Official Records of the World Health Organization, no. 2, p. 100) and entered into force on 7 April 1948."

STATISTICAL MODELS FOR DISTRIBUTIONS OF AMBIENT FINE PARTICULATE MATTER

S. YEH
ORISE Research Fellow, National Risk Management Research Laboratory, Office of Research and Development, U.S. Environmental Protection Agency, 109 TW Alexander Drive, RTP, NC 27711, USA

M. SMALL
Department of Engineering and Public Policy
Carnegie Mellon University, Pittsburgh PA 15213, USA

Abstract

This study evaluates the usefulness of structured non-linear regression models for the prediction of annual ambient fine particulate matter (FPM) concentration distributions. The method developed in this study provides a way to examine and display results for the yearly distribution of FPM when testing emissions control strategy performance. The models are developed using three daily gaseous pollutant concentrations (oxides of nitrogen (NO_x), sulfur dioxide (SO_2), and total hydrocarbons (THC)) and four meteorological measures (wind speed, temperature, relative humidity and precipitation) as explanatory variables. The models are fitted using data from the North Long Beach, Rubidoux (Riverside) and Azusa stations in Los Angeles County and Riverside County, CA for a recent 7-year period (1988-1994). The statistical model is tested for the year 1995 based on precursor concentrations and meteorological conditions in that year, and found to provide reasonably good prediction, though the annual average FPM concentration is overestimated by an average of 26 percent across the three stations. The response surfaces of $PM_{2.5}$ concentrations with respect to all input variables are plotted, and the predicted changes in daily, annual average and annual 98th percentile base-year (1995) $PM_{2.5}$ concentrations are predicted for different precursor reductions. The predicted effects of precursor reductions are further explored by comparing predicted and observed FPM concentrations for 1999 (though the absence of THC data for this year restricts this comparison to plausible ranges). The method developed in this study provides a way to examine and display results for the predicted concentration distributions when evaluating emission control strategy performance. The potential usefulness and limitations of a statistical model of this type are discussed.

1. Introduction

1.1 PARTICULATE MATTER AND HEALTH EFFECTS

B. Morel and I. Linkov (eds.),
Environmental Security and Environmental Management: The Role of Risk Assessment, 127–150.

Airborne particulate matter (PM) is the suspended aerosols of relative stable solid or liquid particles in the air. It is a generic term that refers to a broad class of aerosols that are varying in size, chemical composition and physical characteristics. Particles can be divided into three categories by source and formation mechanisms: coarse, primary fine particles, and secondary fine aerosols. Coarse particles are generated mainly from mechanical processes through crushing or grinding or suspension of dust. Sources of coarse particles include resuspension of dust and soil, biological formation and fly ash from coal and oil burning. Fine particles (aerodynamic particle diameter < 2.5 μm) are largely secondary pollutants formed through chemical reaction, nucleation and condensation of gases onto existing smaller particles or reaction in the liquid phase. Sources of fine particles include the combustion of coal, oil, gasoline, diesel, and wood; and atmospheric transformation of NO_x, SO_2, ammonium and organics. Fine particles are composed of sulfate (SO_4^{2-}), nitrate (NO_3^-), ammonium (NH_4^+), hydrogen ion (H^+), elemental carbon, organic compounds and metals. The atmospheric residence times of fine particles range from days to weeks. (U.S. EPA 1996a)

Over the past three decades, epidemiologists have identified TSP (Total Suspended Particles), PM_{10} (particulate matter less than 10 μm in aerodynamic diameter), $PM_{2.5}$ (particulate matter less than 2.5 μm in aerodynamic diameter), and sulfate as the major (aerosol) pollutant measures with statistically significant, positive associations with acute and chronic health effects. These include premature mortality, increased hospital admissions and emergency room visits (primarily among the elderly and individuals with cardiopulmonary disease), increased respiratory symptoms and disease (in children and individuals with cardiopulmonary disease such as asthma), decreased lung function (particularly in children and individuals with asthma) and alterations in lung tissue, structure and respiratory tract defense mechanisms (U.S. EPA 1996a; b).

1.2 CONTROL STRATEGIES AND EVALUATION APPROACHES IN THE U.S.

On July 18, 1997, the U.S. Environmental Protection Agency updated the new National Ambient Air Quality Standards (NAAQS) for particulate matter (U.S. EPA 1997). The new standards were subsequently challenged in courts and later upheld by the Supreme Court on February 27, 2001. Under the Clean Air Act, states are required to revise their implementation plans to demonstrate that emission control strategies will bring the polluted areas into attainment of the standard.

A key issue in finding effective approaches to meet PM air quality objectives is to understand the relationship between precursor emissions and resulting concentrations of PM. Millions of pieces of emissions and meteorological data can be needed to describe the events leading to local PM formation for a single high pollution episode. Because of this complexity, a common practice has been to model the effect of emission controls on atmosphere PM concentration for only one or a few historically observed high PM events in a few selected areas when evaluating proposed emission control programs. Numerous numerical modeling techniques based on thermoequilibrium of chemical reactions combined with dispersion models have been applied to predict ambient PM concentrations for compliance, such as the Urban

Airshed Model (UAM). However, these applications are significantly limited by the capabilities of available PM models and databases (Seigneur, Pai et al. 1999). Existing PM models require comprehensive databases on meteorology, source information, and emission profiles. Very frequently, site-specific and expensive field experiments are necessary for input to these models for model evaluation (Seigneur, Pai et al. 1999). In most places other than the South Coast of California, none of these databases or models exists for evaluation. Ambient concentrations of pollutants in outdoor air are measured at more than 4000 monitoring stations owned and operated mainly by state environmental agencies. However, only a small number of locations have enough information for detailed mechanistic model assessment. In addition, these applications are very time-consuming and there are great uncertainties associated with even the most expensive and extensive databases, such as emission inventories and PM monitoring and speciation data (Seigneur, Pai et al. 1999; Seigneur 2001).

Another challenge with the episodic model is that, given that secondary pollutants like fine particulate matter are formed in the atmosphere by nonlinear chemical reactions, it is not obvious that the days with the highest observed concentrations historically are the hardest ones to bring below the air quality standard (Lefohn, Shadwick et al. 1998; Winner and Cass 2000). Lefohn and Shadwick et al (1998) suggested that in the case of ozone, the highest values are brought down at a faster rate than values near the annual average. As such, it is important to predict the annual distribution of PM concentrations expected to result from implementation of different control strategies, instead of just the concentration levels associated with a single event. There have been attempts to predict the effects of emission control on the long-term frequency distribution of regional ozone concentrations with photochemical airshed models (Calbo, Pan et al. 1998; Winner and Cass 2000; Hanna and Davis 2002), but no similar work has been done for prediction of the long-term distribution of particulate matter under alternative attainment abatement strategies.

1.3 THE RELEVANCE OF THIS RESEARCH TO THE MIDDLE EAST AND OTHER REGIONS

Around the world, particulate air pollution in urbanized cities has contributed to significant adverse human health effects as we had described in Section 1.1. For example, urban populations in Middle Eastern countries have increased dramatically in recent years: from 48% of the total population in 1980 to 57% in 1995; and they are expected to increase even further (Soubbotina and Sheram 2000). The increasing trend of urban population and higher air pollution levels as a result of economic growth have imposed serious health threats in countries around the world (Soubbotina and Sheram 2000). For example, a report conducted by the European Commission entitled "Air Pollution and Health: A European Information System (APHEIS)," found that Tel Aviv, Israel has the highest mean value of PM_{10} levels (65 $\mu g/m^3$ or 85 $\mu g/m^3$ after correction for the time and location of samples) and 95th percentile PM_{10} concentration (105 $\mu g/m^3$ or 136 $\mu g/m^3$ after correction) among 26 European cities. The annual mean value of PM_{10} pollution in Tel-Aviv increased by 17% in just two years, from 56.4$\mu g/m^3$ to 66$\mu g/m^3$ between 1996 and 1998. It is estimated that the reduction of long-term $PM_{2.5}$ levels in Tel-Aviv from 42.9$\mu g/m^3$ to 20$\mu g/m^3$ would save ~1,300 lives per year, including 750 cardio-pulmonary and 76 lung cancer mortalities per year. In all countries APHEIS investigated, the report suggests that the health risks imposed by urban pollution are on average almost four times the annual death rate from AIDS, 2.6 times the leukemia rate, and 1.5 times the annual rate of traffic fatalities. Similarly, high emission levels of CO, NO_x, hydrocarbons, and lead from urban transportation and the industrial sector pose serious health risks to cities of Lebanon (Lebanon Ministry of Environment 1997). In a study of VOC and PM pollution in Cairo, Egypt, Abu-Allaban et al. (2002) observed very high PM concentrations and, using source apportionment methods, determined that the major sources of PM_{10} were geological material, mobile source emissions, and vegetative burning, while major sources of $PM_{2.5}$ included mobile source emissions, vegetative burning, and secondary formation processes in the atmosphere. El-Metwally and Ramadan (2003) also document the serious damage to cultural materials associated with regional air pollution in the Cairo area.

It is clear that there is a pressing need to provide tools for policy makers to evaluate options for air pollution control and the effectiveness of alternative control strategies in the Middle East and other rapidly urbanizing areas. However, "identifying tools that are appropriate, up to date, and scientifically credible can be a daunting task, especially for those with limited resources" (Bell, Cohen et al. 2002: 2). Furthermore, the efforts required for data collection, database development, and training for the use of sophisticated modeling tools may yet take years to develop.

The objective of this study is to develop a purely data-driven approach that uses input variables that are obtainable from routine daily measurements by the EPA or local government agencies. Thus, the purpose of this work is to provide a simplified methodology that can be used to provide useful information about the general scheme of emission controls that can be compared with sophisticated models should they become available later on. The approach adopted in this study is to relate ambient concentrations of fine particulate matter to its precursor concentrations. It is assumed when using this approach that relationships are available between precursor emissions and precursor ambient concentrations (e.g., that there exists a linear relationship between them). The effect of precursor emission controls on fine particulate matter concentration can then be computed by first applying these emission precursor relationships, followed by application of the precursor-$PM_{2.5}$ relationships developed herein.

The paper is organized as follows. Section 2 describes the data sources and the nature of the air quality data (concentrations, trends) for the three monitoring stations. The structure of the regression models and dependent/independent variables are defined. We fit the models for the period 1988-1994 and validated for the year 1995 for all three stations. The observed response surfaces of independent variables to $PM_{2.5}$ concentrations are presented and plotted for this year (1995) in Section 3. Section 4 uses the statistical models developed in Sections 2 and the response surfaces shown in Section 3 to predict the effects of future emission controls on fine particulate matter concentrations at the three monitoring stations. The predicted results of precursor emission controls are compared to 1999 ambient $PM_{2.5}$. The overall framework of the modeling approach is summarized in Figure 1.

Figure 1: Flowchart of the regression model scheme and its policy application for forecasting the effects of abatement strategies. In Steps 1 and 2, the regression model is fitted and validated using observed ambient concentration data. In Steps 3 and 4, the model is used to project the impact of hypothetical emission control strategies on the distribution of PM and these are validated for 1999 air quality data.

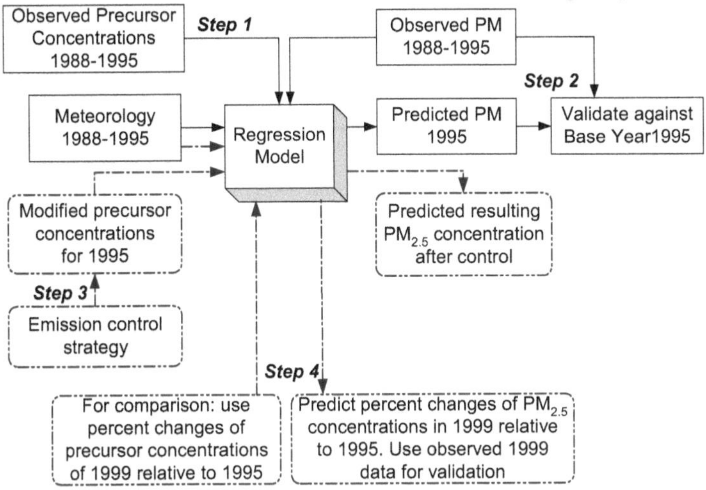

2. Data and model description

2.1 MONITORING STATIONS AND $PM_{2.5}$ AIR QUALITY

The structured statistical regression models evaluated in this study are developed using input parameters that include three daily precursor concentrations and four meteorological variables. The models are applied to the data published for the three FPM monitoring stations available in the California South Coast Air Basin at the time of the study, i.e., Azusa and North Long Beach in Los Angeles County and Rubidoux in Riverside County (Air Resources Board 1999). Data from 1988-1994 are used to estimate the regression coefficients for the regression models and the resulting precursor regression models are validated against the air quality data in 1995.

The three monitoring stations and their annual average concentrations in 1988 and 1995 are shown in Figure 2. There is a significant downward trend in $PM_{2.5}$ concentration levels throughout the study period for all three monitoring stations, partly due to California's continuous effort to reduce ozone and criteria pollutant air pollution during that time. The Long Beach sampling site is located near the upwind edge of the South Coast Air Basin and thus exhibits relatively low concentrations, while the Riverside and Azusa sampling sites are located in relatively polluted downwind areas.

Figure 2: Map of the three fine particle mass monitoring stations in Los Angeles County and Riverside County. The annual average $PM_{2.5}$ concentrations in 1988 and 1995 for the three monitoring stations are also shown on the graph.

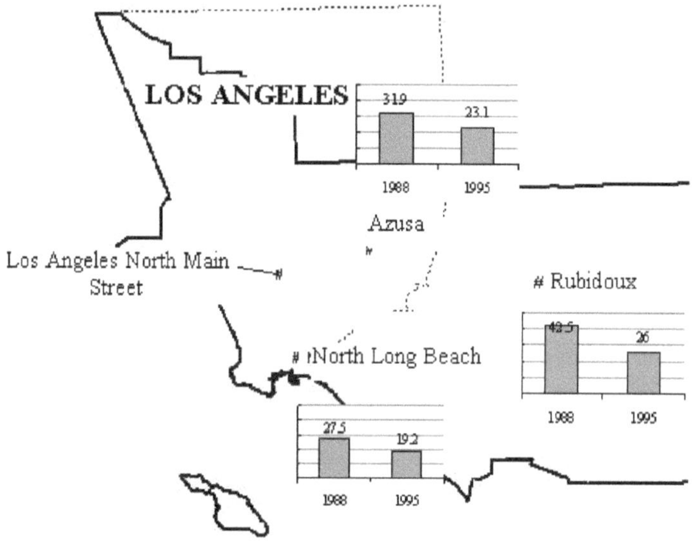

2.2 INPUT VARIABLES

Since the objective of this study is to develop a purely data-driven approach that uses input variables that are obtainable from routine daily measurements, we base our analysis on the data provided by the California Air Resources Board (ARB) which compiles air quality data from monitoring stations throughout California. Two criteria were used to select input independent variables and the sampling period for this study: the completeness and relevance of the data. Measurements of three gaseous pollutants that have the most complete data set and are most relevant to the formation of PM fine particles include NO_x (oxides of nitrogen), THC (total hydrocarbons), and SO_2 (sulfur dioxide). The daily measurements of maximum 1-hour average concentrations are used as the representative variables for these gaseous pollutants. Meteorological data are used to determine the meteorological parameters that most influence $PM_{2.5}$ air quality, and to quantify the relationship between these meteorological parameters and $PM_{2.5}$ concentrations. Daily weather data (average wind speed, temperature, relative humidity and precipitation) were obtained for the study period from the nearest airport weather station at the Los Angeles International Airport. For data analysis and model evaluation, Mathematica® software is used.

Several studies on fine particle concentration and chemical composition in Southern California, including the three monitoring stations in this study, have shown that carbonaceous aerosols (elemental carbon + organic carbon), ammonium sulfate, and ammonium nitrate (NH_4NO_3) are the major contributors to fine particle mass concentrations (Chow, Watson et al. 1994; Christoforou, Salmon et al. 2000; Kleeman and Cass 2001) (Table 1).

Table 1. Particle chemical composition (%) for annual average and days with maximum 24-hr average fine particle mass concentration ($\mu g/m^3$) at Azusa, Long Beach and Rubidoux, CA. Values are the averages of the three years (1982, 1986 and 1993) reported in the study. Source: Christoforou, Salmon et al. (2000).

Monitoring station	Fine mass ($\mu g/m^3$)	Organic material	Elemental carbon	SO_4^{-2}	NO_3^-	NH_4^+	Other
Annual average							
Azusa	26.95	40.5%	8.2%	17.6%	11.4%	8.8%	13.6%
N. Long Beach	26.47	43.1%	10.7%	20.0%	11.4%	8.9%	5.9%
Rubidoux	41.17	26.3%	5.6%	11.5%	27.1%	11.5%	18.0%
Maximum 24-hr average							
Azusa	97.05	14.6%	4.8%	10.3%	23.9%	11.5%	34.9%
N. Long Beach	107.00	28.8%	8.3%	6.5%	27.9%	10.0%	18.5%
Rubidoux	174.73	17.1%	3.4%	5.6%	36.2%	12.0%	25.7%

Particles at Azusa and Riverside contain a significant amount of secondary aerosol which accumulates due to gas-to-particle conversion production as these particles are advected across the urban Los Angeles area (Kleeman, Hughes et al. 1999; Kleeman and Cass 2001; Mysliwiec and Kleeman 2002). By using trajectory models, it is observed that the single largest contribution to regional $PM_{2.5}$ in the South Coast Air Basin surrounding Los Angeles is associated with the accumulation of secondary ammonium nitrate onto combustion particles and onto hygroscopic background marine particles (Kleeman and Cass 1999; Kleeman, Hughes et al. 1999; Kleeman and Cass 2001). Other categories of fine particles are predicted to be associated with: (1) large mineral dust and road dust particles that accumulate only a small amount of secondary products; and (2) primary combustion particles that are released from diesel vehicles, noncatalyst gasoline-powered vehicles, and food processing, and subsequently grow by accumulation of secondary reaction products (Kleeman, Hughes et al. 1999). It is also known that the coastal aerosol at Long Beach contains sea-salt particles and primary carbon particles that are changed significantly as these particles accumulate secondary ammonium nitrate and organics during their travel across the air basin to Riverside (Hughes, Allen et al. 2000).

Nitric oxide (NO) is formed in great quantities during combustion processes and is gradually oxidized to NO_2 once emitted to the atmosphere. The combination of NO and NO_2 — referred to as NO_x — contributes to acid rain and (together with volatile organic compounds) the formation of ground-level ozone. Hydrocarbons are compounds containing various combinations of hydrogen and carbon atoms. They may be emitted into the air by natural sources (e.g., trees, oil or gas seeps), and as a result of fossil and vegetative fuel combustion, fuel volatilization, solvent use, and motor vehicle exhaust emissions. These compounds are a major contributor to secondary aerosols. Sulfur dioxide (SO_2) is an air pollutant emitted to the atmosphere primarily through the burning of sulfur containing materials. It reacts with NH_3 to form aerosol sulfate $(NH_4)_2SO_4$ in the atmosphere. Sulfates can be primary, such as particulate sulfate (SO_3, H_2SO_4) that subsequently adds to existing particles, or nucleated to form new particles. Secondary sulfates are produced as sulfuric acid from the SO_2 oxidation. NO_x is converted to HNO_3 through gas-phase or aqueous-phase reactions, coupled with NH_3 to form NH_4NO_3, ammonium nitrate. Organic compounds are major components of $PM_{2.5}$ in the Los Angeles area (Kleeman and Cass 1999; Kleeman, Hughes et al. 1999; Kleeman and Cass 2001). Hydrocarbons with about 8 or more carbons, when oxidized, lead to products with low volatility that form particles. These hydrocarbons are a subset of "non-methane hydrocarbons" (NMHC), which would be a better predictor of FPM formation. However, the concentrations of NMHC were only measured for a limited portion of the study period and were thus not utilized in the model. The routinely measured parameter "Total Hydrocarbon" includes methane and NMHC, of which approximately 70 percent of THC consists of methane. Motor vehicle exhaust is the main contributor to NMHC concentrations in the South Coast Air Basin, with lesser contributions from liquid gasoline and gasoline vapor.

Meteorological conditions are beyond the control of policy makers; they are included as explanatory variables in the models to avoid specification bias (from omitting relevant variables) and to understand the relationships and interactions between gaseous pollutants in forming particles under different meteorological conditions. Scatter diagrams of $PM_{2.5}$ versus meteorological variables (not shown) indicate that the daily average wind speed has an inverse relationship with the daily $PM_{2.5}$ concentration, while the daily relative humidity seems to be positively associated with $PM_{2.5}$. As the ambient relative humidity increases, the solid particles absorb water, produce a saturated aqueous solution and exhibit an increase in the particle mass as well as particle sizes (Seinfeld and Pandis 1998). In the event of precipitation, the raindrops fall through the air and collide with aerosol particles, sweeping them out as they fall. As a result, PM mass is expected to decrease when precipitation occurs. A scatter plot of 24-h $PM_{2.5}$ mass versus daily precipitation shows that $PM_{2.5}$ concentrations are systematically lower when precipitation occurs, however, there is no significant trend/pattern between the amount of precipitation and the same-day $PM_{2.5}$ mass (not shown). Therefore, a dummy variable is used to represent the precipitation occurrence (0 for a dry day, 1 for a wet day with precipitation less than 0.50 inches, and 2 for a wet day with precipitation equal to or greater than 0.50 inches).

2.3 REGRESSION MODELS

Two multivariate linear regression models were evaluated:

Model I. 2nd-order linear model:

$$\mathrm{PM_i} = \beta_0 + \sum_{k=1}^{4} \beta_{\mathrm{k}} X_{k,i} + \sum_{m=1}^{4} \beta_{\mathrm{m}} X_{m,i} + \sum_{k=1}^{4}\sum_{p=k}^{4} \beta_{\mathrm{k,p}} X_{k,i} X_{p,i}$$

$$+ \sum_{k=1}^{4}\sum_{m=1}^{4} \beta_{\mathrm{k,p}} X_{k,i} X_{m,i} + \sum_{m=1}^{4}\sum_{p=m}^{4} \beta_{\mathrm{k,p}} X_{m,i} X_{p,i} + \varepsilon_i \qquad (1)$$

Model II. 2nd-order log-linear model:

$$\ln \mathrm{PM}_i = \beta_0 + \sum_{k=1}^{4} \beta_k \ln(X_{k,i}) + \sum_{m=1}^{4} \beta_m X_{m,i} +$$
$$+ \sum_{k=1}^{4}\sum_{p=k}^{4} \beta_{k,p} \ln(X_{k,i}) \ln(X_{p,i}) \sum_{k=1}^{4}\sum_{m=1}^{4} \beta_{k,p} \ln(X_{k,i}) X_{m,i}$$
$$+ \sum_{m=1}^{4}\sum_{p=m}^{4} \beta_{k,p} X_{m,i} \ln(X_{p,i}) + \varepsilon_i \qquad (2)$$

where, for a set of observations on day i, the predicted variable FPM (PM fine particle mass concentration) is a linear combination of an intercept β_0, a set of predictor precursor concentration variables X_k and meteorological variables X_m with matching coefficients β, and a residual error ε. The reasons to use a log model are two-fold. First this transforms FPM and precursor distributions from approximate-lognormal to approximate-normal distributions. Secondly log transformation avoids negative concentrations in the prediction. These and other advantages of the log-linear model are discussed in more detail in the next section.

3. Results

The regression models, estimated regression coefficients and regression statistics are summarized in Table 2. SO_2 measurements were not available for the Azusa monitoring station during the study period and therefore could not be included as an explanatory variable in the regression model.

The second-order regression models have non-linear functional forms that can approximate more complex relationships between the dependent variables and the respective independent variables. In most cases, $PM_{2.5}$ formation by gaseous precursors under different meteorological conditions is complex and non-linear behavior is commonly observed. In general, the 2nd-order log-linear model (Model II) has higher R^2 values than Model I and therefore only the results of Model II are reported here. The explanatory variables in the regression models are correlated to some degree for a few reasons. Emission sources usually emit more than one gaseous pollutant at the same time, causing the daily gaseous pollutant concentrations to be correlated. Meteorological conditions also tend to influence the gaseous pollutant levels in the same manner that they affect FPM concentrations, for example, low wind speed resulting in higher pollution concentrations. In general, all pollutants at the three monitors, including fine particulate matter, have higher concentration levels in the winter and lower concentration levels in the summer. Note that multicollinearity violates no regression assumptions.

The regression coefficients are still unbiased and consistent, but possess larger standard errors, i.e. the coefficients cannot be estimated with great precision or accuracy (Gujarati 1995). This is acceptable given that our objective is to estimate a linear combination of these coefficients, the *estimable function*, i.e. the prediction of the expected value of the dependent variable (Gujarati 1995).

Table 2. Estimate of the log-linear multivariable regression model, regression coefficients, and statistical summaries for daily ambient fine particulate matter concentrations at North Long Beach, Rubidoux and Azusa, CA

	North Long Beach		Rubidoux		Azusa	
Explanatory variable	Coefficient	Std error	Coefficient	Std error	Coefficient	Std error
Constant	-6.11E-01	5.76E+00	-1.24E+01	4.47E+00	-1.86E+00	4.17E+00
Ln NO_x	-4.18E-01	1.50E+00	4.18E-01	9.36E-01	-1.00E-01	1.03E+00
Ln SO_2	5.29E-01	3.59E-01	6.24E-01	2.24E-01	NA	NA
Ln THC	3.22E+00	3.19E+00	5.15E+00	2.31E+00	6.50E+00	2.09E+00
Temp	1.66E-01	1.08E-01	2.85E-01	9.61E-02	3.28E-02	8.91E-02
Ws	-6.89E-02	2.71E-02	7.21E-02	2.76E-02	-2.01E-03	2.67E-02
RH	1.43E-02	3.78E-02	6.58E-02	3.34E-02	-1.55E-02	3.11E-02
Prcp	4.40E-01	1.52E+00	-1.83E+00	1.42E+00	4.51E+00	1.48E+00
$(\text{Ln } NO_x)^2$	-1.36E-02	1.63E-01	8.85E-02	8.53E-02	-1.78E-01	1.10E-01
$(\text{Ln } SO_2)^2$	1.81E-02	8.92E-03	2.52E-02	1.20E-02	NA	NA
$(\text{Ln THC})^2$	-2.49E-01	6.14E-01	-8.47E-01	4.61E-01	-9.51E-01	3.62E-01
$(\text{Temp})^2$	-1.32E-03	6.75E-04	-1.85E-03	7.07E-04	-8.30E-04	6.07E-04
$(\text{Ws})^2$	9.99E-05	4.64E-05	1.15E-06	3.86E-05	4.72E-05	3.39E-05
$(\text{RH})^2$	-4.15E-05	1.40E-04	1.90E-04	1.28E-04	2.74E-04	1.08E-04
$(\text{Prcp})^2$	1.46E-01	3.36E-01	2.54E-01	2.24E-01	-9.72E-04	2.18E-01
$(\text{Ln } NO_x)\times(\text{Ln } SO_2)$	5.83E-02	4.41E-02	-1.34E-02	1.99E-02	NA	NA
$(\text{Ln } NO_x)\times(\text{Ln THC})$	4.79E-01	5.60E-01	2.92E-01	3.29E-01	1.16E+00	3.52E-01
$(\text{Ln } NO_x)\times\text{Temp}$	9.45E-03	1.57E-02	-1.19E-02	1.08E-02	-2.87E-02	1.15E-02
$(\text{Ln } NO_x)\times\text{Ws}$	1.01E-03	4.48E-03	6.00E-03	3.16E-03	7.74E-04	3.53E-03
$(\text{Ln } NO_x)\times\text{RH}$	-5.79E-03	8.58E-03	-3.83E-03	5.77E-03	9.50E-05	4.44E-03
$(\text{Ln } NO_x)\times\text{Prcp}$	-7.45E-02	2.78E-01	3.14E-02	1.63E-01	4.79E-01	2.21E-01
$(\text{Ln } SO_2)\times(\text{Ln THC})$	-7.94E-02	7.56E-02	-5.16E-02	5.72E-02	NA	NA
$(\text{Ln } SO_2)\times\text{Temp}$	7.57E-04	3.60E-03	-2.63E-04	1.70E-03	NA	NA
$(\text{Ln } SO_2)\times\text{Ws}$	-1.18E-03	1.02E-03	-7.90E-04	4.98E-04	NA	NA
$(\text{Ln } SO_2)\times\text{RH}$	8.87E-04	1.96E-03	-1.32E-03	7.17E-04	NA	NA
$(\text{Ln } SO_2)\times\text{Prcp}$	-1.51E-02	3.66E-02	5.31E-02	3.34E-02	NA	NA
(Ln THC)×Temp	-3.36E-02	3.05E-02	-6.00E-04	2.70E-02	-8.58E-03	2.67E-02
(Ln THC)×Ws	-9.66E-04	6.22E-03	-1.14E-02	7.77E-03	1.03E-03	8.06E-03
(Ln THC)×RH	4.09E-03	1.37E-02	-1.71E-02	1.32E-02	-1.58E-02	1.15E-02
(Ln THC)×Prcp	5.83E-01	4.89E-01	7.15E-01	4.30E-01	3.39E-01	4.74E-01
Temp×Ws	7.67E-04	3.64E-04	-5.50E-04	3.73E-04	-1.40E-05	3.39E-04
Temp×RH	1.12E-04	4.24E-04	-4.92E-04	4.23E-04	4.56E-04	3.44E-04
Temp×Prcp	-3.28E-02	1.74E-02	1.98E-02	2.03E-02	-2.44E-02	1.77E-02
Ws×RH	-2.11E-04	1.43E-04	-3.56E-04	1.23E-04	-8.43E-05	9.42E-05
Ws×Prcp	6.62E-03	5.03E-03	4.94E-03	4.96E-03	-2.89E-04	3.94E-03
RH×Prcp	-3.03E-03	1.23E-02	-1.28E-02	1.10E-02	-3.11E-02	1.19E-02
N	288		362		333	
R^2	0.68		0.61		0.62	
Adjusted R^2	0.63		0.57		0.59	
Residual variance	0.13		0.24		0.18	

Model I violates the normal variance assumption in the linear regression and exhibits heteroscedasticity (the residuals are larger when FPM concentrations are higher). Heteroscedasticity does not negate the unbiasedness and consistency properties of OLS estimators, but these estimators are no longer minimum variance or efficient.

While heteroscedasticity is not a sufficient reason to throw out an otherwise good model, if an equally good (or superior) model can be achieved without heteroscedasticity residuals, this is preferred, and the log transformation in Model II does eliminate or greatly reduce this problem (Gujarati 1995).

Note that even though our objective is to maximize the R^2 values in this exercise in order to explain the variance in the daily $PM_{2.5}$ concentrations; the variance in our proposed regression model can never be fully explained since:

1. Fine particulate matter's chemical composition and formation processes are complex. Many other pollutants that contribute to the formation of FPM mass are not routinely measured and therefore are not available for inclusion in our model, such as primary fine particle emissions, organic species emissions, ammonium emissions and other pollutants that also contribute to the formation of FPM. A composition analysis of the fine particles can be helpful in speculating about the maximum variability we can explain in the regression analysis based on the explanatory variables included in the model; the higher the fraction of $PM_{2.5}$ formed by the direct conversion of the input precursor gases, the higher the R^2 likely to be obtained in the regression analysis.
2. While the effects of meteorological conditions on ambient FPM concentrations are included in the model, factors not included, such as ground level pressure, and the height of the inversion layer also affect ambient particulate matter concentrations.
3. Model specification error. The regression model is a way to approximate the response relationships between variables, but does not capture the full complexity of the chemical reactions and physics of the formation, fate and transport in the atmosphere.

3.1 MODEL CROSS VALIDATION

Comparisons of the observed fine particulate matter concentrations and simulated $PM_{2.5}$ concentrations by Model II based on the measured precursor gaseous concentrations and daily meteorological conditions in 1995 are presented in Figure 3.

The regression models in general make reasonable, if somewhat imprecise, predictions of the daily $PM_{2.5}$ concentrations for the three monitoring stations. In general, the predictions display a higher degree of variance at higher concentrations than at lower concentrations. The log-transformed models also result in a slight overestimation of the simulated concentrations and therefore the simulated annual average concentrations are slightly higher than the average of the values observed at the three stations. For example, the mean observed $PM_{2.5}$ concentrations are 23.1 $\mu g/m^3$, 19.4 $\mu g/m^3$, and 26.0 $\mu g/m^3$ for Azusa, North Long Beach and Rubidoux respectively, and the respective mean predicted concentrations are 29.2 $\mu g/m^3$ (26% overestimate), 22.4 $\mu g/m^3$ (15% overestimate), and 35.6 $\mu g/m^3$ (37% overestimate) (Figure 3). This overestimation may result from some systematic change in meteorological or air chemistry conditions, not included in the model, in 1995 vs. the calibration years. However, differences in predicted vs. observed concentrations of this magnitude are common in the independent validation stage of model development (Seigneur 2001).

3.2 RESPONSE SURFACES OF STATISTICAL REGRESSION MODELS

Figure 4 A-C show the regression lines (based on Model II) of the mean $PM_{2.5}$ predictions as functions of the relevant variables, while keeping other variables at their mean values and precipitation equal to zero. Note that when the response functions depend critically on the level of other pollutants, the predicted response functions with respect to the selected precursor concentrations are shown for the low, mean, and high levels of the other precursor concentrations. The regression models predict non-linear (and sometimes non-monotonic) relationships with the increases of NO_x, SO_2 and THC concentrations for all three stations.

Figure 3: Scatter plots of observed $PM_{2.5}$ concentrations at three monitoring sites ((a) North Long Beach, (b) Rubidoux, and (c) Azusa, CA), in 1995 versus simulated concentration by log-linear multivariable regression model (Model II) based on the measured precursor gaseous concentrations and daily meteorological conditions in 1995.

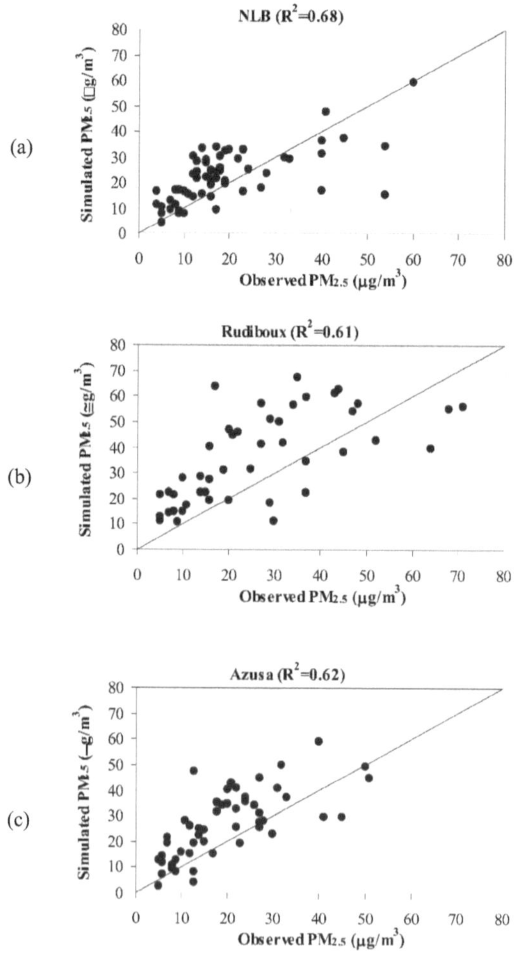

Figure 4A: Predicted $PM_{2.5}$ concentrations at North Long Beach as functions of the explanatory variables in the x-axis while keeping other variables at their mean values and precipitation equal to zero. The mean concentrations for NO_x, SO_2, and THC are 0.07 ppm, 0.012 ppm, and 3.30 ppm respectively. In the case of NO_x, the response functions for low and high concentrations of THC are also tested. In the cases of SO_2 and THC, the response functions for low and high NO_x concentrations are presented here. The relationship between wind speed and $PM_{2.5}$ is shown under both dry and wet conditions. The mean values of temperature = 64°F, wind speed = 7.8 mile per hr and mean relative humidity = 70%

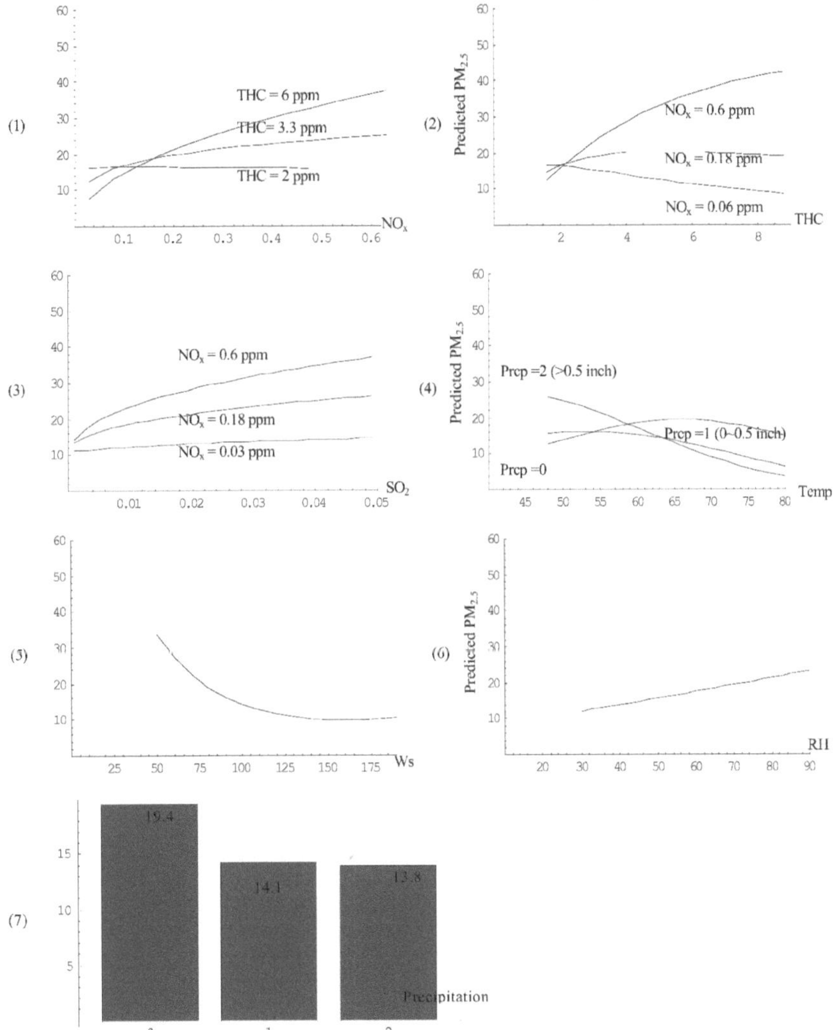

Figure 4B: Predicted $PM_{2.5}$ concentrations at Rubidoux as functions of the explanatory variables in the x-axis while keeping other variables at their mean values and precipitation equal to zero. The mean concentrations for NO_x, SO_2, and THC are 0.06 ppm, 0.004 ppm, and 3.8 ppm respectively. In the case of NO_x, the response functions for low and high concentrations of THC are also tested. In the cases of SO_2 and THC, the response functions for low and high NO_x concentrations are presented here.

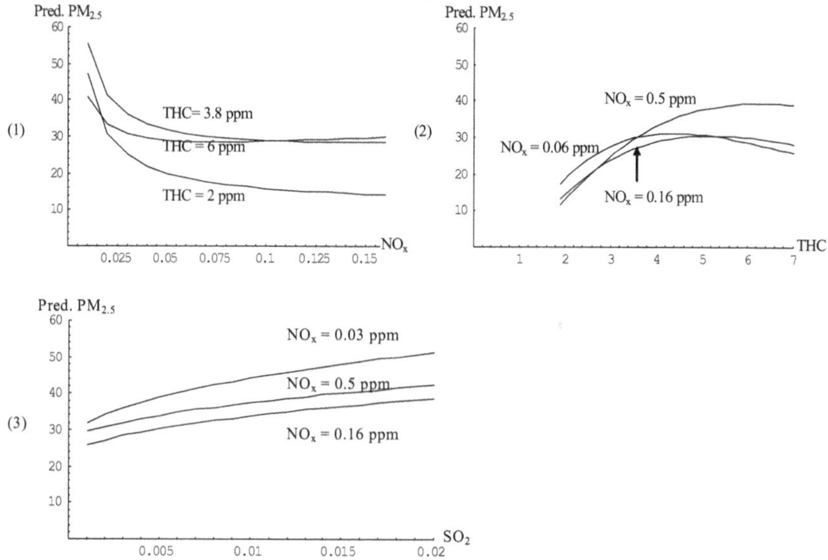

Figure 4C: Predicted $PM_{2.5}$ concentrations at Azusa as functions of the explanatory variables in the x-axis while keeping other variables at their mean values and precipitation equal to zero. The mean concentrations for NO_x, and THC are 0.07 ppm and 2.9 ppm respectively. In the case of THC, the response functions for low and high NO_x concentrations are also tested.

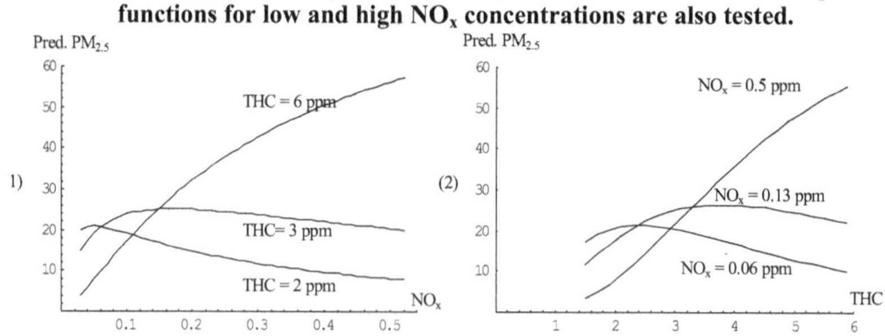

At Azusa and North Long Beach, lower NO_x concentrations are associated with lower $PM_{2.5}$ concentrations at high THC concentration levels. As THC concentrations are lowered below 3 ppm, however, the reductions of NO_x concentrations are associated with constant or even slightly higher levels of $PM_{2.5}$

(Figures 4A(1) and 4C(1)). At Rubidoux, very low concentrations of NO_x have a negative relationship with $PM_{2.5}$ concentration regardless of the levels of THC and SO_2 (Figure 4B(1)). As Figures 4A-C(2) show, lower levels of THC are associated with lower concentrations of $PM_{2.5}$. The relationship is stronger for higher NO_x concentrations. However, when NO_x concentrations are at their lowest (less than 0.06 ppm), lower levels of THC are associated with an increase in $PM_{2.5}$ concentrations at the Azusa and North Long Beach stations (Figures 4A(2) and 4C(2)).

In both North Long Beach and Rubidoux where SO_2 measurements were available, the regression models show that, when the values of other explanatory variables are kept at their mean, the concentrations of $PM_{2.5}$ are slightly lower with lower concentrations of SO_2 (Figures 4A(3) and 4B(3)). The relationship remains the same, though different in magnitude, for different levels of NO_x. However, note that at Rubidoux (Figure 4B(3)) lower concentrations of NO_x are associated with higher levels of $PM_{2.5}$. This observation is consistent with what is found in Figure 4B(1).

Wind speed and temperature (under wet conditions) are negatively associated with fine particle concentrations, while RH is positively correlated (Figures 4A(4)-(6)). Under dry conditions, temperature has an inverted U shape relationship with $PM_{2.5}$ concentrations, with median temperatures associated with higher levels of $PM_{2.5}$. Precipitation under our specified meteorological conditions lowers the concentrations of $PM_{2.5}$, with the largest reductions for the initial rain. The observations above are consistent with the fact that in the southern California air basin, high ambient fine particle levels are mostly observed in the winter season when temperature and wind speed are typically low and relative humidity is high. The winter season in Los Angeles is usually accompanied by low temperature and high relative humidity. The conversions of oxides of sulfur or NO_x to sulfate and nitrate aerosols are generally a cool-wet (e.g., winter) phenomenon and these reactions are accelerated under humid conditions, when the conversion can occur inside water droplets. The temperature phenomena we observed in this analysis may be accentuated by the surrogate effects of the mixing height. It is known that in Los Angeles the mixing heights are lower during cold weather and higher when the temperature is high. Since the mixing height is not included in the model, it is very likely that the temperature variable in the regression model also acts as a surrogate for the effect of mixing height: lower mixing heights (lower temperature) being associated with higher FPM concentrations.

4. Predicted potential effects of emission controls

The method developed in this study provides a way to examine and display results for the predicted concentration distributions when evaluating emission control strategy performance. This is done by using new precursor variable distributions assumed to occur under the emission control strategies and predicting the resulting $PM_{2.5}$ daily concentrations using the regression model (Step 3 of Figure 1).

4.1 SIMULATION OF EMISSION CONTROLS

Figure 5 shows the predicted percent change of $PM_{2.5}$ annual mean concentrations relative to 1995 mean concentrations as a function of the percent reduction of the selected gaseous pollutant(s) (zero percent reduction corresponds to the model predicted concentrations under current precursor conditions).

Figure 5: The predicted percent change of $PM_{2.5}$ annual mean concentrations relative to 1995 mean concentrations as a function of the percent reduction of the selected gaseous pollutant(s). Zero percent reduction corresponds to the model predicted concentrations under current precursor conditions.

(a)

(b)

(c)

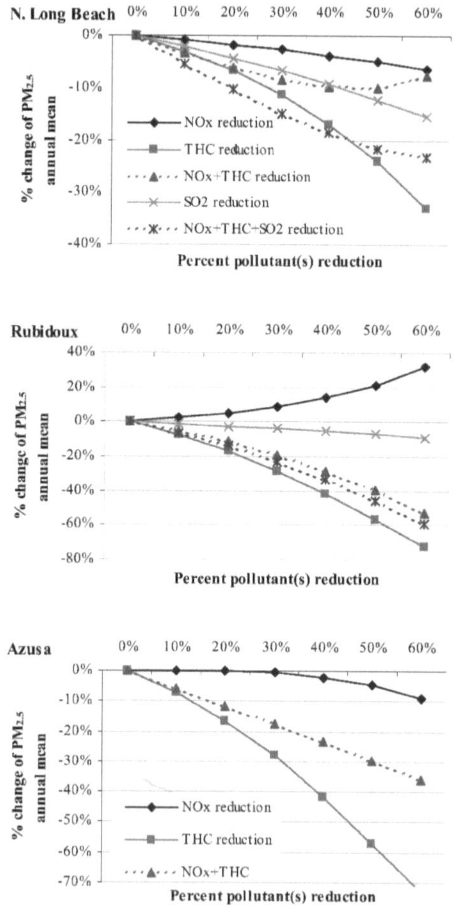

The results show that all three stations are sensitive to reductions in THC concentrations, which in general give the greatest reductions in all individual/combined precursor reduction scenarios. Reductions in NO_x emissions yield little predicted reductions in FPM annual concentrations at North Long Beach and Azusa stations. However, the model predicts a nonlinear increase in FPM annual average concentrations with reductions in NO_x emissions at the Rudiboux station.

Co-reductions in both THC and in NO_x emissions produces results that lie somewhere between the individual effects of THC and NO_x emissions reductions in all three stations. Controlling SO_2 emissions leads to small predicted reductions of $PM_{2.5}$ annual average concentrations at North Long Beach and Rubidoux. Combing reductions in NO_x, THC and SO_2 seems to be the second best control strategy next to the THC-only emission control.

A more careful look at the predicted percent changes in daily concentrations indicates that the percent changes in daily concentration vary significantly depending on the daily meteorological conditions. Table 3 shows the results for NO_x and THC emissions control by season (summer and winter). In all three stations, NO_x emission control contributes the greatest increase in FPM daily concentrations at Rubidoux in the summer season, and the greatest reduction in FPM daily concentration at Azusa in the winter season. The effects of THC emission reductions have statistically significant seasonal differences in all three stations, with more reductions of $PM_{2.5}$ daily concentrations predicted in the winter than in the summer.

Table 3. Emission reduction of precursor pollutants and the averages of the predicted percent changes of *daily* concentrations by season. Numbers in the bracket correspond to the minimum and maximum changes in daily concentrations relative to the base year. Summer is defined from the month of April to September. Test of significance is for difference in summer and winter response at each station.

Emission Control (% reduction from the base case, 1995)		NLB		Rubidoux		Azusa	
		Summer	Winter	Summer	Winter	Summer	Winter
NO_x	20%	-0.9% (-7.8%, 7.4%)	-1.0% (-10.7%, 9.5%)	5.4%$^{++}$ (-13.1%, 14.0%)	2.3%$^{++}$ (-7.4%, 10.5%)	0.7%$^{+}$ (-18.7%, 16.4%)	-3.2%$^{+}$ (-24.7%, 8.6%)
	40%	-2.0% (-17.1%, 17.5%)	-2.1% (-22.9%, 22.7%)	14.7%$^{++}$ (-26.6%, 36.7%)	6.9%$^{++}$ (-15.1%, 27.3%)	-0.5%$^{+}$ (-39.3%, 38.0%)	-8.4%$^{+}$ (-49.2%, 17.7%)
	60%	-3.6% (-28.9%, 32.8%)	-3.2% (-37.6%, 43.7%)	33.6%$^{++}$ (-40.6%, 80.9%)	17.3%$^{++}$ (-22.9%, 59.3%)	-6.0%$^{+}$ (-61.8%, 66.7%)	-17.6%$^{+}$ (-72.2%, 25.4%)
THC	20%	-3.0%$^{+++}$ (-29.0%, 9.4%)	-12.9%$^{+++}$ (-32.2%, 3.5%)	-17.0%$^{++}$ (-39.6%, -7.0%)	-21.7%$^{++}$ (-34.6%, 0.5%)	-14.3%$^{+++}$ (-26.7%, 4.6%)	-20.1%$^{+++}$ (-37.4%, -5.2%)
	40%	-9.5%$^{+++}$ (-56.0%, 18.3%)	-28.6%$^{+++}$ (-60.4%, 4.3%)	-41.8%$^{+}$ (-72.1%, -25.1%)	-48.6%$^{+}$ (-66.6%, -10.6%)	-38.1%$^{+++}$ (-57.3%, -3.7%)	-47.4%$^{+++}$ (-70.2%, -23.0%)
	60%	-22.6%$^{+++}$ (-79.1%, 23.3%)	-48.4%$^{+++}$ (-82.7%, -1.7%)	-71.9%$^{+}$ (-92.6%, -56.6%)	-76.8%$^{+}$ (-89.8%, -40.3%)	-69.3%$^{+++}$ (-84.8%, -34.4%)	-77.2%$^{+++}$ (-92.0%, -56.0%)

$^{+}$ Statistically significant summer-winter difference at less than 10% level.
$^{++}$ Statistically significant summer-winter difference at less than 5% level.
$^{+++}$ Statistically significant summer-winter difference at less than 1% level.

4.2 COMPARISON WITH OTHER STUDIES

In California, due to high concentration of NH_4NO_3, high NH_3 and low SO_2 emissions, it has been suggested that controlling SO_2 has little effect on the reduction of FPM concentrations (Ansari and Pandis 1998; Kleeman and Cass 1999). A reduction in SO_2 emissions, especially without a reduction in NO_x emissions, could lead to an increase in NH_4NO_3 concentrations, and thus the total FPM mass, in the Los Angeles region (Ansari and Pandis 1998; Kleeman and Cass 1999; West, Ansari et al. 1999). In the western United States, where higher NH_3 and lower SO_2 emissions permit complete neutralization of SO_2, the concentration of nitrate is higher than that of sulfate. As SO_2

concentrations in the atmosphere are reduced, the NH_3 left in the atmosphere after neutralization of H_2SO_4 is able to react with HNO_3 to form NH_4NO_3 (Table 4). In addition, Ansari et al. provides detailed mechanistic explanations for the chemical processes that affect these results under different meteorological conditions (Ansari and Pandis 1998; Kleeman and Cass 1999; West, Ansari et al. 1999). This is consistent with our findings here (Figures 4(A)(B) and Figures 5(a)(b)). The counterproductive effects of SO_2 reductions are predicted by our response-surface analysis for conditions occurring on only two days out of 59 measured concentrations at the North Long Beach station and two days out of 48 at the Rubidoux stations in the year 1995. Three of these four days are in April and one day is in December.

The atmospheric processes that form secondary particulate matter from precursor NO_x emissions are found to be nonlinear. Kleeman and Cass (1999), Pai, Vijayaraghavan et al. (2000), and Meng, Dabdub et al. (1997) have predicted nonlinear, positive relationships with the control of emissions of NO_x and $PM_{2.5}$ concentrations in CA (Table 4). However, several recent studies have found the reduction in precursor NO_x emissions may actually increase the "conversion efficiency" of the atmospheric chemistry, leading to an increase in particulate NO_3^- concentrations (Ansari and Pandis 1998; Pun and Seigneur 2001; Mysliwiec and Kleeman 2002).

Conversion efficiency refers to the fraction of NO_x emitted to the atmosphere that is eventually converted to nitrate. Conversion efficiency is a function of meteorological variables such as temperature, relative humidity, and solar radiation (Ansari and Pandis 1998; Mysliwiec and Kleeman 2002). Pun and Seigneur (2001) uses a box model to simulate the sensitivity of PM nitrate formation in San Joaquin Valley (SJV), CA during January 4-6, 1996. Their study showed that the concentration of PM nitrate is sensitive to reductions in VOC emissions, and a decrease in NO_x emissions may have the counter-intuitive effect of increasing PM nitrate. It also showed that PM formation in the SJV during the winter (low temperature, high RH, and sufficient free ammonium) is HNO_3 sensitive, that HNO_3 formation is oxidant-sensitive, and that oxidant formation is sensitive to reductions in VOC emissions. The reduction in PM as a result of VOC emission reductions is caused by the limited availability of the oxidants OH and O_3. On the contrary, a decrease in NO_x emissions leads to an increase in PM due primarily to an increase in O_3 concentrations, and therefore increases the radical concentrations disproportionally. As shown in Table 4, a 50% reduction in VOC results in a 24-hr average $PM_{2.5}$ nitrate concentrations that is reduced by 18% and 36% on days 2 and 3; while a 50% reduction in NO_x emission actually increases PM nitrate concentration by as much as 30% on both day 2 and 3. The isopleths of 24-hour average PM nitrate concentration in the paper shows that 50% VOC reduction and 50% NO_x reduction are associated with respective 35% and –30% changes in PM nitrate concentration. When VOC and NO_x are both reduced by 50%, PM nitrate concentration is predicted to be unchanged. This is consistent with the Ansari and Pandis (1998) generic analysis that suggests that wintertime PM concentrations in the SJV should be very sensitive to a change in HNO_3 concentration, but should not be sensitive to NH_3 concentrations.

Table 4. Predicted effect of precursor emission reductions on $PM_{2.5}$ concentration from previous modeling studies (values are presented as percent changes with respect to base cases).

Emission Control (%)					$\Delta PM_{2.5}$ (%)
NO_x	SO_2	NH_3	VOC	Others	
Kleeman and Cass (1999), 24-h average $PM_{2.5}$ concentration at Claremont CA, August 28, 1987					
				$PM_{2.5}$ -41%	-24%
	-72%				-1.4%
		-49%			-13%
-56%				ROG -70%	-10%
-56%	-72%	-49%		$PM_{2.5}$ -41% ROG -70%	-46%
Pai, Vijayaraghavan et al. (2000), Max 1-h $PM_{2.5}$ concentration at Riverside CA, August 28, 1987					
-50%					-24%
			-50%		< -1%
-50%			-50%		-22%
Meng, Dabdub et al. (1997), Max 1-h $PM_{2.5}$ concentration at Riverside CA, August 28, 1987					
-50%					-18%
			-50%		> -19%
-50%			-50%		-5%
Pun and Seigneur (2001), 24-h average $PM_{2.5}$ concentration in the Fresno Area, San Joaquin Valley CA, January 4-6, 1996					
-50%					$PM_{2.5}$ nitrate: day 2 +30% day 3 +27%
			-50%		$PM_{2.5}$ nitrate: day 2 –18% day 3 –36%
					$PM_{2.5}$ ammonium: day 2 –15% day 3 –29%

4.3 COMPARISON TO 1999 DATA

To test our predictions of the effects of precursor emission reductions on the change of annual $PM_{2.5}$ concentrations assuming future meteorology is unknown, but assumed to follow that of the base year, we match the percent changes of annual mean precursor concentrations in 1999 to the base year 1995 and make predictions for the 1999 annual $PM_{2.5}$ concentrations based on our statistical model (Step 4, Figure 1). The statistics of precursor concentrations and observed $PM_{2.5}$ concentrations are summarized in Table 5.

However, the data we obtained from the Air Resources Board does not have any information on the THC levels in 1999 for the three monitoring stations in this study. Therefore, we make predictions of $PM_{2.5}$ annual mean and 98th percentile concentrations (computed as the 98th percentile from the sorted series of daily values which is ordered from the lowest to the highest number) based on the percent changes of annual mean NO_x and SO_2 (Table 5) and we vary the percent changes of THC annual concentration from –30% to +30%. In 1999, the mean NO_x ambient concentrations decrease while SO_2 ambient concentrations and annual mean $PM_{2.5}$ concentrations increase in all three stations compared to the year of 1995 (Table 5).

Implicit in this is the assumption that the meteorological conditions are the same in 1995 and 1999. The results are shown in Figure 6. With the assumed THC levels ranging from –30% to +30% of the 1995 levels, the predicted $PM_{2.5}$ annual mean and 98th percentile concentrations generally fall within the range of our prediction.

Table 5. Annual mean concentrations of observed precursor pollutants and observed $PM_{2.5}$ and the percent changes in 1999 relative to the base year 1995

		NLB	Rubidoux	Azusa
1995	Mean NO_x (ppm)	0.144	0.124	0.127
	Mean SO_2 (ppm)	0.010	0.003	NA
	Mean $PM_{2.5}$ ($\mu g/m^3$)	19.2	26.0	23.1
1999	Mean NO_x (ppm)	0.128	0.121	0.083
	Mean SO_2 (ppm)	0.011	0.004	NA
	Mean $PM_{2.5}$ ($\mu g/m^3$)	21.47	31.62	25.53
	98th percentile $PM_{2.5}$	50.5	78.7	66.3
Δ(%)	Mean NO_x (%)	-10.8 %	-2.3 %	-34.2 %
	Mean SO_2 (%)	8.1 %	30.6 %	NA
	Mean $PM_{2.5}$ (%)	11.7 %	21.5 %	10.4 %

5. Discussion

The purpose of the regression model developed in this study is to analyze the precision with which FPM concentrations can be calculated when emission inventories and chemical input data are unknown or too limited to apply advanced physicochemical models successfully. While the regression model has low demand for input parameters, it can be used to predict long-term probability distributions of daily FPM concentration based on an assumed distribution of precursor gaseous concentrations and historic meteorological variables. Although the application of simple models might not be necessary or desirable for regions where emission inventories and atmospheric conditions are well characterized both physically and chemically, such as Southern California, such a model could have importance in less intensely studied areas, as most other regions fall under this category. A chemical equilibrium model may be a better choice for understanding the physical and chemical processes involved in the formation of ambient particles. However, statistical models, such as the one we develop here, can provide a useful complementary tool for policy makers making air pollution forecasts for input into decisions regarding abatement measures. Chemical equilibrium models are theoretically sophisticated and desirable for forecasting, however, they are not practical choices in many locations and circumstances.

In the regression model predictions, we note that reductions in ambient levels of SO_2 concentration could lead to an increase of $PM_{2.5}$ concentrations under very special conditions. We also predict that a decrease in NO_x ambient concentration has the counter-intuitive effect of increasing FPM concentrations, especially at Rudiboux. In general, the concentration of fine particulate mater was suggested by this study to be sensitive to reductions in THC concentrations at high levels of NO_x.

Figure 6: The predicted annual mean (solid box) and 98th percentile (dash box) $PM_{2.5}$ concentrations in 1999 as the results of base year (1995) precursor emission changes, assuming the meteorological conditions follow that of the base year. The predictions are based on percent changes of annual mean precursor concentrations (NO_x and SO_2) and percent changes of annual THC concentrations varying from –30% to +30%. The long-dash and short-dash lines correspond to the observed annual mean and 98th percentile $PM_{2.5}$ concentrations respectively in 1999 at the three stations

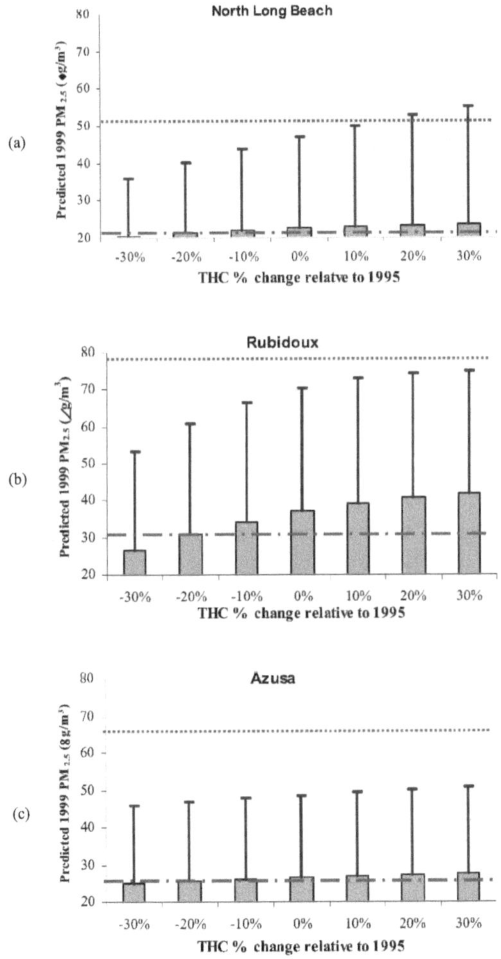

THC is used in this paper as a surrogate for VOC since it is measured routinely and published by the Air Resources Board. In the highly polluted atmosphere of Los Angeles, approximately 70% of THC consists of methane. Therefore only a small portion of THC is likely to imitate the chemical relationship of ambient concentrations of VOC on concentrations of ambient $PM_{2.5}$. The interpretation of the results should thus be cautious: we assume that the reductions of methane and NMHC are proportional to the simulated reduction in THC in this study, but only the reduction in NMHC is associated with a reduction in ambient $PM_{2.5}$ concentration.

The important role of ammonia is increasingly recognized in the atmospheric science community. Unfortunately this parameter is not measured at most monitoring stations across the United States, including the three that we studied here. In this study, we assume that the level of ammonia concentration stays constant for all the control strategies proposed herein. The atmospheric conditions in Los Angeles and Riverside County are ammonium-rich, which permit complete neutralization of SO_2 and the amount of NH_4NO_3 is limited by the amount of inorganic nitrate (e.g., HNO_3) available to form particulate mass (Hughes, Allen et al. 2002). Most comprehensive, dynamic models predict that reductions in ammonia emissions will have a very significant impact on total PM levels (Ansari and Pandis 1998; Kleeman and Cass 1999; Blanchard and Hidy 2003). It is not clear whether there will be drastic changes to the response surfaces predicted herein when ammonia concentrations are reduced significantly. Further research is needed to explore this issue when ammonia data became more readily available. Furthermore, the model developed in this study may be sensitive to different emission-weather-PM regimes. However, the methodology for model identification, fitting, and cross-validation can serve as guide for application in other areas.

Given the potentially serious health effects associated with particulate air pollution in the Los Angeles area and urban cities around the world (Soubbotina and Sheram 2000, U.S. EPA 1996b; Yeh and Small 2002), the results of the current study suggest that the best strategy to improve the air quality needs to be evaluated very carefully. Detailed air quality models that include a description of nonlinear chemical reaction pathways must be used to understand the chemical processes. However, a statistical model such as the one developed here can also help to provide practical information about the general direction of the effects likely to occur as a result of emission reductions. Following verification with a mechanistic (but typically episodic) air quality model, the statistical model can be a useful tool for initial prediction of the long-term distribution of particulate matter concentrations.

6. Acknowledgement

The authors thank Professors Luis Cifuentes, Benoit Morel, Spyros Pandis, and an anonymous reviewer for their helpful comments to significantly improve the quality of the manuscript. This research was made possible through support from the Center for Integrated Study of the Human Dimensions of Global Change. This Center has been created through a cooperative agreement between the National Science Foundation (SBR-9521914) and Carnegie Mellon University. Support was also provided by the Vira Heinz Endowment through the H. John Heinz III Professorship of Environmental Engineering. Views or opinions expressed in this paper are those of the authors alone and do not represent those of their organizations or the sponsors of this work.

7. References

1. Abu-Allaban, M., Gertler, A. W., and Lowenthal, D. H. (2002). A Preliminary Apportionment Of The Sources Of Ambient PM_{10}, $PM_{2.5}$, And VOCs in Cairo. Atmospheric Environment **36**(35), 5549-5557.
2. Air Resources Board (1999). California Ambient Air Quality Data, 1980 - 1998. CD Number: PTSD-99-012-CD. December, 1999: Planning and Technical Support Division, Air Resources Board. California Environmental Protection Agency.
3. Ansari, A. S. and Pandis, S. N. (1998). Response Of Inorganic Pm To Precursor Concentrations. Environmental Science & Technology **32**, 2706-2714.
4. APHEIS (2004). Air Pollution and Health: A European Information System (APHEIS). Health Impact Assessment of Air Pollution and Communication Strategy. http://www.apheis.net/.
5. Bell, M. L., Cohen, A., Davis, D. L., Grant L., and Sun, G. (2002). Discussion Summary. ISEE Workshop on Public Health Impacts of Fossil Fuels, Vancouver, Canada.
6. Blanchard, C. and Hidy, G. (2003). Effects Of Changes In Sulfate, Ammonia, and Nitric Acid On Particulate Nitrate Concentrations In The Southeastern United States. J. Air & Waste Manage. Assoc. **53**(3), 283-290.
7. Calbo, J., Pan, W., Webster, M., Prinn, R. and Mcrae, G. (1998). Parameterization Of Urban Subgrid Scale Processes In Global Atmospheric Chemistry Models. Journal Of Geophysical Research - Atmospheres **103**(D3), 3437-3451.
8. Chow, J. C., Watson, J. G., Fujita, E. M., Lu, Z. and Lawson, D. R. (1994). Temporal and Spatial Variations Of $Pm_{2.5}$ and Pm_{10} Aerosol In The Southern California Air Quality Study. Atmospheric Environment **28**(12), 2061-2080.
9. Christoforou, C. S., Salmon, L. G., Hannigan, M. P., Solomon, P. A. and Cass, G. R. (2000). Trends In Fine Particle Concentration and Chemical Composition In Southern California. J. Air & Waste Manage. Assoc. **50**, 43-53.
10. El-Metwally, A. A. and Ramadan, A. B. (2003). The Role of Air Pollutants and Sewage Waste on Acceleration of Degradation of Islamic Cultural Heritage of Cairo, in Linkov, I. and Ramadan, A. eds., Comparative Risk Assessment and Environmental Decision Making. Kluwer Academic Publishers, Dordrecht, pp. 367-374.
11. Gujarati, D. N. (1995) Basic Econometrics. New York: Mcgraw-Hill, Inc.
12. Hanna, S. R. and Davis, J. (2002). Evaluation Of A Photochemical Grid Model Using Estimates Of Concentration Probability Density Functions. Atmospheric Environment **36**(11), 1793-1798.
13. Hughes, L., Allen, J., Bhave, P., Kleeman, M., Cass, G., Liu, D., Fergenson, D., Morrical, B. and Prather, K. (2000). Evolution Of Atmospheric Particles Along Trajectories Crossing The Los Angeles Basin. Environmental Science & Technology **34**(15), 3058-3068.
14. Hughes, L. S., Allen, J. O., Salmon, L. G., Mayo, P. R., Johnson, R. J. and Cass, G. R. (2002). Evolution Of Nitrogen Species Air Pollutants Along Trajectories Crossing The Los Angeles Area. Environmental Science & Technology **36**(18), 3928-3935.
15. Kleeman, M. J. and Cass, G. R. (1999). Effect Of Emissions Control Strategies On The Size and Composition Distribution Of Urban Particulate Air Pollution. Environmental Science & Technology **33**(1), 177-189.
16. Kleeman, M. J. and Cass, G. R. (2001). A 3D Eulerian Source-Oriented Model For An Externally Mixed Aerosol. Environmental Science & Technology **35**, 4843-4848.
17. Kleeman, M. J., Hughes, L. S., Allen, J. O. and Cass, G. R. (1999). Source Contributions To The Size and Composition Of Atmospheric Particles: Southern California In September 1996. Environmental Science & Technology **33**, 4331-4341.
18. Lebanon Ministry of Environment (1997). Automotive Fuel Strategies For Clean Air In Lebanon: Lead Phasing Out and Diesel Fuel Policy.
19. Lefohn, A. S., Shadwick, D. S. and Ziman, S. D. (1998). The Difficult Challenge Of Attaining EPA's New Ozone Standard. Environmental Science & Technology **32**(11), 276a-282a.
20. Meng, Z., Dabdub, D. and Seinfeld, J. H. (1997). Chemical Coupling Between Atmospheric Ozone and Particulate Matter. Science **277**(4), 116-119.
21. Mysliwiec, M. and Kleeman, M. J. (2002). Source Apportionment Of Secondary Airborne Particulate Matter In A Polluted Atmosphere. Environmental Science & Technology **36**(24), 5376-5384.
22. Pai, P., Vijayaraghavan, K. and Seigneur, C. (2000) Particulate Matter Modeling In The Los Angeles Basin Using Saqm-Aero. J. Air & Waste Manage. Assoc. **50**(1), 32-42.

23. Pun, B. and Seigneur, C. (2001). Sensitivity Of Particulate Matter Nitrate Formation To Precursor Emissions In The California San Joaquin Valley. Environmental Science & Technology **35**(14), 2979-2987.
24. Seigneur, C. (2001). Current Status Of Air Quality Models For Particulate Matter. Journal Of Air & Waste Management Association **51**(11), 1508-1521.
25. Seigneur, C., Pai, P., Hopke, P. K. and Grosjean, D. (1999). Modeling Atmospheric Particulate Matter. Environmental Science & Technology **33**(3), 80a-86a.
26. Seinfeld, J. and Pandis, S. N. (1998) Atmospheric Chemistry and Physics. New York: John Wiley and Son, Inc.
27. Soubbotina, T. P. and Sheram, K. A. (2000). Beyond Economic Growth: Meeting the Challenge of Global Development. Washington, D. C., The World Bank.
28. U.S. EPA (1996a). Air Quality Criteria for Particulate Matter. Vols. I - III, Washington Dc: Office Of Research and Development, U.S. Environmental Protection Agency.
29. U.S. EPA (1996b). Review Of The National Ambient Air Quality Standards For Particulate Matter: Policy Assessment Of Scientific and Technical Information, OAQPS Staff Paper, Research Triangle Park, NC: Office Of Air Quality Planning and Standards, U.S. Environmental Protection Agency.
30. U.S. EPA (1997). National Ambient Air Quality Standards For Particulate Matter. Final Rule: Federal Registry.
31. West, J. J., Ansari, A. S. and Pandis, S. N. (1999). Marginal $PM_{2.5}$: Nonlinear Aerosol Mass Response To Sulfate Reductions In The Eastern United States. J. Air & Waste Manage. Assoc. **49**, 1415-1424.
32. Winner, D. A. and Cass, G. R. (2000). Effect Of Emission Control On The Long-Term Frequency Distribution Of Regional Ozone Concentrations. Environmental Science & Technology **34**, 2612-2617.
33. Yeh, S. and Small, M. J. (2002). Incorporating Exposure Models In Probabilistic Assessment Of The Risks Of Premature Mortality From Particulate Matter. Journal Of Exposure Analysis and Environmental Epidemiology **12**(6), 389-403.

QND: A SCENARIO-BASED GAMING SYSTEM FOR MODELING ENVIRONMENTAL PROCESSES AND MANAGEMENT DECISIONS

G.A. KIKER
Dept of Agricultural and Biological Engineering, University of Florida, PO Box 110570, Gainesville, FL 32611-0570, USA

N.A. RIVERS-MOORE
Institute for Water Research, Old Geology Building
Rhodes University, P.O.Box 94, Grahamstown, 6140, SOUTH AFRICA

M.K. KIKER
Zama Environmental, PO Box 142752, Gainesville, FL 32614-2752, USA

I. LINKOV
Cambridge Environmental Inc., 58 Charles Street, Cambridge, MA 02141, USA

Abstract

In this chapter, we introduce a generic environmental modeling system that has been developed using an object-oriented approach. The Questions and Decisions™ (QnD™) model system combines both number-based calculations and value-style judgments. It can integrate ideas and data that are well-studied with concepts that are estimated from expert knowledge and experience. A specific QnD version is constructed through conversations with stake-holders and decision-makers. The wishes of the stakeholders are created within the model using configurable objects designed to be quickly made and quickly altered through subsequent learning and iteration.

QnD-simulated ecosystems are represented by combinations of component, process, and data objects that are constructed through the use of XML-based input files. This design allows different ecosystem/habitat/organism/chemical combinations to be efficiently formed, simulated, and documented. The flexibility of the model is demonstrated through its non-spatial application to a terrestrial ecosystem (Kruger National Park, South Africa) and a spatial risk assessment application within an idealized US river system (as a demonstration for the US Army Corps of Engineers). Unlike traditional decision support systems that direct outputs at discipline-specific management, the model has been created as a game to stimulate discussions and analysis among managers, scientists, and stakeholders who are working increasingly closely within an adaptive management context.

B. Morel and I. Linkov (eds.),
Environmental Security and Environmental Management: The Role of Risk Assessment, 151–185.

1. Introduction

Ecosystems exist within matrices of human landscapes. These ecosystems have value (Costanza *et al.*, 1997) and must be managed appropriately in order to sustain the benefits they provide. The growing problem is that the scale at which human activities are occurring is approaching the scale at which ecosystem dynamics occur (O'Neill et al., 1998), such that ecological processes and economic activity can become part of the same dynamic system. Consequently, ecosystems become less resilient and the appropriate management of these systems becomes increasingly complex, especially with differing human values and expectations of the system. The interplay of various social, technical and environmental forces at differing time and spatial scales has been termed "*Panarchy*" (Gunderson *et al.*, 1995; Gunderson and Holling, 2002), and attempts to integrate the various social-, technical-, or ecological-based solutions often highlight one viewpoint while shortchanging the others (Gunderson and Holling, 2002).

Interdisciplinary collaboration is the key to resolving natural resource problems of the 21st century (Holling, 1999; O'Neill *et al.*, 1998), wherein greater sharing and collaboration between scientists is implicit as a way of achieving "bigger" science (Houlahan, 1998) to better manage ecosystems. This does not necessarily mean "large-scale science" (Walters, 1997), nor does it mean that results from small-scale experiments can be scaled up without a clear understanding of scaling rules (Rastetter *et al.*, 1992). We feel that large-scale science should not be an endpoint in itself, but that interdisciplinary research should seek the appropriate scale of the problem, where the scales of observation and management match the scale at which the problem is occurring (Jewitt and Görgens, 2000a).

"The goal of conservation management is shifting from managing species for their intrinsic value, to managing them for their interactive roles in ecosystem functioning" (Rogers, 1997). Management of natural resources occurs under uncertainty, but the use of resources and the need for management will continue (Johnson, 1999a). The management challenge is to gain insight into change in complex natural systems. One approach, known as strategic adaptive management, is based on the concept of managing natural systems through a process of careful testing rather than trial and error (Walters, 1997). Institutional barriers and inertia pose the greatest threat to the successful implementation of adaptive management (Walters, 1997; Walters *et al.*, 2000). Additional problems include too much focus on the models while ignoring the problems, scale linkages between different models (Walters, 1997; Jewitt and Görgens, 2000b), and the definition of appropriate goals (Johnson, 1999b).

Often management decisions must be made in the absence of adequate data, which is where modeling becomes a useful management tool. Thus a model's development should be driven by the objectives of the management program, rather than the available data (Starfield & Bleloch, 1991). "In a decision-making context, the ultimate test of a model is not how accurate or truthful it is, but only whether one is likely to make a better decision with it than without it" (Starfield, 1997). Scenario modeling is a useful tool for envisaging future situations in an unknown future (Breen, 1998). Models help to expose gaps in data and understanding, and help to screen policy options, especially under conditions where time is limited and systems are sensitive (Walters *et al.*, 2000).

In this chapter, we introduce a generic environmental modeling system that has been developed using an object-oriented approach. The flexibility of the model is demonstrated through its non-spatial application to a terrestrial ecosystem (Kruger National Park, South Africa) and a spatial application within a generalized river system (as a demonstration for the US Army Corps of Engineers). Unlike traditional decision support systems that direct outputs at a discipline-specific management, the model has been created as a game to stimulate discussions between managers, scientists, and stakeholders who are working increasingly closely within an adaptive management framework (Rogers, 1997).

2. Problem Statement and Objectives

Natural systems are complex webs of synergies with species interactions providing a reflection of fast and slow processes (Vannote *et al.*, 1980; Frissel *et al.*, 1986). A question that often arises is, "How do we understand seemingly random or stochastic patterns, and how do we manage so that system variability is maintained?"

Typically, ecologists have adopted a reductionist approach to understanding ecosystems. There is a quest for the ecological "Holy Grail" – a set of mathematical equations that explain system behavior, obtained by breaking systems down to their basic units. Certainly there is merit in this idea. For example, Stone and Ezrati (1996) provides numerous examples where apparently random oscillations can be explained by deterministic chaos, where a small number of equations can model a system precisely. While this approach may be the most manageable one, it ignores the idea that through synergies the sum of the individual parts is greater than the whole. Consequently, deterministic equations are not able to adequately explain system behavior. Very often neither the data nor the ability to represent natural processes as equations exists (Matsinos *et al.*, 1994). Furthermore, natural systems do not seek equilibrium through successions, but rather exist in a state of non-equilibrium where disturbance is important.

The use of simple, pragmatic models that require relatively fewer parameters than complex models is useful in ecological studies (Jeppesen and Iversen, 1987). This approach was useful in highlighting certain management issues within the Colorado ecosystem (Walters, 2000), where a suit of small models at multiple scales of time and space were used to assist scientists and managers. Our approach is to make use of simple "rules" and relationships to model complex systems; simple game models are an effective way of exploring this complexity. The interaction of simple sets of rules are able to create dynamic "systems," such as John Conway's "Game of Life" (Green, 1998), and complex patterns if we view natural objects as simple computers with their own sets of rules (i.e. cellular automata) (Green, 1998). This approach can be applied at the landscape level if the landscape is seen as a collection of cells. These ideas can in turn form a useful basis for more complex models that are designed to promote management. However, bringing together different models at different scales is a daunting task, and it involves finding common units, such as fish habitats (biotic models) and geomorphological units (abiotic models) (Jewitt *et al.*, 1998).

The idea of management as a "game" involving different roleplayers can reveal important general patterns of system behavior, as illustrated by the "Nonpoint" model developed by Carpenter *et al.* (1999). This is essentially a simple system with few roleplayers, and it serves to show the interaction between fast and slow variables (multiple time scales) and illustrates the point that continual learning is crucial for sustainability. However, this model lacks a spatial component and is specific to a single system. Starfield (1993) presented a frame-based modeling approach which consists of collections of smaller models representing different states (frames) within a single system. Different frames were invoked according to certain sets of conditions and rules. Frame-based models could be made more powerful to management through the inclusion of a spatial component. Spatially explicit modelling is useful in quantifying patterns and linking them to ecological processes and mechanisms (Matsinos *et al.*, 1994).

While models have been successfully used for both early system understanding and for later system optimization, we believe there is a significant need for a quickly configurable, spatial model to efficiently represent both formal and informal knowledge in an iterative, interactive format for further management exploration. This model would take initial ecosystem understanding and allow scenarios to be played out to both further refine ecosystem understanding as well as sketch out potential management responses for further, systematic exploration. This model would operate efficiently in between simplified systems understanding models and more complex numerical/optimization approaches by adopting elements of each approach and being very adaptable to either stakeholder goals/preferences or scientific understanding.

This chapter has the following objectives:

- Introduce the design and structure of the Questions and Decisions™ (QnD™) model system;
- Demonstrate the use of the QnD model for adaptive management scenarios in two different ecosystems with multiple drivers and stressors;
- Highlight the lessons learned and next steps for the modeling system.

This chapter is divided into four sections. The first section introduces the QnD model, its design and construction. The second section describes the strategy of developing and using QnD with various stakeholders and scientists in addressing environmental challenges. The third section shows two QnD applications in ecosystem management, and the fourth section highlights the overall lessons learned from QnD applications.

3. QnD: An Object-Oriented Management "Game"

QnD is an acronym for "Questions and Decisions," or alternatively "Quick 'n Dirty," as both of these phrases emphasize the ideas we incorporated into the model from the outset: that our model would not only be appropriate to general environmental management problems for specific areas, but also that the model would be generic enough to be readily convertible between different ecosystems with different sets of drivers and problems. However, it is important to understand the modeling context of the QnD model before further details of the model itself are given.

Object-oriented models have been increasingly used to model ecological systems at various scales (for example, Matsinos *et al.*, 1994; Mooij and Boersma, 1996; Railsback, 1999; Railsback and Harvey, 2001; Sekine *et al.*, 1996). The object-oriented modeling approach is useful in modeling natural systems, since the "attributes of inheritance, polymorphism, data protection and modularity, provides a natural framework for simulating real-world phenomena involving individual organisms" (Matsinos *et al.*, 1994). Mooij and Boersma (1996) found that "the object-oriented programming paradigm is well suited for the creation of simulation models of ecological systems." Objects (fish, elephant, habitat, *etc.*) interact with each other according to sets of procedures and rules. An object-oriented modelling approach has the advantages that each model object falls within a hierarchy of other objects, so that inheritance relationships avoid unnecessary coding (Mooij and Boersma, 1996). Objects lower down the hierarchy inherit all the attributes of the objects higher up the hierarchy (Budd, 1991; Silvert, 1993). This approach means that it is considerably easier to add new objects to a model, and redundancy in programming code is minimized (Silvert, 1993). The Object-Oriented Programming (OOP) approach means that one can develop models that are simpler and closer to natural ecosystem structure than with procedural languages; it is also possible to modify and refine these models more efficiently. Ecological processes can be modelled at different scales within the same model, depending on the purpose of the model (Matsinos *et al.*, 1994; Mooij and Boersma, 1996). By using rules within an OOP, animals are able to interact with their habitat and with other animals (Mueller, 1991). Animal objects are assigned characteristics and behaviours; the animal objects live within habitat objects that carry relevant information such as vegetation and soil type (Mueller, 1991).

Fishwick (1995) describes the ATLSS (Across-Trophic-Level System Simulation) modeling system for the Florida Everglades. This was a collection of different models at different scales, depending on the trophic level being studied. For example, individual-based models were used for higher-trophic organisms, while general population models were used for organisms from lower trophic levels. Collectively, the models formed a "multimodel" (Fishwick, 1995) to provide a landscape-scale ecosystem model.

In designing the QnD model, we have developed an intermediate-scale management "game" model that draws on many of the ideas described in the previous section. The aim is to present the model as a game which involves both managers and scientists. Such a modeling system will add to the existing abiotic-biotic models already developed for wildlife areas such as the Kruger National Park, South Africa (Weeks *et al.*, 1999; Mackenzie *et al.*, 2000). The model links abiotic drivers to biotic responses using simple rules and cause-and-effect relationships. The object-oriented framework of the QnD model provides flexibility in the code, where additional objects and methods can be added with ease. Thus our model was developed with the following principles in mind:

- **One design, many ecosystems**: to provide a generic object-oriented modeling framework that can be adapted for different ecological systems;

- **Single and double loop learning**: to provide adaptive management support by scenario playing to view tactical, strategic and system-wide issues;
- **Ecology meets engineering**: to allow complex ecological situations to be constructed from relatively simple model designs;
- **Right problem, right scale**: to incorporate a degree of scalability (small and big time steps in both spatial and temporal scales);
- **Ecosystems *have value and are valued***: to some broadly definable extent, the concept of existence value and political issues/popularity should be taken into consideration;
- **Precise and vague, together**: to allow some things to be known or valued at precise levels while other things within the same ecosystem may be vaguely known or valued. These two aspects must be included and embraced within an iterative framework;
- **Fast development and continued iteration**: to allow interested people to quickly set up model simulations and just as easily change them when further learning occurs.

The entire QnD system is coded in the Java language and is a combination of original code and open source libraries/application programming interfaces (APIs). The QnD system is divided into two parts: the *Simulation Engine* and the *Game View* as shown in Figure 1. Each part has a primary objective to either ***create information*** or to ***communicate information*** to the users. The Simulation Engine works to synthesize the various data and systems concepts into useful constructs that can provide systematic calculations and information in a modular and quickly-altered platform. The Game View allows users to see the system in more graphical (less number-based) methods and to implement management options in a simplified way.

3.1 QND MODEL DESIGN: THE "GAME VIEW"

The Game View constitutes what a "player" sees and reacts to with the various management options in the player's world. Each game view is made of a map viewer (GeoToolsLite API), scrolling time series charts (Chart2D API), warning lights, and management selection widgets.

The game view has several types of outputs that can be configured by the user via XML (e**X**tensible **M**arkup **L**anguage) file inputs. By presenting the outputs in a selectable form, the QnD system allows users to choose how they want to see their output, including the following output options as described in Figure 2 and listed below:

- Geographic Information System (GIS) maps that are updated on each simulated time step;
- Mouse-activated charts and text for individual spatial areas (pie charts and text line descriptions);
- Warning lights that change at user-selected critical levels;
- Scrolling time-series charts (listed on user-defined, tabbed pages);
- User-defined, text output files in comma separated format.

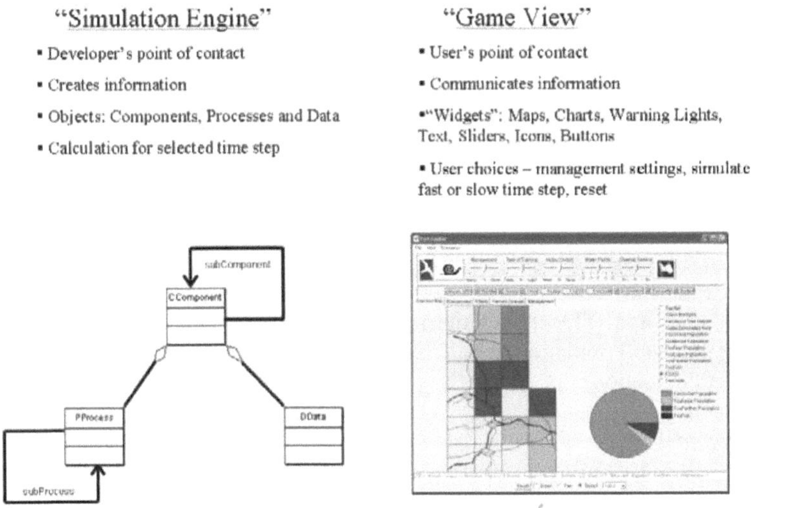

Figure 1. QnD model main parts: simulation engine and game view.

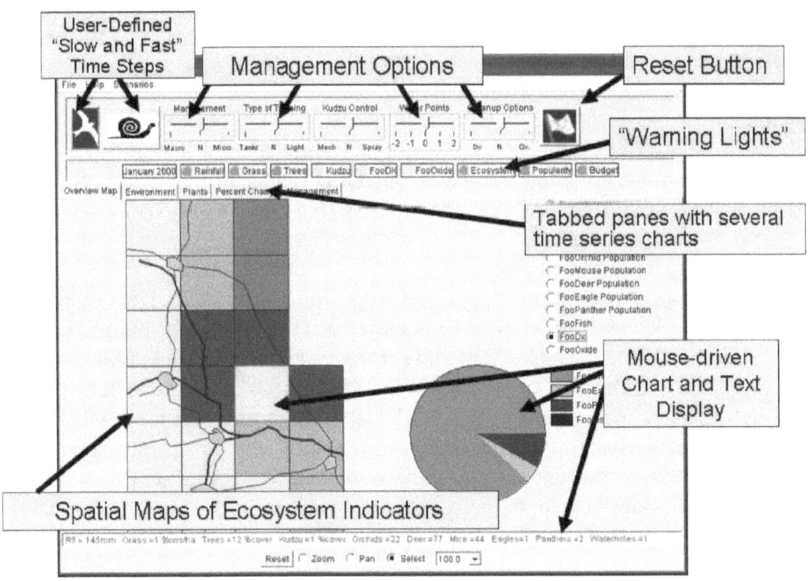

Figure 2. QnD game view features.

3.2 QND MODEL DESIGN: THE "SIMULATION ENGINE"

The QnD simulation engine is made of a few basic objects linked together into simple or complex designs, determined by the needs of decision participants. The most elemental objects of QnD are Components, Processes, and Data as shown in Figure 1. A Component is an object that is of interest to the user. Processes are the actions that involve Components. Data are the descriptive objects assigned to Components. If one uses parts of grammar as an analogy, Components are the nouns. Processes are the verbs. Data objects are the adjectives or adverbs. For clarification, a **"C"** prefixes Components, a **"P"** prefixes Processes, and a **"D"** prefixes Data objects. For example, the statement "*An elephant will trample two trees per day*" could be interpreted as the Components (**C**Elephant and **C**Tree) with a Process ("**P**Trample") and Data (**D**ElephantPopulation and **D**TreePopulation). In this case, the Process "**P**Trample" would use the **D**ElephantPopulation to calculate the reduction in the **D**TreePopulation (by 2 x **D**ElephantPopulation).

3.2.1 QnD Component Objects

The most fundamental building block components in QnD include CWorld, CSpatialUnits, CHabitats, Organisms, and Chemicals, and their relationships are described in Figure 3. The CWorld object contains all the objects and serves to define the spatial limits of the simulated system. A CSpatialUnit is the basic spatial unit of the QnD system. CSpatialUnits can be linked to one another and have a specific location. A CSpatialUnit can have either zero or any number of CSpatialUnits connected to them. In addition, these connections can be labeled with useful words to group similar types of connections. For example, a riverine description may be "UPSTREAM" to describe all connections that move against a prevailing current. CHabitats exist within CSpatialUnits and are not spatially defined. CHabitats make up a certain percent area of a CSpatialUnit. At least one default habitat exists (and occupies 100% of the CSpatialUnit) if the user does not set up any other CHabitats. A CHabitat can hold any number of COrganisms or CChemicals. With the QnD object framework, both simple and complex designs are possible.

3.2.2 QnD Data Objects

DData objects store all the relevant information for a simulation. All DData objects are created in the input files and represent a composite variable as seen in Figure 4. Each DData has several attribute variables that allow for various calculations. All attributes are not used for each DData as some data object definitions may use other attribute features while others do not. For example, a DData object that is linked with a time series file (through its DriverLink attribute) may constantly change current values over time while another may represent a static variable in the simulation and may not use any other attributes besides current value.

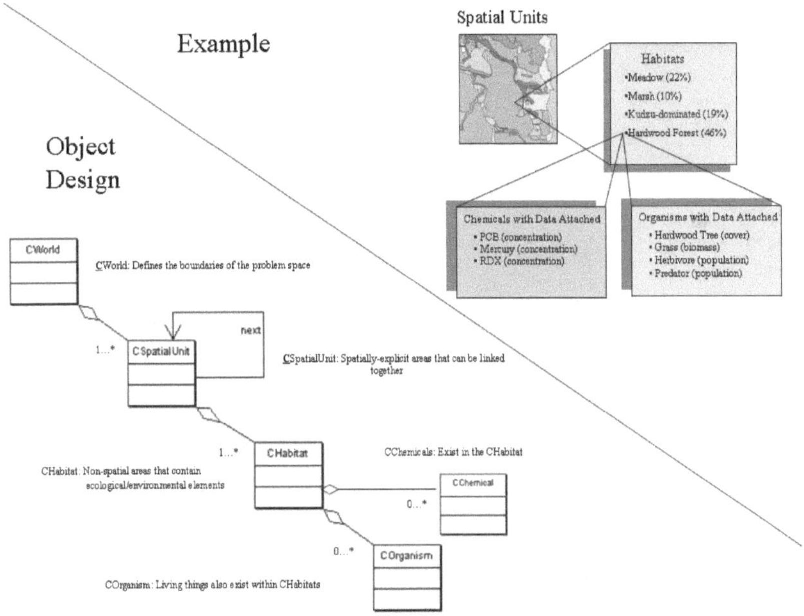

Figure 3. QnD component design and example.

3.2.3 QnD Process Objects

Processes provide the action within QnD. Process objects use DData objects as inputs, provide a calculation or series of calculations, and then write the resulting products into output DData objects. Processes can be used individually as described in Figure 5. In addition, processes can be designed with constituent sub-processes within them to create a series of processes for more complex interactions, as described in Figure 6. Table 1 shows the different types of processes currently available in QnD. The example in Figure 6 shows how sub-processes can be linked with interactions between DData objects and two processes. The current values of DGrassBiomass and DShrubBiomass are added together with a PAdd process to supply the current value of a DTotalBiomass data object. This same DTotalBiomass current value is the input to a PRelationship object that creates an output that is placed into the cumulative effect of a DElephantPopulation data object.

Table 1. PProcess objects for the QnD model.

Process Type	Purpose
PAdd	Input1 + Input2 + Input3… + Input_n = Output
PSubtract	Input1 – Input2 - … … - Input_n = Output
PMultiply	Input1 x Input2 x Input3… Input_n = Output
PDivide	Input1 / Input2 / … … Input_n = Output
PTransfer	(Input – TransferAmount) & (Output + TransferAmount)
PRelationship	Two dimensional input/cause (x axis) is used to interpolate an output/effect (y axis) value.
PSimpleLookUpTable	Uses two input data values to choose another output value from user-defined table

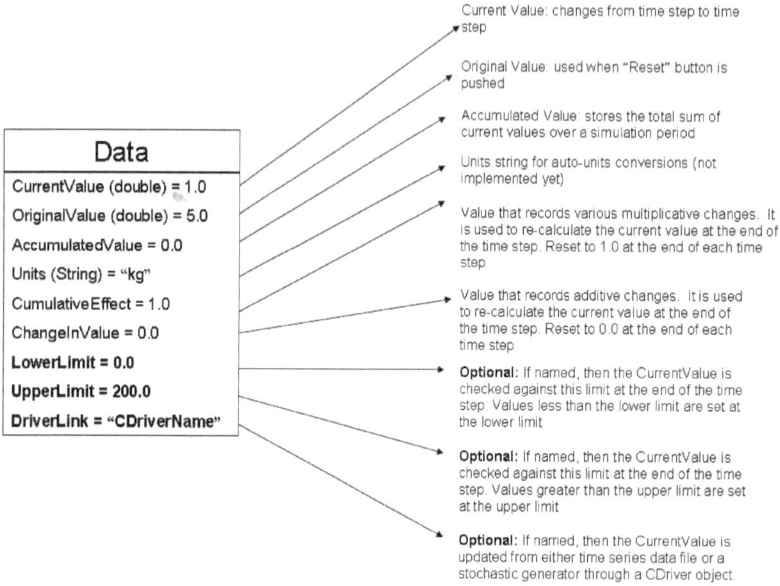

Figure 4. Data object design and attribute descriptions.

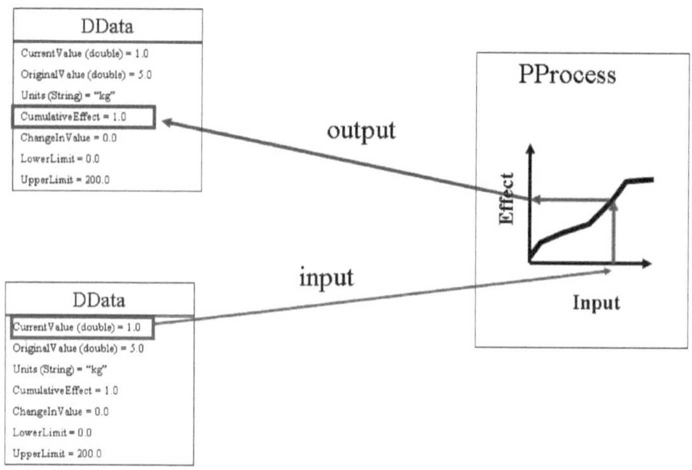

Figure 5. Example PRelationship process object.

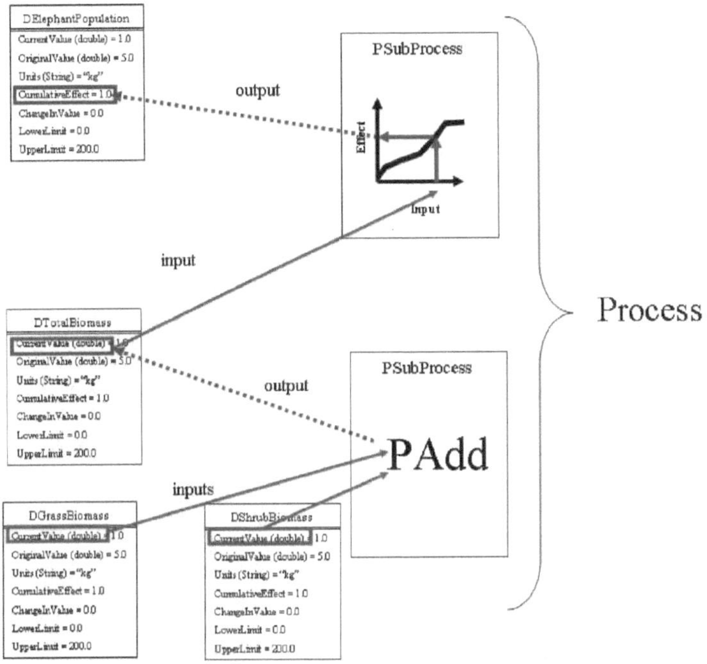

Figure 6. Process objects can be created with multiple "sub-process" objects to form more complex calculations.

3.3 ASSEMBLING QND OBJECTS

Each version of QnD is created entirely from the XML file inputs. Figure 7 shows the seven input files and their use to create the QnD Simulation Engine and Game View. The QnDStartHere file is read by QnD to find the input and output file paths as well as the exact filenames of the other XML files. The QnDWorld file is read to construct the various spatial units along their constituent habitat, organism, and chemical objects. Any DData objects that represent spatially unique properties such as local concentrations or population levels are included in the QnDWorld file. The QnDTopology file is used to link various spatial units with each other. The QnDOrganism and QnDChemical files are used to create DData and PProcess objects that all occur in all instances. The QnDOutput.xml file is read to create user-defined files of DData values in a comma-separated text format.

The QnDManagement file is used to define the various parts of the game view, including the map layers and user-selected maps, scrolling time series charts, warning lights, and management options. In addition, certain simulation engine components such as scenarios and their time series files or stochastic generation settings are set in this file.

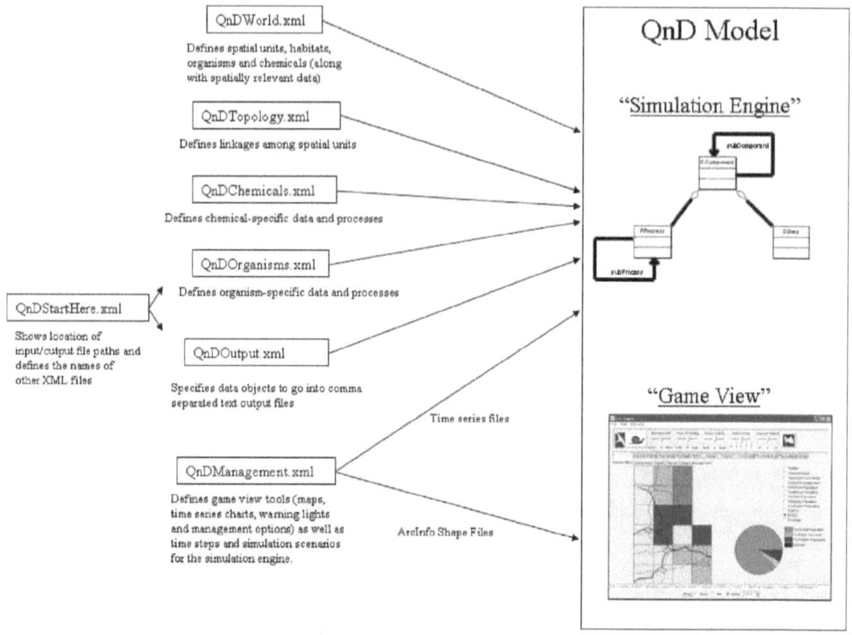

Figure 7. QnD input files are used to generate both the simulation engine and the game view.

After all input files have been read and the objects instantiated, the QnD system waits for user input including the following options:

- Look over the simulation information
    - Switch Map View (with radio buttons)
    - Switch Chart View (with tabbed pages)
    - Review specific spatial unit data values (by pointing with the mouse)
- Set some management options by interacting with the map and setting management sliders
- Simulate a short or long-term time step
- Restart the simulation to its initial settings by clicking on the reset (white flag) icon.

3.4 PEOPLE AND QND

There are 3 stereotypical groups of people (or actors) involved in QnD simulations, as shown in Figure 8. These three example groups are named *players*, *developers*, and *coders*. Players interact mostly with the Game View while playing and exploring the ecosystem, potential management responses, and trade-offs. Many players are stakeholders but can be anyone who has an overall interest in the system. They see the simulated world as a larger, integrated ecosystem and have broad, varying interests. While players may have some interest in certain areas of the QnD Simulation Engine, they mostly interact with the Game View, the management options, functional information, and QnD operations that approach some level of reality as they understand it. In this fashion, players provide an important reality check to the overall design and function of the QnD system.

We divided the traditional model/code developer role into two separate roles (*developer* and *coder*) to include specialists that are not well-versed in computer science and formalized modelling to functionally interact with the QnD system. *Developers* design and implement the game view and simulation engine objects using the XML input files. While a developer might be a player as well, their primary role is to translate the broader ideas of the players into functional object designs that are represented in the input files. Another fundamental role of developers is to provide any formalized calibration or validation of the simulation engine/game view that is desired by the players. This confidence building aspect is an important function in building trust and interest into any simulation results that are seen by the overall group as critical.

While developers may have interest and/or access to QnD java source code, they should not be spending much time altering source code to achieve their modeling objectives. This role is assigned to smaller, more technical group of *coders*. Coders interact mostly with the java source code and concern themselves with the overall applicability and expansion of the game view and simulation engine parts, as well as the functional deployment of the QnD models. Coders have control of and responsibility for the overall design and evolution of the QnD system for all groups of players and developers. Coders may take specialized suggestions from players and developers and implement them at a broader, more abstract level within the source code to take advantage of new developments in the Java language, computer science concepts, or internet technologies.

4. QnD: Development and Gaming Strategy

The QnD model has been developed as a useful tool embedded in a larger process of stakeholder and public participation when utilized to generate questions and decisions for complex environmental management. Development of the QnD game and its application was inspired by some of the principles described by Gunderson *et al.* (1995), Gunderson and Holling (2002), Miller (1999) and Checkland (1999) as a way to view a complex environmental problem situation from a variety of technical, social and cultural perspectives. This section reviews the theory and practice that contextualizes the development of the QnD model.

4.1 QND: MODELING AS LEARNING

The QnD development process is typically embedded in a context of environmental management where information is uncertain and decisions regarding improvement need

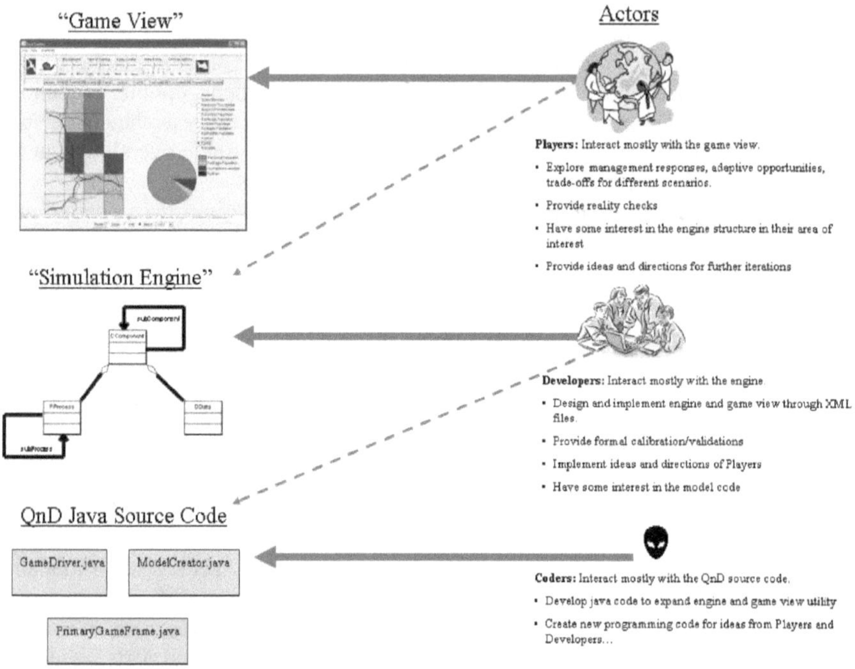

Figure 8. Diverse groups of people will interact the QnD system in different areas.

to be made. Collaborative building of a QnD model creates a critical dialogue amongst stakeholders, simultaneously gathering the technical data that is available and also clarifying values and beliefs about the environmental system. The modeling development process as a learning process is based on a soft systems understanding of problem exploration and problem understanding.

Traditional problem solving approaches based on a mechanistic and reductionist view of the world have shown their inadequacy in the face of the vast scale of modern problems. Capra (1996) suggests that major problems cannot be understood in isolation, but instead must be viewed in terms of interconnectivity and interdependence. He calls for not just a holistic way of viewing problems, but an "ecological" mindset that recognizes how each aspect of a problem is imbedded in a natural and social environment. Checkland (1976, 1981, 1999), whose primary concern is with the social world, makes a similar observation when he suggests that integration, rather than further fragmentation, is needed to think about complex problem situations. He argues that scientific inquiry can be described as a particular kind of "learning system." To this end, the QnD model can be utilized to assist dialogue and learning within a problem situation.

Soft systems methodologies are essentially learning models, and they can be related to Kolb's cognitive/action cycle (Kolb, 1984) representing four different knowledge forms: diverging (what is there?), assimilating (what can we do?), converging (what is important?), and accommodating (what does it mean?). Figure 9 shows soft systems considerations in terms of Kolb's knowledge forms (after Bawden *et al.*, 1984). The QnD model can be used as a facilitative device to take participants iteratively through the four stages of the learning cycle. The model presented as a game generates discussion amongst the stakeholders about their understanding of the problem situation and provides an interactive way of testing and debating a variety of actions that could be taken to improve the situation. Instead of talking about the implication of various actions in the abstract, the QnD game allows participants to try out different management alternatives and investigate possible repercussions of those decisions. Additionally, the game can be structured around different scenarios from different perspectives, allowing participants to test their assumptions within different future worlds, thus revealing the biases in differing perspectives. QnD has been designed to be used within more formalized scenario generation processes, such as the planning approach developed at Royal Dutch/Shell (van der Heijden, 1999).

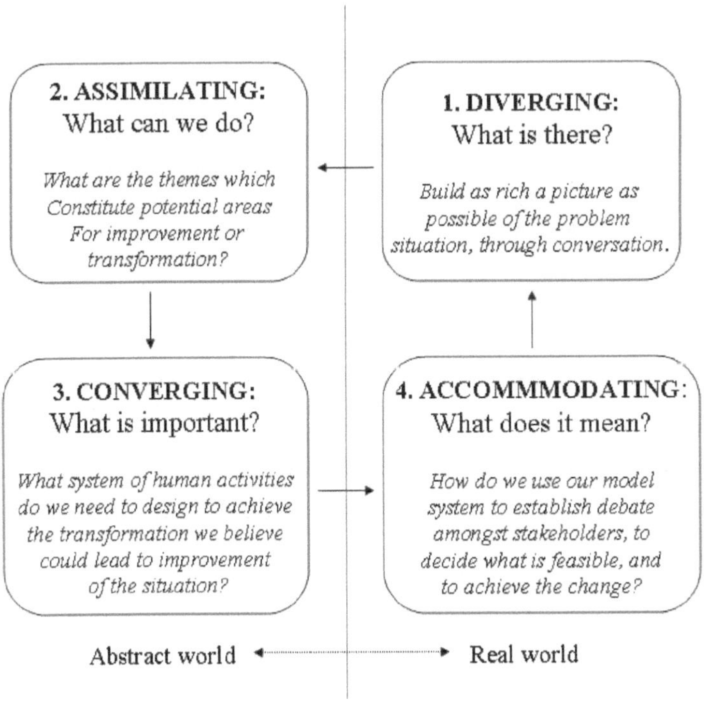

Figure 9. Soft systems considerations represented in terms of Kolb's Knowledge Forms (after Bawden *et al.*, 1984).

The notion of modelling as learning differentiates itself from a traditional scientific hypothesis-testing approach which seeks to establish a firm problem definition early on in the research process. Using Kolb's terms presented above, this traditional approach rushes past diverging and assimilating in an attempt to converge on a definition of the problem as early as possible, in order that research may begin and solutions may be found. One disadvantage of this early convergence is that formalizing the problem definition too soon in the planning process can establish a faulty foundation with a biased starting point; as a result, the solutions that are generated are solutions to the "wrong" problem. If the problem is defined without a full consideration of all possible opinions, then the solution will be off target. Churchman (1979) recommends a process of "sweeping in" pros and cons, friends and enemies in order to reveal the range of possible assumptions about the problem situation and what would constitute improvement. Using QnD in the early stages of a decision making process enables participants to "sweep in" this bigger picture through debating the problem situation, building the model, and playing the game. Divergence is an uncomfortable process, as those who are schooled in a scientific way of thought desire to formalize the problem mess as soon as possible. The QnD model development process is a tangible way to help participants make sense of this mess as they learn more about the many possible perspectives and dynamics in the problem situation, equipping them to make decisions and to continue exploring possible actions that may improve the situation.

Model development traditionally forces the problem situation into a structured form, and in the process loses important features of the situation, such as the human component. Most technical models have in the past been built around mathematical equations which give the impression of precision and reliability in an attempt to find the one best way to solve problems. This deterministic use of models and computer technology has created suspicion in the general public who were not convinced that the computer could process the data and print out a neat and tidy solution for their messy problem situation. The general public could see through the enthusiastic use of technology, recognizing that the model was often working on the "wrong" problem from a mechanistic and reductionist point of view, able to input only quantitative information that is statistically reliable. Many decision makers have realized that the complex problems of today are not only about finding technical solutions, but also about understanding stakeholder experience, knowledge, and values. Because of QnD's design, both qualitative and quantitative information fit side by side. Both hard data, such as field-measured experiments, and soft data, such as experiential learning or general impressions, are valid model inputs.

Model development has been used elsewhere as a facilitative device, usually to generate very simple models. One potential limitation of these models is that they are primarily a means for uniting stakeholders around a single systems viewpoint and are less relevant for detailed ecosystem management exploration. QnD is differentiated in that a little more complexity is desired. QnD is not created in a one-time meeting. Instead, participants interact at least two times within a QnD learning process as shown graphically in Figure 10. The first meeting is an initial "genesis" session to elicit the key features of concern in the problem situation from potential players. Then developers build an initial game, returning one to two weeks later to play the game with participants and to test scenarios. After playing the game, revisions to the model can be made, and the game played and revised repeatedly as needed during the decision-making process.

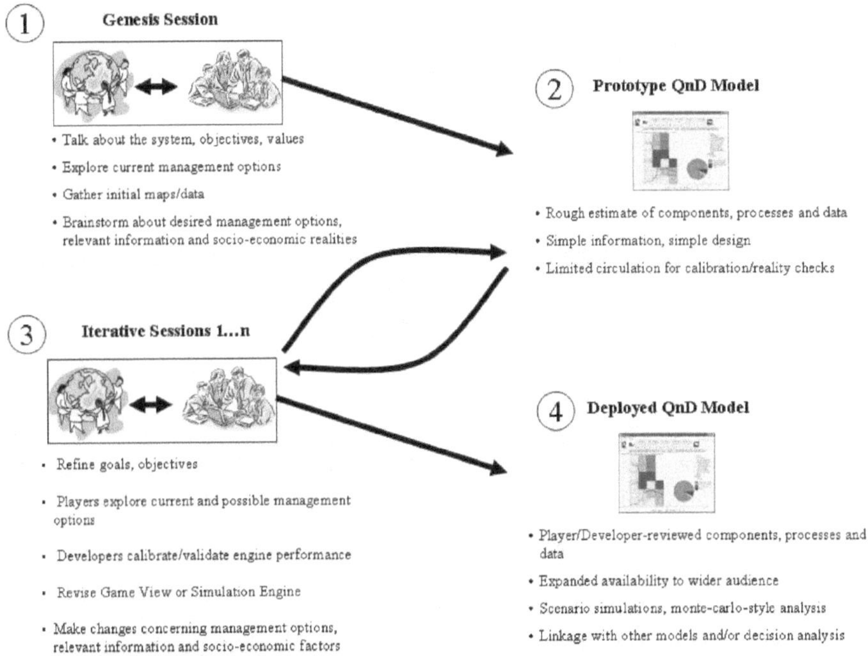

Figure 10. QnD development is iterative and allows group learning to be incorporated into the model.

Three primary activities are used to develop a QnD model/game. First, participants describe the problem and its elements in words and pictures. Through conversations with stakeholders, a series of pictures, stories, experiences, simple diagrams, or equations are recorded to get an overall view of the problem. Secondly, words and pictures are interpreted into QnD objects. The various system descriptions from the initial meeting are used by the QnD developer to fashion the initial engine and game view sections. An essential element of the QnD model is that the game view should be constructed as much as possible from the user's perspective while the engine can be a combination of technical and subjective relationships. The third primary activity takes place during the second meeting with stakeholders when they discuss and debate the problem situation using the QnD model scenarios in order to identify desirable and feasible actions and changes that would improve the problem situation. This discussion in which stakeholders interact with various QnD elements may highlight three resulting activities: (1) changing the QnD engine to provide a more adequate simulation of measured events; (2) changing the QnD game view to better represent management information requirements or potential actions; or (3) identifying new aspects of the problem situation that were previously hidden from scrutiny. By playing QnD scenarios, users find that they are able to explore the positive and negative repercussions related to each potential management option. Participants are able to discuss both informal "rules of thumb" and technical aspects of management decisions. In addition, QnD enables stakeholders to explore from a variety of perspectives how a decision might impact ecosystem components as well as socio-political and economic factors.

5. QnD Case Studies

A good model should be useful to managers while being founded on data and assumptions that can be justified scientifically. While model outputs may approximate the real world situation (*i.e.* there is significant correlation between observed and modeled data), it is important that the mechanisms underlying the model output are the right ones (Snowling and Kramer, 2001). Hereafter, a model only becomes useful to natural resource managers if the model inputs can be coupled with different scenarios and the outputs compared against some kind of meaningful threshold.

Management inevitably occurs under situations of imperfect data and incomplete assumptions, but invariably takes the form of "what happens if...?" questions, such as "What will happen to this system if we have a dry season?"; "What happens if a severe flood occurs?", and "What happens if I do nothing?". Thus natural resource managers operate within a spectrum that ranges from "do nothing" to micro-management, which is possible in our model by applying management actions on a cell-by-cell basis, to macro-management, where a blanket management policy is applied to the entire area of interest. Furthermore, it is useful for managers to experience the effects on the system of large and random disturbances that are beyond their control. This section highlights two applications of QnD with non-spatial and spatial simulations. Both case studies highlight QnD development and game/scenario playing processes within differing ecological contexts.

In the two case studies briefly summarized in this chapter, we have tried to describe the model structure and function in simplified terms without large detailed object designs or mathematical equations to communicate the basic purpose, behavior, and lessons learned from each of these simulations. More specialized, technical documentation is available through the QnD web page (www.risktrace.com).

5.1 QND:NPR – ADAPTIVE MANAGEMENT WITHIN THE NORTHERN PLAINS OF THE KRUGER NATIONAL PARK

The Kruger National Park (KNP) is located in the northeast corner of South Africa, as seen in Figure 11. The park covers a wide variety of climates, ecosystems, and soils. Both infertile (deep sandy) and fertile (basalt-based) soils are present, and annual precipitation ranges from 400 mm to 750 mm, with high spatial and temporal variability. These varying environments provide a high species diversity in both plants and animals (Joubert, 1986). Rogers and Biggs (1999) outlined an adaptive management framework that explores the consequences of management decisions by measuring the model outputs against critical thresholds. In the KNP, these are defined as "thresholds of potential concern" or "TPCs." Successful TPCs are based on research that has identified agents of change, as well as suitable indicators of this change. Systems are variable, and the TPCs for indicators within these systems need to reflect this variability. Furthermore, the TPCs need to be objectively defined and defendable and exist within an iterative cycle of monitoring (Rogers et al., 1999). The TPCs are not an end in themselves, but rather assist in achieving pre-defined management goals, which in turn are constantly reassessed. Part of the goal maintenance system of Rogers and Bestbier (1997) is that the consequences of management actions are examined within a goal-oriented framework Models become useful to management if they have the capacity for evaluating different scenarios and testing the consequences of these against objective yardsticks (TPCs). The object-oriented approach of QnD enables managers and modelers to add indicators for agents of change and measure their response against TPCs under different management scenarios.

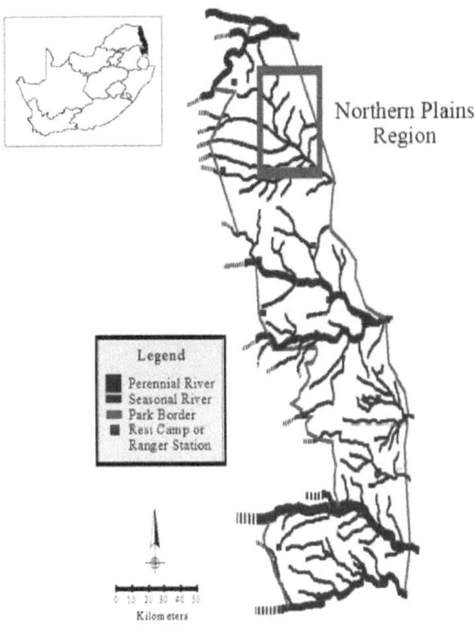

Figure 11. The Northern Plains Region of the Kruger National Park (after Kiker, 1998).

Given the large size of the KNP (approximately 20,000 km^2), a smaller site was chosen to focus the QnD modeling effort into an area of specific interest to both scientists and park managers. The Northern Plains Region (NPR), as shown in Figure 11, is ecologically defined by KNP scientists: *Colophospermum Mopane* Shrubveld on Basalt

(Gertenbach, 1983) comprises very flat plains of slopes averaging 0.62°. The mean elevation of the NPR is 343 meters above sea level with a standard deviation of 41 meters. The mean annual rainfall varies between 450 and 500 mm per year. The woody vegetation of the NPR is almost completely dominated by dense *Colophospermum mopane* (Mopane) shrub trees with some scattered trees of other species such as *Lonchocarpus capassa*, *Combretum imberbe* and *Dalbergia melanoxylon*, (Venter *et al.*, 2003).

The NPR has been of interest to KNP scientists and managers because of its role in supporting rare antelope species including roan (*Hippotragus equinas*) and sable (*Hippotragus niger*) antelope, along with Lichtenstein's Hartebeest (*Sigmoceros lichtensteinii*). Declines in those rare species have focused attention on the NPR's role in maintaining biodiversity within the KNP (Mills *et al.*, 1995; Grant, 1999; Grant and Van der Walt, 2000; Grant *et al.*, 2002). Grant *et al.* (2002), Grant, 2003 describe the management goals in terms of a command and control-style paradigm and the emergence of a newer adaptive management paradigm using the resilience concepts found in Gunderson and Holling (1995). More interventionist management was used in the form of water provision, burning, and elephant culling.

Management actions within the Northern Plains occur from a combination of individual expert knowledge and consensus opinion from population and vegetation monitoring data. Historically roan antelope has not responded to various management actions (Grant et al., 2002; citing Pienaar, 1963). Populations have ranged from 150 to 300 since the 1930s. Since the 1960s, KNP management has focused specifically on creating and maintaining viable rare antelope populations. The consensus opinion of the KNP scientists and managers from the early 1960s was a need for more suitable habitat. The proposed mechanism of the problem was that excess water from the artificial water sources was sustaining higher populations of zebra for longer periods than their traditional seasonal visit. The combined effects of an extended drought, higher zebra populations, and higher associated predators caused a population drop. As a management "experiment," water points were closed in roan habitat with continued assessment and monitoring. These management experiments were conducted within a climate with high annual rainfall.

The primary purpose of this QnD version was to not to necessarily predict future ecological state variables as much as it was to inform and educate interested parties of the ecosystem management issues within the KNP. Playing the QnD:NPR software allowed non-KNP professionals to test various management responses within an operational framework against simplified TPCs to judge the success within ecosystem, financial and public values.

The long-term objectives of the QnD:NPR game were the following:

- **Ecosystem Viability**: maintain a balance between vegetation, herbivore and carnivore populations;
- **Political/Management Success**: maintain a balance between public perceptions and opposition to certain management options (*i.e.* elephant culling);
- **Financial Management Success**: maintain financial reserves under limited budgetary resources.

Figures 12 and 13 show the basic structure of the QnD:NPR model. The QnD components, processes and data objects were built from previous model and ecosystem studies (Kiker, 1998; du Toit and Biggs, 2003; and O'Connor and Kiker, 2004). Rainfall is the primary driver and varies randomly from 300 to 800 mm/yr. We assume that there is only one large spatial unit with a default bushveld habitat. All species of grass are represented by a simple COrganism Grass with a corresponding data object DBiomass (metric tons/ha). All tree species are represented by a CTree object with a related DPercentCover data object. Animal species such as elephant, roan antelope, zebra and lion are represented by individual objects, each with a corresponding DPopulation data object. No chemical objects were used in this QnD version.

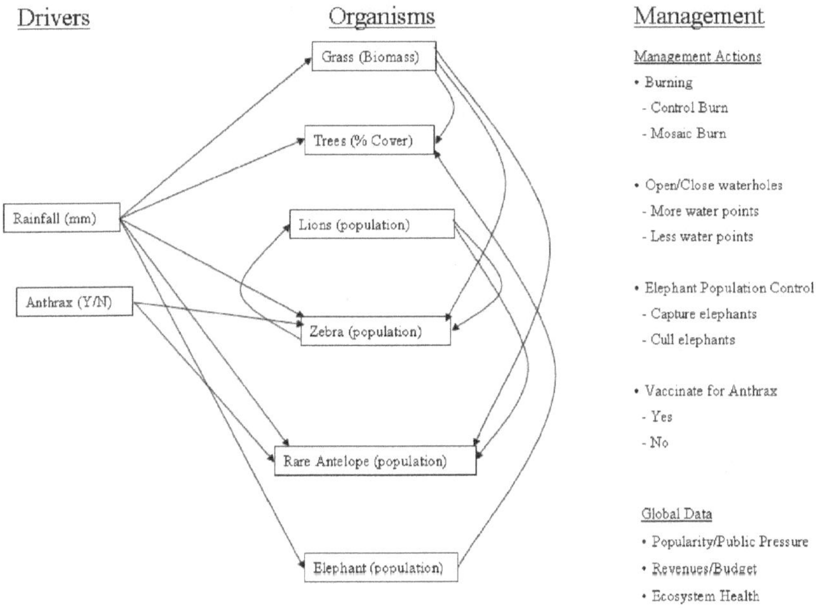

Figure 12. Driver and organism interactions in the QnD:NPR model.

Processes that interlink between organisms are shown in Figure 12. In this version of QnD, all process objects are simple linear relationships in the form:

$$\text{Annual Change in Data Value} = f(\text{Data Value}_1, \text{Data Value}_2, \ldots \text{DataValue}_n)$$

where each f(Data Value) is a PRelationship object as described in Table 1 and Figure 5. Multiple effects are combined in a multiplicative fashion. For example, the effect of rainfall upon tree cover may increase the percent cover by 5% while the effects of elephant populations may decrease cover by 8%. These two annual effects would be incorporated as:

$$\text{Tree Cover}_{n+1} = \text{Tree Cover}_n *(1.05) *(0.92)$$

Figure 12 graphically describes the following annual processes:

Grass Biomass (metric tons/ha) = f(Rainfall, mm/yr)

Tree Cover change (percent) = f(Rainfall, Grass Biomass, Burning Policy, Elephant Population)

Lion Population change = f(Zebra Population)

Zebra Population change = f(Rainfall, Grass Biomass, Lion Population, Anthrax Occurrence, Number of Water Points)

Rare Antelope Population change = f(Rainfall, Grass Biomass, Lion Population, Zebra Population, Anthrax Occurrence, Number of Water Points)

Elephant Population change = f(Rainfall, Elephant Population Control Policy)

Anthrax occurrence is set at a base level of 0.05 probability. For each consecutive dry year (less than 400 mm), the outbreak probability increases by 0.2 up to a maximum probability of 0.95. Any wet year resets the outbreak probability to the base level. If an outbreak occurs on an unvaccinated population, then population reductions of 20% in zebra and 40% in rare antelope populations occur.

Management interactions are described in Figure 13. The following four management actions were explored:

- ❑ Water Management: Adding or Closing Water Points
- ❑ Elephant Population Management: None, Culling, or Live Capture
- ❑ Vaccination for Anthrax
- ❑ Fire Management: Patch Burning, Lightning Fires, Plot Burning

Management success is related to three primary indices: Ecosystem Health, Available Budget, and Management Popularity. These measures were established to give players an appreciation for the issues that ecosystem managers were being judged.

Ecosystem Health change = f(Tree Cover, Lion Population, Elephant Population, Rare Antelope Population, Zebra Population, Fire Management Policy)

Change in Budget = f(Ecosystem Health, Number of Elephants Captured, Lion Population, Anthrax Vaccination Policy)

Change in Management Rating = f(Ecosystem Health, Lion Population, Elephant Population, Number of Elephants Culled)

The primary scenario to be explored was random annual rainfall between 300 and 800mm. Outputs were recorded on a single time series graph and included percent change in various herbivore and vegetation species from a baseline starting year of

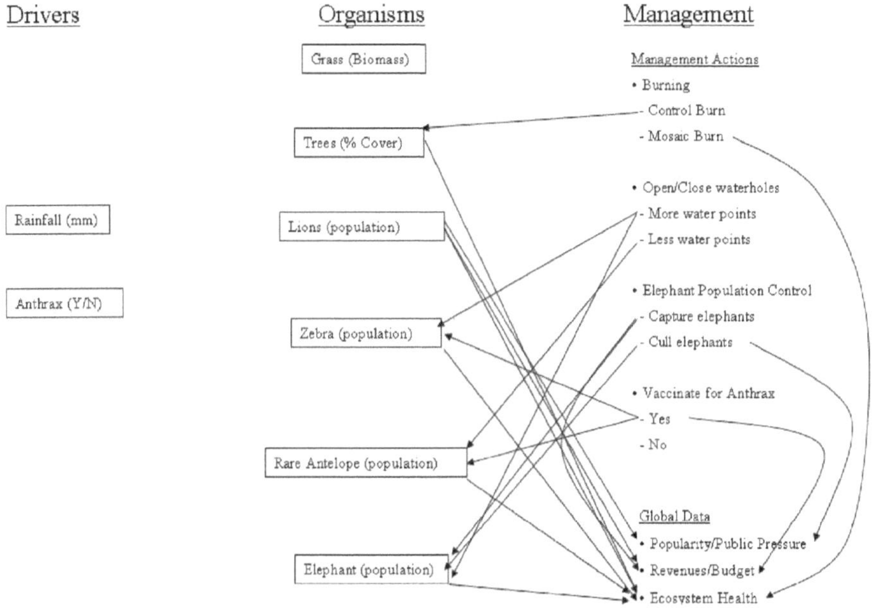

Figure 13. Management-organism interactions in the QnD:NPR model.

2000. Outputs as warning lights included rainfall, grass biomass, anthrax warning level, ecosystem health, management popularity and park revenue. The object of the game was to manage the area for as long as one could before having an ecosystem crash (the ecosystem health level becoming critical), public outcry for a resignation (the management popularity becoming critically low), or exhaustion of park budgets (the budget levels becoming critically low).

This first version of QnD was created to show users the various tradeoffs in ecosystem management and to learn how to manage adaptively an ecosystem with limited resources and options. Simulated ecosystem responses were for the most part accurate to various conditions within the Northern Plains Region. Figure 14 shows a typical QnD:NPR simulation.

Of the various students and interested players who interacted with the game, the longest game length was approximately 30 years. This success was due to a combination of good adaptive management and simple luck of having few drought seasons. The general average first response was approximately 7 to 15 years. Upon restarting the game, almost all players tended to improve their management scores from their first attempt. This occurred even when the players received less favorable climate conditions on the second attempt.

As players played the game, they tended to improve their management skills by learning what interventions worked under what general situations. This adaptive learning went up to a point where the simple "luck of the draw" factor of yearly rainfall began to determine players' longer term success.

Over the simulations, several emergent strategies were suggested and debated among players in the scenario:

"Rainfall trumps everything, even elephants" – almost all management was in response to or in expectation of different rainfall levels. Under some scenarios, droughts provided some temporary population control of elephants. Players often would become more skilled at setting up the ecosystem conditions to mitigate small-scale droughts, although droughts longer than 4 consecutive years were often non-recoverable.

"Micro-managing with water?" – given that players could open or close water points with no financial cost, the first choice of action was usually to modify water point numbers. Given the current QnD:NPR structure, this had a larger positive influence on rare antelopes, zebra and elephants, but often led to overpopulation of both species and large population crashes once elephant populations increased to levels 25 to 50% above baseline. Population crashes in rare antelopes were usually instigated by droughts, competition from zebras, or anthrax outbreaks.

"Manage elephants early and in smaller numbers" – the social and financial cost of elephant control dictated that frequent control of fewer elephants is less costly (politically and financially) than few large-scale population control actions. In almost all neophyte simulations by players, control of elephants was delayed because of the high cost of the decision. The less politically charged (but financially draining) action of live capture was almost always adopted first, with culling as an option of last resort. In addition, some discussion as to whether the simplistic, linear popularity penalties for elephant culling were accurate to reality, as they tended to dominate the use of this option.

"Vaccinate for anthrax only when you have to?" – player groups were more mixed over the best strategy for vaccination. Some favored vaccination whenever more than one year of drought occurred, and others favored a strategy of building up rare antelope and zebra populations as a buffer against large scale die-offs. This strategy worked in some rainfall scenarios, while Figure 14 shows this strategy failing under a combination of an anthrax outbreak and a multi-year drought.

Overall, these strategy discussions were not meant to find an optimal management scheme for the KNP. More realistically, QnD:NPR formed what was jokingly called a "decision sympathy system" (as opposed to the traditional decision support system) to allow non-KNP players to appreciate some of the pressures and limitations faced by KNP managers. The model was quite useful for initiating discussions on basic ecosystem dynamics and whether certain management options would be effective over the long term. Further QnD research and development into more spatial applications within the KNP are underway.

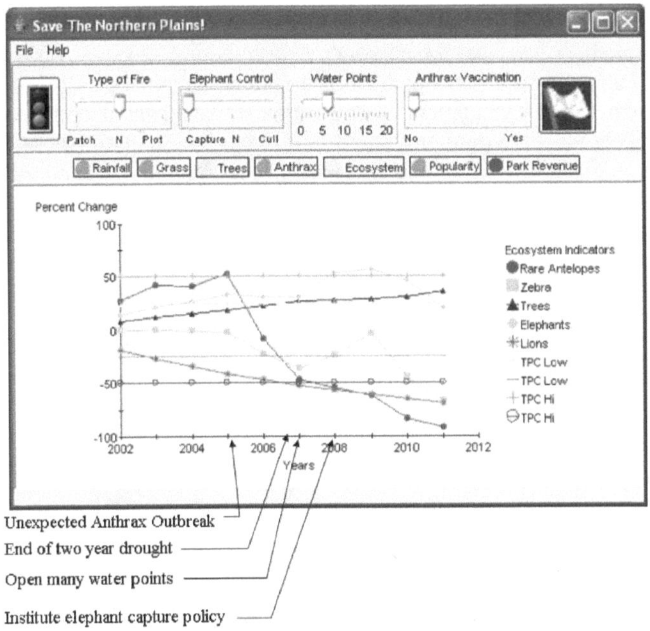

Figure 14. Example QnD:NPR results showing a player's management responses to an unexpected anthrax outbreak and two year drought.

5.2 QND:FOORIVER – INTEGRATING SEDIMENT AND ECOSYSTEM MANAGEMENT FOR RISK MINIMIZATION

A riverine ecosystem/dredging/risk version of QnD was developed as a demonstration project for the US Army Corps of Engineers – Engineer Research and Development Center (USACE-ERDC). The FooRiver demonstration was created to show how QnD can be used in a stakeholder setting with problem definition, model design, and scenario generation. Functionally, this demonstration version was built from comparative risk assessment concepts and data used in Kane Driscoll *et al.* (2002). A functional management map of the FooRiver system is illustrated in Figure 15. Sections of the FooRiver are divided into 5 river reach management areas (Shaka's Rapids, Petronella Reach, Mandela Straights, Joe's Bend, and Bobville stretch) and two reservoirs (FooLockDam1 and FooLockDam2). The FooRiver flows into an estuary as the practical edge of the management area.

Stakeholders in the FooRiver basin have environmental challenges that center around the dredging and disposal of contaminated sediments. FooCB (a fictitious non-metabolized organic, similar to PCBs and Dioxins) is present in river sediments at varying concentrations throughout the FooRiver system. Significant trophic transfer of

FooCB can occur from sediments, through benthic invertebrates and into fish populations. These fish populations are consumed by local recreational anglers.

FooRiver stakeholders have expressed concern over both ecosystem features (abundance levels of benthic invertebrates and fish) and contaminant levels (FooCB levels in fish and subsequent risk to anglers). The primary management options of interest to FooRiver stakeholders are dredging within the river reaches and disposal of the contaminated dredged material. Economic and social concerns are also important for stakeholders in that management budgets are limited and some management choices are more popular than others. In addition, stakeholders have questions about management responses under different climate scenarios – including the southern oscillation index ("El Niño," Neutral/Normal, "La Niña") – which tend to have different flow and sediment fluxes.

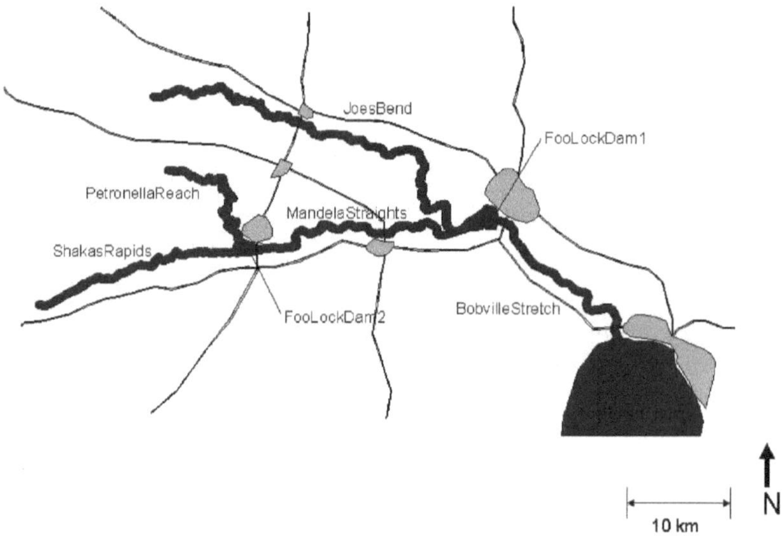

Figure 15. Management map of the FooRiver system.

The primary interactions in each river reach between CDriver, COrganism, and CChemical objects are described graphically in Figure 16. Each reach has sediment, water flow, and contaminant inputs that are used to determine the flux in sediment FooCB concentration:

Change in Sediment FooCB (ppm) = ƒ(Flow Level, Sediment Influx, FooCB Influx)

Abundance of benthic invertebrates is influenced by simplified logistic growth function and reduced by FooCB concentration and dredging effort.

> Benthic Invertebrate Abundance change = f(Current Benthic Abundance, FooCB Concentration, Dredge Effort)

The FooCB concentration of benthic invertebrates is a derived by a biota-sediment-accumulation factor ($BSAF_{inv}$).

> Benthic Invertebrate FooCB Concentration = Sediment FooCB Concentration x $BSAF_{inv}$

Abundance of fish is positively influenced by simplified logistic growth function and reduced by fish and invertebrate FooCB concentration.

> Fish Abundance change = f(Current Fish Abundance, Current Fish FooCB Concentration, Current Fish FooCB Concentration)

The FooCB concentration of fish is a derived by a bio-sediment-accumulation factor ($BSAF_{fish}$).

> Fish FooCB Concentration = Benthic Invertebrate FooCB Concentration x $BSAF_{fish}$

Risk levels in fish consumed by recreational anglers are simulated by a simple step-wise relationship from fish fooCB concentrations.

> Human Risk Level = f(Fish FooCB Concentration)

The management-organism-chemical interactions are described in Figure 17. As an initial construct for management, dredging is described at four levels:

- None
- Low
  - Removes 20% of FooCB from sediments
  - Reduces benthic invertebrate abundance by 10%
  - Reduces total budget resources by 20%
- Medium
  - Removes 40% of FooCB from sediments
  - Reduces benthic invertebrate abundance by 25%
  - Reduces public satisfaction level by 15%
  - Reduces total budget resources by 30%
- High
  - Removes 60% of FooCB from sediments
  - Reduces benthic invertebrate abundance by 80%
  - Reduces public satisfaction level by 35%
  - Reduces total budget resources by 35%

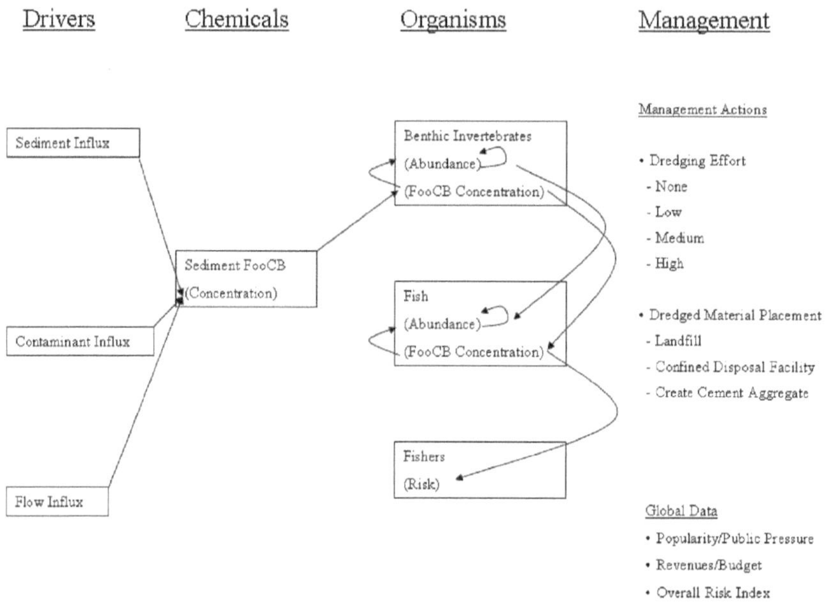

Figure 16. Driver-chemical-organism interactions in the QnD:FooRiver model.

These dredging effects are defined to show both the positive and negative ecosystem effects of dredging as well as the socio-economic cost of implementing different management levels. Once dredging has occurred then the material must be placed in one of three potential locations:

- Landfill
  - Reduces total budget resources by 10% per unit dredge effort
  - Reduces public satisfaction level by 5%
- Confined Disposal Facility
  - Reduces total budget resources by 5% per unit dredge effort
  - Reduces public satisfaction level by 20%
- Create Cement Aggregate Material
  - Reduces total budget resources by 15% per unit dredge effort
  - Increases public satisfaction level by 5%

This initial demonstration version of QnD:FooRiver was constructed to show the utility of QnD in integrating different ecosystem, management, and socio-economic data.

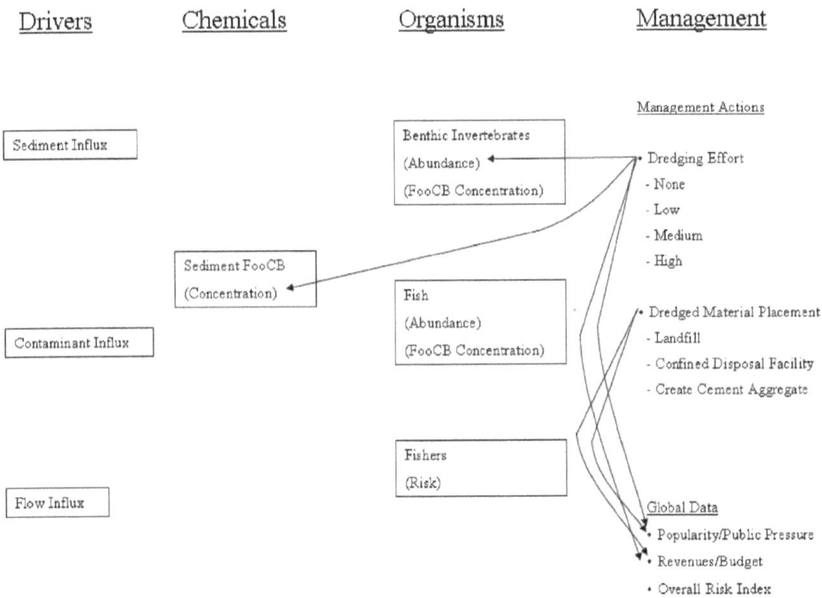

Figure 17. Chemical-organism-management interactions in the QnD:FooRiver model.

Figure 18 shows a simulation for one of the upper reaches (Shaka's Rapids) over a fifty year period under three climate scenarios (La Niña, El Niño and Neutral). This reach had a high initial FooCB concentration in its sediments. Figure 18 shows the progression of sediment concentrations, fish abundance, and fish concentrations under no-dredging management policies. In the uppermost time series chart, sediment FooCB concentrations decline under Neutral and El Niño scenarios but increase sharply under the La Niña scenario due to its tendency for lower reach flows and sediment loads. The middle time series chart shows fish abundances declining sharply in the Neutral and La Niña scenarios while rebounding in the El Niño scenario. The lowermost time series shows the FooCB concentrations in fish declining in the Neutral and El Niño scenarios while increasing in the La Niña scenario.

Figure 18 shows the dynamics of one management area within eight of the entire river system. The decisions on where and how much to dredge can be quite complex when integrating local complexity (in both time and spatial scales) into a cohesive system-wide management plan. In almost all spatial versions of QnD tested, players find that the addition of separate spatial areas tends to complicate both understanding and management of the overall ecosystem. Thus, once the spatial concerns are added into the model, it is much harder to formulate adaptive management strategies, especially when social/economic factors are included. The temptation to "deal with hotspots" (either spatial or temporal ones) and ignore all other areas tends to dominate thinking to the detriment of the entire system, while in some cases the decisions allow some areas to decline substantially while attempting to "save the whole."

In summary, the initial QnD:FooRiver model construct represents the first iteration/round of communication between Developers and Players. All aspects of the FooRiver simulation engine are simplified representations that can be expanded in detail to allow more elaborate calculations of bioaccumulation, abundance, risk, or even fish or FooCB movement between reaches. With these simplified processes in hand, players and developers can explore the processes and their interaction with the overall ecosystem response and potential management issues.

6. Discussion/Conclusions

The QnD modeling software and its associated development methodology was created to quickly and efficiently construct a management/stakeholder-relevant model that integrates both explicit scientifically-derived data and expert/anecdotal knowledge. Given QnD's object-oriented design and XML-based input files, systematic iteration with stakeholders is encouraged and promoted. New and novel ideas about the problem and potential solutions can be explored, adopted, or discarded to promote greater learning about the system.

Development of the QnD model is undertaken within a larger context of stakeholder engagement and public participation. When eliciting information to build QnD scenarios, many different perspectives are expressed, each with its own assumptions about cause-effect relationships and beliefs about what potential interventions would constitute ecosystem improvement. The development process involves working with stakeholders to build the model, play the game, and revise the model, and it is undertaken within a soft systems approach. The soft systems approach distinguishes the QnD gaming and scenario-building process from the more traditional use of models as system predictors. The QnD development process can accommodate both hard data, such as field-measured experiments, and soft data, such as experiential learning, impressions, or general "rules of thumb." The model is used to facilitate dialogue and learning about the factors that influence the environmental system under consideration, and also to explore potential management actions.

While QnD has been used in a traditional simulation modeling context of simulating processes, reproducing measured field conditions, and predicting future conditions (Best *et al.*, 2004; Kiker and Linkov, 2005), the KNP and FooRiver case studies provided in this chapter show how QnD is configured to the wishes of different player and developer groups. The primary goal of these two versions is not necessarily to predict future ecosystem events with high precision, but to show the complexity of ecosystem management choices within a scenario context. Even under simplified object systems and spatial scales, adaptive management is a complex problem. The temptation to go against self-proclaimed policies at localized levels for short-term gains is constantly tempting, especially when some level of political pressure is placed on management choices.

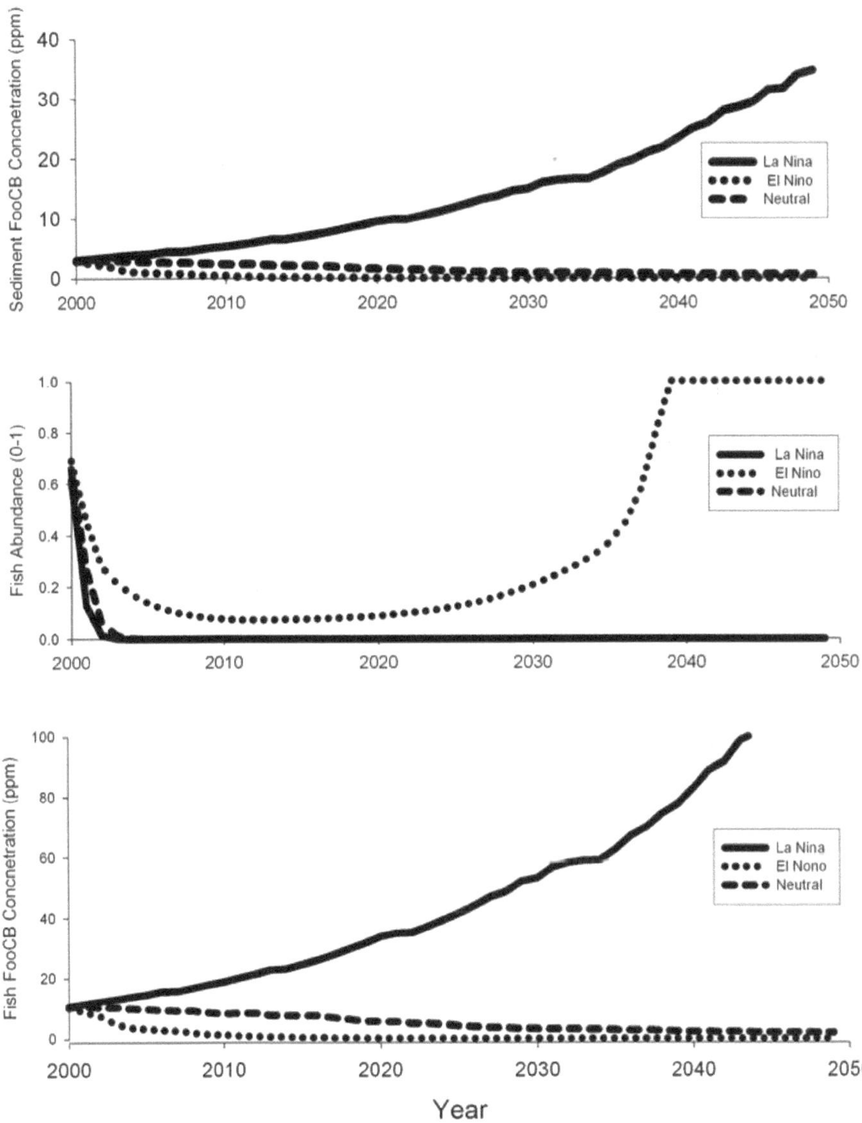

Figure 18. Example results from the QnD:FooRiver model (Shaka's Reach, three climate scenarios).

The role of external agents on management (*i.e.* "back-seat management") can influence and limit options quite directly. It may be beneficial to include these influences in QnD model design with the level of attention that one might give to ecosystem dynamics and detail.

As mirrored within the QnD development methodology explored in this chapter, further development of the model system is ongoing. A more direct linkage with multi-criteria decision analysis is being developed to allow scenario-based exploration of various policies (collections of management actions that function under user set rules). One primary advantage of the player/developer/coder roles is that each group is able to innovate according to their function within the model development and game playing arenas. Players can expect a model that conforms more to their understanding of their world and future worlds with scenario development. Developers can design and implement objects to create a modular system that allows for testing and for changes to be quickly undertaken. Coders can work toward implementing technical advances that further the ease and power of model deployment within the internet-connected world.

An essential objective of QnD has been to actively involve many types of interested people in both model development and its subsequent execution to explore management scenarios. The model system strives to use different experiences and skill sets to an overall advantage in the decision-making process instead of limiting some stakeholders to outside roles as reviewers and/or critics. QnD is an evolving system that continues to develop as new groups of people interact with it and attempt to effectively manage and adaptively address environmental challenges.

7. Acknowledgements

We thank the Kruger National Park, Scientific Services Section, including Dr C.C. Grant, Dr H Biggs and Dr N Zambatis for their input to the QnD:NPR model. The South African Water Research Council deserves thanks for its support of Dr N Rivers-Moore. We acknowledge the Dredging Operations Environmental Research (DOER) program of the Engineering Research and Development Center (ERDC), United States Army Corps of Engineers (USACE) for support in creating the FooRiver demonstration software. We also thank Dr Todd Bridges, Director of the Center for Contaminated Sediments, Environmental Laboratory-ERDC-USACE, for comments and directional guidance.

8. References

1. Bawden, R.J., Macadam, R.D., Packham, R.G., and Valentine, I. (1984). Systems thinking and practices in the education of agriculturalists. *Agricultural Systems* **13**: 205-225.
2. Best, E.P.H., Fredrickson, H.L., MacFarland, V.A., Hintelmann, H., Jones, R.P., Lutz, C.H., Kiker, G.A., Bednar, A.J., Millward, R.N., Price, R.A., Lotufo, G.R., and Ray, G.L. (2004). Pre-construction biogeochemical analysis of mercury in wetlands bordering the Hamilton Army Airfield (HAAF) Wetlands Restoration Site. Interim Report to USACE District, San Francisco,

by USACE Engineer Research and Development Center, Waterways Experiment Station, Vicksburg, MS, November, 2004.
3. Breen, C. (1998). Change: A strategic leadership issue for river management. http://www.ccwr.ac.za/knprrp/cbre(web).htm
4. Budd, T. (1991). *An Introduction to Object-Oriented Programming*. Reading, Massachusetts: Addison-Wesley.
5. Capra, F. (1996). *The Web of Life*. New York: Anchor.
6. Carpenter, S., Brock, W. and Hanson, P. (1999). Ecological and social dynamics in simple models of ecosystem management. *Conservation Ecology* **3**(2): 4. http://www.consecol.org/vol3/iss2/art7/
7. Checkland, P.B. (1976). Science and the systems paradigm. *International Journal of General Systems* **3**(127): 259-269.
8. Checkland, P.B. (1981). *Systems Thinking, Systems Practice*. Chichester, UK: John Wiley & Sons.
9. Checkland, P.B. (1999). *Systems Thinking, Systems Practice: A 30-Year Retrospective*. New York: John Wiley & Sons.
10. Churchman, C.W. (1979). *The Systems Approach*. New York: Basic Books.
11. Costanza, R., d'Arge, R., de Groot, R., Farber, S., Grasso, M., Hannon, B., Naeem, S., Limburg, K., Paruelo, J., O'Neill, R.V., Raskin, R., Sutton, P., and ven den Belt, M. (1997). The value of the world's ecosystem services and natural capital. *Nature* **387**: 253-260.
12. du Toit, J., Rogers, K.H., and Biggs, H.C. (2003). *The Kruger Experience: Ecology and Management of Savanna Heterogeneity*. Washington, DC: Island Press.
13. Fishwick, P.A. (1995). *Simulation Model Design and Execution: Building Digital Worlds*. London: Prentice Hall.
14. Frissel, C.A., Liss, W.J., Warren, C.E., and Hurley, M.D. (1986). A hierarchical framework for stream habitat classification: viewing streams in a watershed context. *Environmental Management* **10**(2): 199-214.
15. Gertenbach W.P.D. (1983). Landscapes of the Kruger National Park. *Koedoe* **26**: 9-121.
16. Green, D.G. (1993). Cellular automata. http://www.resonancepub.com/cellular.htm
17. Grant, C.C. (2003). Northern plains programme. http://www.sanparks.org/parks/kruger/conservation/scientific/programmes/North_Plains_Prog.pdf
18. Grant, C.C. and van der Walt, J.L. (2000). Towards an adaptive management approach for the conservation of rare antelope in the Kruger National Park – outcome of a workshop held in May 2000. *Koedoe* **43**: 103-112.
19. Grant, C.C., Davidson, T., Funston, P.J., and Pienaar, D.J. (2002). Challenges faced by in the conservation of rare antelope: a case study from the northern basalt plains of the Kruger National Park. *Koedoe* **45**: 45-66.
20. Gunderson, L. H., Holling, C. S. and Light, S. S. (1995). Barriers and Bridges to the Renewal of Ecosystems and Institutions. Columbia University Press. New York.
21. Gunderson, L. H. and Holling, C. S. 2002. Panarchy: Understanding Transformation in Human and Natural Systems. Island Press. Washington.
22. Holling, C. S. (1999). Visions: A personal essay. *Conservation Ecology* **3**(1): 12. http://www.consecol.org/vol3/iss1/art12/
23. Houlahan, J. (1998). Big problems, small science. *Conservation Ecology* **2**(1): R1. http://www.consecol.org/vol2/iss1/resp1/
24. Jeppesen, E. I. and Iversen, T.M. (1987). Two simple models for estimating daily mean water temperatures and diel variations in a Danish low gradient stream. *Oikos* **49**: 149-155.
25. Jewitt, G.P.W., Heritage, G.L., Weeks, D.C., Mackenzie, J.A., van Niekerk, A, Görgens, A.H.M., O'Keeffe, J., Rogers, K., and Horn, M. (1998). *Modelling Abiotic-Biotic Links in the Sabie River*. WRC report no. 777/1/98. Water Research Commission, Pretoria.
26. Jewitt, G.P.W. and Görgens, A.H.M. (2000a). Facilitation of interdisciplinary collaboration in research projects: lessons from a Kruger National Park Rivers Research Programme project. *South African Journal of Science* **96**(8): 410-414.
27. Jewitt, G.P.W. and Görgens, A.H.M. (2000b). *Scale and model interfaces in the context of integrated water resources management for the rivers of the Kruger National Park*. WRC report no. 627/1/00. Water Research Commission, Pretoria.
28. Johnson, B.L. (1999a). Introduction to the special feature: Adaptive management – scientifically sound, socially challenged? *Conservation Ecology* **3**(1): 10. http://www.consecol.org/vol3/iss1/art10/
29. Johnson, B.L. (1999b). The role of adaptive management as an operational approach for resource management agencies. *Conservation Ecology* **3**(2): 8. http://www.consecol.org/vol3/iss2/art8/

30. Joubert, S.C.J. (1986). The Kruger National Park – an introduction. *Koedoe* **29**: 1-11.
31. Kane Driscoll, S.B., Wickwire, W.T., Cura, J.J., Vorhees, D.J., Butler, C.L., Moore, D.W., and Bridges, T.S. (2002). A comparative screening-level ecological and human health risk assessment for dredged material management alternatives in New York/New Jersey Harbor. *International Journal of Human and Ecological Risk Assessment* **8**: 603-626.
32. Kiker, G.A. (1998). *Development and Comparison of Savanna Ecosystem Models to Explore the Concept of Carrying Capacity.* PhD Dissertation. Cornell University, USA.
33. Kiker, G.A. and Linkov, I. (2005). The QnD model/game system: Integrating questions and decisions for multiple stressors. In: Goncharova, N., Arapis, G., eds, *Ecotoxicology, Ecological Risk Assessment and Multiple Stressors.* Amsterdam: Kluwer, 2005.
34. Kolb, D.A. (1984). *Experiential Learning: Experience as a Source of Learning Development.* New Jersey: Prentice Hall.
35. Mackenzie, J.A., van Coller, A.L., and Rogers, K.H. (2000). *Rule Based Modelling for Management of Riparian Systems.* Pretoria, Water Research Commission.
36. Matsinos, Y.G., DeAngelis, D.L., and Wolff, W.F. (1994). Using an object-oriented model for an ecological risk assessment on a great blue heron colony. *Mathematical and Computer Modelling* **20**(8): 75-82.
37. Mills, M.G.L., Biggs, H., Funston, P., Pienaar, D.J., van der Walt, J.L., Viljoen, A., and Viljoen, P.C. (1995). *Annual Report on the Northern Plains Project – June 1995.* Internal Report. Scientific Services Section. National Parks Board of South Africa 9/95.
38. Mooij, W. M. and Boersma, M. (1996). An object-oriented simulation for individual-based simulations (OSIRIS): Daphnia population dynamics as an example. *Ecological Modelling* **93**: 139-153.
39. Mueller, H.E. (1991). Object-oriented simulation: a new ecological tool emerges. *GIS World* **4**(5): 42-44.
40. O'Connor, T. and Kiker, GA. (2004). Collapse of the Mapungubwe society: Vulnerability of pastoralism to increasing aridity. *Climatic Change* **66**:49-66.
41. O'Neill, R.V., Kahn, J.R., and Russell, C.S. (1998). Economics and ecology: the need for detente in conservation ecology. *Conservation Ecology* **2**(1): 4. http://www.consecol.org/vol2/iss1/art4/
42. Packham, R., Roberts, R., and Bawden, R. (1989). Our faculty goes experiential. In: Warner-Weiland, S., McGill, I., eds, *Making Sense of Experiential Learning: Diversity in Theory and Practice.* Buckingham: SRHE and Open University Press, pp. 129-149.
43. Railsback, S. and Harvey, B. (2001). *Individual-based Model Formulation for Cutthroat Trout, Little Jones Creek, California.* General Technical Report PSW-GTR-182, Pacific Southwest Research Station, Forest Service, U. S. Department of Agriculture, Albany, California.
44. Railsback, S.F., Lamberson, R.H., Harvey, B.C., and Duffy, W.E. (1999). Movement rules for individual-based models of stream fish. *Ecological Modelling* **123**: 73-89.
45. Railsback, S.F. (2001). Concepts from complex adaptive systems as a framework for individual-based modelling. *Ecological Modelling* **139**: 47-62.
46. Rastetter, E. B., King, A.W., Cosby, B.J., Hornberger, G.M., O'Neill, R.V., and Hobbie, J.E. (1992). Aggregating fine-scale ecological knowledge to model coarser-scale attributes of ecosystems. *Ecological Applications* **2**(1): 55-70.
47. Rogers, K. (1998). Managing science/management partnerships: a challenge of adaptive management. *Conservation Ecology* **2**(2): R1. http://www.consecol.org/vol2/iss2/resp1/
48. Rogers, K., Roux, D., and Biggs, H. (2000). The value of visions and the art of visionaries. *Conservation Ecology* **4**(1): R1. http://www.consecol.org/vol4/iss1/resp1/
49. Rogers, K. and Bestbier, R. (1997). Development of a protocol for the definition of the desired state of riverine systems in South Africa. http://www.ccwr.ac.za/knprrp/riverine/index.html: 93.
50. Rogers, K. and Biggs, H. (1999). Integrating indicators, endpoints and value systems in strategic management of rivers of the Kruger National Park. *Freshwater Biology* **41**: 439-451.
51. Rogers, K. (1997). Operationalizing ecology under a new paradigm. In: Pickett, S.T.A., Ostfield, R.S., Shachak, M., and Likens, G.E., eds., *Enhancing the Ecological Basis of Conservation: Heterogeneity, Ecosystem Function and Biodiversity.* New York: Chapman and Hall, pp. 60-77.
52. Sekine, M., Nakanishi, H., and Ukita, M. (1996). Study on fish mortality caused by the combined effects of pesticides and changes in environmental conditions. *Ecological Modelling* **86**: 259-264.
53. Silvert, W. (1993). Object-oriented ecosystem modelling. *Ecological Modelling* **68**: 91-118.

54. Snowling, S.D. and Kramer, J.R. (2001). Evaluating modelling uncertainty for model selection. *Ecological Modelling* **138**: 17-30.
55. Starfield, A. M., Cumming, D.H.M., Taylor, R.D., and Quadling, M.S. (1993). A frame-based paradigm for dynamic ecosystem models. *AI Applications* **7**(2): 1-13.
56. Starfield, A. M. (1997). A pragmatic approach to modeling for wildlife management. *Journal of Wildlife Management* **61**(2): 261-270.
57. Starfield, A. M. and Bleloch, A.L. (1991). *Building Models for Conservation and Wildlife Management*. Edina, MN: Burgess International Group.
58. Starfield, A. M., Smith, K.A. and Bleloch, A.L. (1990). *How to Model It: Problem Solving for the Computer Age*. New York: McGraw-Hill, Inc.
59. Stone, L. and Ezrati, S. (1996). Chaos, cycles and spatiotemporal dynamics in plant ecology. *Journal of Ecology* **84**: 279-291.
60. Van der Heijden, K. (1999). *Scenarios: The Art of Strategic Conversation*. West Sussex: John Wiley and Sons.
61. Vannote, R.L., Minshall, G.W., Cummins, K.W., Sedell, J.R., and Cushing, C.E. (1980). The river continuum concept. *Canadian Journal of Fisheries and Aquatic Sciences* **37**: 130-137.
62. Venter, F.J., Scholes, R.J., and Eckhardt, H. (2003). The abiotic template and its associated vegetation pattern. In: du Toit, J., Rogers, K.H., and Biggs, H.C., eds, *The Kruger Experience: Ecology and Management of Savanna Heterogeneity*. Washington, DC: Island Press.
63. Walters, C. (1997). Challenges in adaptive management of riparian and coastal ecosystems. *Conservation Ecology* **1**(2): 1. http://www.consecol.org/vol1/iss2/art1/
64. Walters, C., Korman, J., Stevens, L.E., and Gold, B. (2000). Ecosystem modeling for evaluation of adaptive management policies in the Grand Canyon. *Conservation Ecology* **4**(2): 1. http://www.consecol.org/vol4/iss2/art1/
65. Weeks, D. C., Jewitt, G.P.W., and Heritage, G.L. (1999). *Planning the Future of the KNPRRP Abiotic-Biotic Links Knowledge-Based Models*. Pretoria, Water Research Commission.
66. Wilson, A.M. (1992). *An Introduction: Systems Thinking at Hawkesbury: 1983-1993*. Agpak SA-4, Hawkesbury, Australia: School of Agriculture and Rural Development, University of Western Sydney.

MAPPING OF RELATIVE RISK

Based On District-Wise Aggregated Data

U. SCHLINK, O. HERBARTH, A. KINDLER, P. KRUMBIEGEL
Department of Human Exposure Research & Epidemiology, UFZ-Centre for Environmental Research Leipzig-Halle, GERMANY.

K. STREBEL, B. ENGELMANN
University of Applied Sciences (HTWK) Leipzig, GERMANY

Abstract

Environmental security involves the geo-referenced assessment of health risks. A straightforward estimation of relative risks can be based on aggregated health data, which are often routinely collected in administrative districts and, therefore, cheap and easily available. Disease maps based on raw data can be misleading, however. The value of the maps can be improved by applying a statistical model that, firstly, accounts for the spatial autocorrelation and, secondly, admits the inclusion of exposure data (e.g. from monitoring networks). A third important task is the inclusion of socioeconomic and further individual-specific confounders.

The paper presents a hierarchical model that facilitates the mapping of the relative risk and the adjustment for socio-demographic and individual-specific variables (ecological adjustment). While the basis of the model is the conditional autoregressive (CAR) approach, we extend this approach by including a model for the individuals at risk. Bayesian inference is used for the risk estimation.

As an example, the statistical model is applied to the analysis of the spatial epidemiology of *Helicobacter pylori* in Leipzig, Germany. So far, the epidemiology of this infection is not fully understood. The *Helicobacter pylori* [^{13}C]urea breath test was offered to all school beginners in the urban area of Leipzig to determine the colonization prevalence and potential transmission pathways of the bacterium. A total of 2888 school-starters participated in the test and their parents completed detailed, self-administered questionnaires. Calculating the spatial distribution of *H. pylori* among school-starters in the urban area of Leipzig, we present maps of district-wise relative risks and demonstrate the usefulness of the ecological adjustment.

1. Introduction

Preventing and adverting a danger threatening the public demands for a geo-referenced assessment of risk. For this purpose, risk mapping may be a valuable tool. Appropriate risk maps can, firstly, provide a sound image of the prevalent situation, enabling the

B. Morel and I. Linkov (eds.),
Environmental Security and Environmental Management: The Role of Risk Assessment, 187–199.

identification of hot spots and, thus, providing decision support to the authorities. Secondly, the maps can help explain the complex and multi-causal formation of the considered risk.

Unfortunately, there are some technical problems with risk mapping procedures that were intensely discussed in the recent literature. We summarize the basic ideas of the established techniques and apply them to the following example of environmental health data. The epidemiology of the *Helicobacter pylori* infection is so far not fully understood and there are several factors accused to have an important effect on the spread of this infection across the population. Because the organism appears to be ubiquitous and may be picked up anywhere, we hypothesized that indirect fecal-oral transmission could possibly be a dominant pathway in its acquisition.

Maps of the geographical distribution of the *H. pylori* infection may help to represent the current situation and to control this complex epidemic, in which the quality of drinking water probably plays a particular role as well as socioeconomic parameters. Such maps present, e.g. for the administrative districts of an urban area, measures of relative risk. A very crude measure is the so-called standardised morbidity ratio (SMR), which is the ratio between observed and expected cases per local district. Though the SMR takes into account the population density varying over the map, imprecisely estimated SMRs, based on only a few cases, may be the extremes of the map, and hence dominate the pattern. Not only is the SMR an unsatisfactory measure as it ignores the statistical aspect of the data, but also there is no chance to adjust for confounding effects, such as socioeconomic parameters and environmental agents.

The search for an appropriate geo-referenced risk measure has been described in particular in the literature on mortality statistics [5]. Currently, successful approaches to risk mapping are facilitated by statistical models, which allow the estimation of relative risks. As there is an infectious agent as well as spatially varying risk-factors and confounders, spatial autocorrelation or clustering of cases can be assumed. In the statistical model, the former property is represented by conditional auto-regression [2,3,8,9]. Additionally, the model can control for exposure and confounder data by means of ecological regression. The exposure data may be available at different spatial scales and this requires the application of a hierarchical model [20].

The main aim of the present paper is, based on data of an epidemiological study, to calculate district-wise relative risks to *Helicobacter pylori* infection in the urban area of Leipzig, Germany. In an attempt to comprehensively model the individual-specific as well as the region-specific impacts to the infection and to calculate spatial risks adjusted for exposure and confounding, the paper presents a statistical approach on the basis of the conditional auto-regressive (CAR) model. The basic spatial units are the districts of the urban area of Leipzig.

The inclusion of a confounder requires stratification of data. This procedure yields in missing counts due to the limited data base. To handle missing observations we set up a hierarchical approach containing a model for the number of individuals at risk.

The presented method, which is described here for an individual-based epidemiological case-control study, can be generalised for use with district-wise aggregated health data that are often available from official governmental statistics.

2. Epidemiological data

The epidemiological data were gathered in frame of the Leipzig *Helicobacter pylori* study [11, 13] during two periods, one in 1998 and one in 2000, in the Leipzig area comprising the city and the rural county. Subject of the present paper are only the data from 1998 of those children, living in the city, with a completed questionnaire and participation in the *H. pylori* test. The study involved the administration of the gastric *H. pylori* colonization test using the stable-isotope-acid *in vivo* [^{13}C]urea breath test and a detailed, parent-completed questionnaire. The [^{13}C]urea breath test involved two breath samples, one taken before and another 30 min after drinking 75 mL orange juice with 75 mg [^{13}C]urea (99.3 atom% 13C; chemical purity according to U.S. Pharmacopeia assay: 99.8%; Chemotrade Leipzig). A child was considered infected with *H. pylori* (positive) when the 13C values of the two exhaled carbon dioxide test samples (measured by a 13C isotope analyzer, FANci; Dr. Fischer Analysen, Leipzig, Germany) differed significantly. The underlying biochemical principle is that [^{13}C]urea is split in the stomach only in the presence of *H. pylori*.

We used the epidemiologic questionnaire to elicit information on education, employment, and medical history of the proband's parents and siblings, past and present home address, living conditions, leisure-time garden plots, and a detailed medical history of the child. Information was sought on the family's nationality and travels to foreign countries. Further questions focused on the child's past and present contacts to various pets and different sources of drinking water.

3. Statistical model

For a population, health data are often available in good terms as summary counts or rates for a formally defined region such as a county or district. In fact we assume that within a region i, having n_i individuals at risk, counts (O_i) or rates ($\frac{O_i}{n_i}$) are observed for subgroups of the population that are susceptible and relevant for investigations of environmental security. In our example, we have subgroups arising through classification by gender (male or female), pet ownership (yes or no), and education level (low, average, high). In general notation, an observation can be denoted by O_{il}, where $i = 1,...,g$ indexes the districts, and $l = 1,...,L$ indexes the subgroups. In application, subgroups are defined through factor levels, so that if we use gender (s), pet ownership (p), education level (e), etc. subscript l would be replaced accordingly.

Basically, we assume that O_{il} is an observed count arising from an associated relative risk ψ_{il}. Utilizing Bayesian inference, the construction of a statistical model comprises two stages: firstly, the specification of a likelihood model for the vector of observed counts O given the vector of relative risks ψ, and, secondly, the specification of a prior model over the space of possible ψ's. By means of Markov

Chain Monte Carlo (MCMC) computational algorithms we calculate a posterior for ψ given the observations O. The set of posterior means or medians of the ψ_{il} is then used to create a risk map. Typically, the objective of the prior specification is to stabilize the risk estimates by smoothing the crude map, although environmental security concerns imply additional interest in explaining the risks.

Our likelihood model assumes that, given the relative risks ψ_{il}, the counts O_{il} are conditionally independent Poisson variables. Note, that the Poisson serves as an approximation to a binomial distribution, say $O_{il} \sim Bin(n_{il}, p_{il})$, where n_{il} is the known number of individuals at risk in district i within subgroup l and p_{il} is the associated disease rate. Often the counts are sufficiently large so that they can be assumed to approximately follow a normal density. For the consideration of small or sparsely populated areas this is not the case. Partitioning into subgroups results in many small values of O_{il} (including several zeroes), so we confine ourselves to the Poisson model, which is a valid approximation of the Binomial model when $p_{il}n_{il}$ can be considered constant while n_{il} becomes large.

We define the relative risk ψ by $n_{il}p_{il} = E_{il}\psi_{il}$, where $E_{il} = n_{il}\overline{p}$ is the expected count in district i for subgroup l. Here the overall disease rate $\overline{p} = \frac{\sum_{il} O_{il}}{\sum_{il} n_{il}}$ is used for internal standardisation of risk. If the expected counts are defined with respect to some external reference, then the model is externally standardised [1]. The E_{il} are the expected counts under a null, constant relative risk model and are assumed to be known in that their sampling variation is ignored. This approach is used by Clayton and Kaldor [5], but models can be generalised to accommodate statistical uncertainty.

Thus the Poisson model reads

$O_{il} \sim Poi(n_{il}p_{il}) = Poi(E_{il}\psi_{il})$, that is $$\Pr(O_{il} = z) = e^{-E_{il}\psi_{il}} \frac{(E_{il}\psi_{il})^z}{z!}, \tag{1}$$

and the log Likelihood Function is

$$\ln L = \sum_{il} \left[-E_{il}\psi_{il} + O_{il}\ln(E_{il}\psi_{il}) - \ln(O_{il}!) \right]. \tag{2}$$

A very crude map arises from the Likelihood model (2) alone, using the maximum likelihood (ML) estimates of ψ_{il} based only on O_{il}. For the specified Poisson model the ML estimates of the relative risk (i.e. $\left.\frac{\partial \ln L}{\partial \psi_{il}}\right|_{\hat{\psi}_{il,ML}} = 0$) are the so-called standardized morbidity ratios (SMR): $\hat{\psi}_{il,ML} = SMR_{il} = \frac{O_{il}}{E_{il}} = \frac{O_{il}}{\overline{p}n_{il}}$. From (2) we also find that $\text{var}(SMR_{il}) = \frac{O_{il}}{(E_{il})^2} = \frac{O_{il}}{\bar{p}^2 n_{il}^2}$, which expresses that such maps, based on SMR's of the raw data, often feature large outlying relative risks in sparsely populated regions (n_{il} small), so that the map is visually dominated by the rates with the highest uncertainty. There are at least three shortcomings of the SMR measure: Firstly, their unstable behaviour, which is caused by the fact that they are calculated from data of the considered district only and do not make use of observations in other areas. The simple likelihood approach fails to account for an anticipated similarity of relative risks in nearby or adjacent regions. Secondly, often the data exhibit extra-Poisson variation (over-dispersion) due to dependence on unmeasured factors and/or data errors [19]. Thirdly, SMR s do not account for socio-demographic and individual-specific confounders (ecological adjustment).

Statistical smoothing methods can be used to overcome these problems. To stabilise the risk estimates, the idea is that a smoothed estimate for each area 'borrows strength' (i.e. precision) from data in other areas, by an amount depending on the precision of the raw estimate for each area [20]. Such spatial modelling introduces association across regions by means of weighted averaging. To account for confounders, a regression term is included into the statistical model. With the purpose to simultaneously estimate all parameters we integrate all analyses into a single model by assuming a hierarchical structure. This leads to narrower confidence intervals than independent (non-hierarchical) analyses, but with the toll that the estimates are biased towards the mean response. Estimation is carried out using Bayesian methods and an appropriate prior distribution has to be specified for the relative risk.

The log relative risk is composed by the following parts:

$$\ln(\psi_{il}) = \alpha x_l + \theta_i + \phi_i \,. \tag{3}$$

The first term includes the effects of standardised covariates (x_{il}), i.e. $mean(x) = 0$, $\text{var}(x) = 1$, with an uninformative prior (large variance)

$$\alpha_l \sim Norm(0, \sigma_\alpha), \quad \sigma_\alpha \approx 10^3 \,. \tag{4}$$

The θ_i are iid random effects that produce an exchangeable model for the extra-Poisson variation, and the ϕ_i are random effects that induce spatial correlation. As the αx_l component of the model adjusts for covariates and potential confounders, the θ_i and ϕ_i are adjusted random effects for region-specific log relative risks. Their variation can be interpreted as compensation for model misspecification, for example failure to include important covariates. Including additional, so far unknown covariates might reduce the magnitude of these random effects. The regional random effects θ_i capture heterogeneity among the districts [1]. Because i indexes the districts arbitrarily, an exchangeable prior for the θ_i is appropriate, that is

$$\theta_i \overset{iid}{\sim} Norm(\kappa, 1/\tau) \,. \tag{5}$$

We then add a flat hyper-prior for κ but require a proper hyper-prior (typically a gamma distribution) for τ.

$$\kappa \sim Norm(0,10^8), \qquad \tau \sim Gamma(0.001,0.001) \qquad (6)$$

For district i the spatial effect, reflecting geographic proximity and resulting in correlated counts, is ϕ_i. In the spirit of the Gaussian Markov Random Field approach of Cressie and Chan [7], who extend Besag [2], we model the ϕ_i using a conditional autoregressive (CAR) model. It builds the full joint distribution from complete conditional distributions for each ϕ_i given all others. For weights $w_{ij} \geq 0$ reflecting the influence of ϕ_j on the expectation of ϕ_i we have

$$\phi_i \mid \phi_{j \neq i} \sim Norm\left(\frac{\sum_{i \neq j} w_{ij}\phi_j}{\sum_{i \neq j} w_{ij}}, \frac{1}{\lambda \sum_{i \neq j} w_{ij}}\right), \ \lambda \sim Gamma(0.001,0.001) \qquad (7)$$

Equation (7) is called the 'intrinsic CAR model'.

The hyper-parameter λ controls the strength of the spatial similarity induced by the CAR prior; larger values of λ indicate stronger spatial correlations between neighbouring districts. As part of the model, the proximity (weight) parameters w_{ij} need to be specified. In our adjacency model we define a set ∂_i of neighbors of district i comprising regions contiguous to district i and let $w_{ij} = 1$ if $j \in \partial_i$ and 0 otherwise; that means w_{ij} is the adjacency matrix. It is easy to establish that the intrinsic CAR model provides a locally weighted averaging of risk, the variance inversely proportional to the number of neighbours. A global average is not fixed, wherefore (7) represents an improper prior.

An alternative specification was the proper CAR model, using

$$\phi_i \mid \phi_{j \neq i} \sim Norm\left((1-\gamma)\alpha + \gamma\frac{\sum_{i \neq j} w_{ij}\phi_j}{\sum_{i \neq j} w_{ij}}, \frac{1}{\lambda \sum_{i \neq j} w_{ij}}\right), \qquad (8)$$

$\lambda \sim Gamma(0.001,0.001)$, $\alpha \sim Norm(0,10^6)$, and $\gamma \sim Uniform(\gamma_{\min}, \gamma_{\max})$ with the global average (α). The weighting (autocorrelation) parameter $0 < \gamma < 1$ represents the relative influence of local mean over the global mean.

To allow for flexibility, Besag, York and Mollie [4] recommend combining the intrinsic CAR prior and an exchangeable normal prior into the so called convolution model, which is specified in our approach (3+5+7). Only the sum $\theta_i + \phi_i$ is uniquely identified by the likelihood (2), but the structure of the priors (5+7) allows the possibility of separating θ_i and ϕ_i because the data can inform about each. We are interested in making inferences on region-specific, covariate-adjusted relative risks, $e^{\phi_i+\theta_i}$, which are smoothed towards a combination of global and local risks, with the relative contribution of each determined by the data. For implementation we used the car.normal procedure of the WinBUGS 1.4 program [18].

A particular problem with epidemiological data is the occurrence of missing values. In the present study for some urban districts there are no *H. pylori* cases observed in some subgroups and, therefore, the corresponding count O_{il} was set to zero.

A much more problematic situation arises when, for a subgroup in an urban district, there are no individuals at risk $n_{il} = 0$. Consequently, both E_{il} and O_{il} vanish and the risk is indeterminate. In order to avoid the situation of indeterminate risks and to enable the estimation process, we set up a model for the expected individuals at risk n_{il} and replaced the missing figures (i.e. if $n_{il} = 0$) by the estimated figures. In the present study the individuals at risk are children living in the urban area of Leipzig, Germany. We assume that, at least approximately, in each urban district the number of children n_{il} follows a Poisson distribution and is linearly related to the number of inhabitants ρ_i, which is well known from yearly census data [15]. Additionally, we assume an overall effect of the considered subgroup x_l to the number of individuals at risk. The complete specification of the model is:

$$n_{il} \sim Poi(\kappa_{il}), \text{ with } \qquad \log(\kappa_{il}) = \gamma_1 + \gamma_2 x_l + \gamma_3 \rho_i, \qquad (9)$$

4. Model fitting

It is obvious from the high complexity of our model discussed in the previous section that some form of the MCMC algorithm will be needed to obtain estimates of the posterior quantities of interest. While models of a straightforward structure can be estimated using the Gibbs sampler [9], the present complex hierarchical model requires the Metropolis algorithm [6,16] as a way of obtaining the necessary samples.

In the sampling algorithm we ran two parallel MCMC chains for 1.000 iterations. Graphical monitoring of the chains for a representative subset of the parameters, along with sample autocorrelations and Gelman & Rubin [10] diagnostics, indicated an acceptable degree of convergence by around the 100th iteration. Using every 10th iteration from both chains, we obtained the 95% posterior credible sets. Model fit was assessed using the DIC criterion [17].

5. Results and Discussion

The raw estimates (SMRs) of the relative risk to *Helicobacter pylori*, presented in Figure 1, indicate highest risks (i.e. 'hot spots') for the urban districts #70 (Lindenau), #80 (Möckern), #43 (Lößnig), and #50 (Schleußig).

Such maps may, however, be misleading to authorities and to the public. As SMRs are merely point estimates, there is no information about the statistical significance of the differences in the presented risk. The few observations per district (8 in #70 (Lindenau), 37 in #80 (Möckern), 36 in #43 (Lößnig), and 29 in #50 (Schleußig)) might have produced large fluctuations in the risk estimates. Another cause for spatial heterogeneities in risk might be a heterogeneous distribution of known risk factors, such as contact to pets. Therefore, without a proper statistical model it is not clear if, in a district, an elevated risk represents the effect of unknown district-specific risk factors or is just the consequence of statistical fluctuations or confounding.

The applied CAR model accounts for the statistical fluctuations. While the results presented in Figure 2 indicate correspondence between SMRs and risk estimates of the CAR model, the latter are more useful as they provide confidence intervals. Our results confirm the significantly elevated risk only for two urban districts (#80 (Möckern): 1.40…3.64…7.43, #43 (Lößnig): 1.69…4.32…8.46), while the low number of observations makes the 95% confidence interval of the risk too wide for the other two districts (#70 (Lindenau): 1.00…4.73…13.46, and #50 (Schleußig): 0.97…3.12…7.00), so that here relative risk is not significant larger than 1.

In a second step we adjusted for the effects of individual-specific covariates. Table 1 presents the estimated risks attributable to some selected risk factors and comparatively lists the results of Herbarth and colleagues [11], who used an individual based approach not considering the spatial distribution.

Table 1 indicates that all the studied risk factors significantly contribute to the overall risk to a *Helicobacter pylori* infection; and our results are in agreement with the literature. Note that the Odds Ratio (OR) approximately equals the Relative Risk (ψ) in a case-control study of a rare disease [12], which is valid as $\Pr(H.pylori) \approx 0.04$ in our study.

Table 1: Relative risks to *H. pylori* infection attributable to selected factors. Estimates are based on the CAR+regression model. For comparison, we present the odds ratio (OR) from individual-specific logistic regression [11].

Risk factor	ψ [95%CI]	OR [95%CI]
Travel to Asia	2.40…3.40…4.81	1.6…3.7…8.7
Contact with pet hamsters	2.12…2.86…3.87	1.2…2.4…4.7
Low education	1.10…1.39…1.75	-
Single father/mother	1.22…1.51…1.88	-

As these risk factors/confounders are by no means homogeneously distributed in the urban area due to social segregation effects, they might explain the main part of spatial variation in risk. This is indeed true for the risk factor 'Contact with pet hamsters': adjusting for the contact with pets all significantly elevated risks disappeared. In contrast, adjusting for the risk factor 'Travel to Asia', the risk of district #43 (Lößnig) remains significantly increased (see Figure 2). Obviously only the risk factor 'Contact with pet hamsters' is able to explain the total spatial variation.

Fig. 1. Standardized Morbidity Ratio (SMR) for Helicobacter pylori in the urban districts of Leipzig

Fig. 2. Ranking of urban districts as to their relative risk (ψ) estimated (a) as SMR, (b) using the CAR model (ψ with 95% CI), (c) using CAR with additionally adjusting for 'Travel to Asia' (ψ with 95% CI).

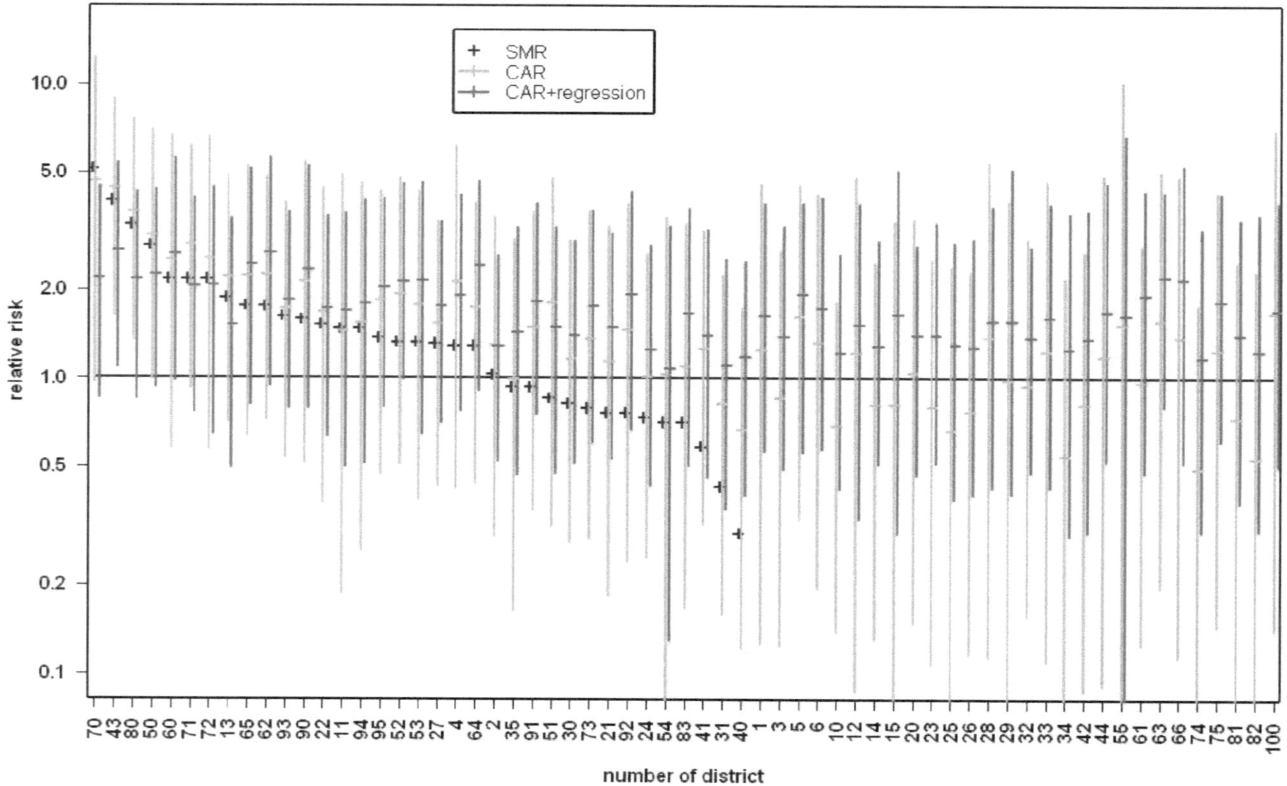

Model (9) for the number of individuals at risk allows for the estimation of additional parameters (Table 2). These estimates mean that, taking the statistical variation into account, the expectation value of the percentage of children of age 6 in an urban district is a function of the inhabitants: $E\left(\frac{n_i}{\rho_i}\right) \approx \frac{e^{2.1+0.1\rho_i/1000}}{\rho_i} \approx 0.01 - 0.008(1-\frac{1000}{\rho_i})$. Approximation by Taylor expansion in terms of $\frac{1000}{\rho_i}$ at $\rho_i = 1000$ shows that the number of school starters is a constant fraction of ~1% of the total number of inhabitants (varying in Leipzig between ~1000 and ~17000 inhabitants per urban district). The percentage of school starters slightly decreases with increasing population. For children exposed to a risk factor the expected frequency per district is reduced and depends on the considered risk factor. For example, the fraction of children being in contact with pet hamsters is $e^{\gamma_2} = e^{-2.5} \approx 0.08$ times the total number of children per urban district.

TABLE 2. Estimated parameters for eqn. (9) modelling the individuals at risk.

Risk factor	$\hat{\gamma}_1$	$sd(\gamma_1)$	$\hat{\gamma}_2$	$sd(\gamma_2)$	$\hat{\gamma}_3 10^{-3}$	$sd(\gamma_3 10^{-3})$
Travel to Asia	2.12	0.07	-2.70	0.17	0.11	0.01
Contact with pet hamsters	2.10	0.07	-2.46	0.12	0.11	0.01
Low education	2.09	0.06	-1.74	0.07	0.11	0.01
Single father/mother	1.98	0.06	-1.17	0.06	0.11	0.01

Finally, we consider the relationship between the district-specific risk (ϕ_i) characterised by spatial correlation (eqn. 7), and the random effects (θ_i, eqn. 5). The latter capture heterogeneity among the urban districts that is caused by differing numbers of individuals at risk in the subgroups in the urban districts. This unbalanced design results in factor-dependent mean values κ (see Table 3). In contrast, Table 3 indicates that the variance (λ) of the autoregressive part does not depend on the investigated risk factor.

TABLE 3. Parameters for random ($\theta_i \sim Norm(\kappa, 1/\tau)$) and autocorrelated ($\phi_i$, CAR) district-wise risks.

Risk factor	κ	$sd(\kappa)$	$\sigma = 1/\tau$	$sd(\sigma)$	λ	$sd(\lambda)$
Travel to asia	0.38	0.15	0.17	0.12	0.027	0.004
Contact with pet hamsters	0.16	0.14	0.21	0.17	0.027	0.005
Low education	-0.21	0.15	0.21	0.16	0.028	0.005
Single father/mother	-0.14	0.13	0.19	0.17	0.028	0.005

6. Conclusions

In our study we mapped the spread of the *Helicobacter pylori* infection among school starters in the urban area of Leipzig. A map created from a naïve plot of the morbidity ratio displays four 'hot spot' urban districts with elevated *H. pylori* prevalence. We demonstrated that this map is misleading because both spatial statistical fluctuations and confounding have to be taken into account.

The presented approach incorporates spatial autoregression as well as the effect of individual-specific risk factors. Some of the known risk factors explain all the spatial variation of the risk to *Helicobacter pylori* infection. No statistically significant hot spots remain after adjustment for known risk factors. This adjustment procedure (ecological regression) is a necessary condition to make district-specific risks comparable.

Though, for each risk factor, a separate analysis was made, we faced the problem of missing data in several subgroups and urban districts. We handled this problem by including a model for the number of individuals at risk, which replaced the missing counts by estimates. The extended approach was therefore quite time-consuming. A model involving several risk factors and their interaction simultaneously will suffer much more from missing counts in certain districts and subgroups. As this type of model is necessary for a comprehensive and integrated assessment of the risks, much effort will have to be devoted to a suitable specification of the statistical model.

The adjusted district-specific relative risks suggest that naïve maps may display seeming hot spots, which do not exist. Rather, the risk to *H. pylori* is homogeneously distributed in the whole urban area. This emphasises the prevalence of uniform sanitary conditions throughout the city of Leipzig; in particular, a constant quality of drinking water. Due to a central water supply, such homogeneous high quality of drinking water has to be expected and, therefore, our result is not surprising. As in the rural surroundings of Leipzig there are some alternative sources of drinking water, such as wells in the garden, a comparative study between urban districts and rural districts will be interesting and is currently under analysis. Moreover, the presented techniques may help assessing the environmental security in regions that do not benefit from a central high quality water supply.

7. Acknowledgements

We thank all of the children and their parents for their interest and participation in the study. The participation of K.S. in a course on spatial Bayesian modelling was funded by the "UFZ-Umweltforschungszentrum" and the "Förderverein der HTWK Leipzig", which is gratefully acknowledged.

8. References

1. Bernardinelli L, Montomoli M (1992) Empirical Bayes versus fully Bayesian analysis of geographical variation in disease risk. Statistics in Medicine 11, 983-1007.
2. Besag J (1974) Spatial interaction and the statistical analysis of lattice systems (with discussion). Journal of the Royal Statistical Society B36, 192-236.
3. Besag J, Kooperberg C (1995) On conditional and intrinsic autoregressions. Biometrica 82(4), 733-746.
4. Besag J, York J, Mollie A (1991) Bayesian image restoration with two applications in spatial statistics. Annals of the Institute of Statistical Mathematics 43, 1-59.
5. Calyton D, Kaldor J (1987) Epirical Bayes estimates of age-standardized relative risks for use in disease mapping. Biometrics 43, 671-681.
6. Chib S, Greenberg E (1995) Understanding the Metropolis-Hastings Algorithm. The American Statistician 49, 327-335.
7. Cressie NAC, Chan NH (1989) Spatial modelling of regional variables. Journal of the American Statistical Association 84, 393-401.
8. Dorling D (1998) Mapping disease patterns. In: Armitage P, Colton T (eds.) Encyclopedia of Biostatistics, vol 6, 2391-2401.
9. Gelfand AE, Smith AFM (1990) Sampling-based approaches to calculating marginal densities. Journal of the American Statistical Association 85, 398-409.
10. Gelman A, Rubin DB (1992) Inference from iterative simulation using multiple sequences (with discussion). Statistical Science 7, 457-511.
11. Herbarth O, Krumbiegel P, Fritz GJ, Richter M, Schlink U, Müller DM, Richter T (2001) Helicobacter pylori prevalences and risk factors among school beginners in a German urban center and its rural county. Environmental Health Perspectives 109(6), 573-577.
12. Kleinbaum DG, Kupper LL, Morgenstern H (1982) Epidemiologic research – principles and quantitative methods, Van Nostrand Reinhold, New York.
13. Krumbiegel P, Herbarth O, Fritz G, Schlink U, Gutsmuths F-J, Kindler A, Richter T (2000) Helicobacter pylori prevalence in Leipzig's 1998 school entries: methodology and first results. International Journal of Hygiene and Environmental Health 203, 11-16.
14. Lawson AB (1998) Statistical map. In: Armitage P, Colton T (eds.) Encyclopedia of Biostatistics, vol 6, 4267-4271.
15. Leipzig (1999) Statistisches Jahrbuch, Amt für Statistik und Wahlen.
16. Metropolis N, Rosenbluth AW, Rosenbluth MN, Teller AH, Teller E (1953) Equations of state calculations by fast computing machines. Journal of Chemical Physics 21, 1087-1091.
17. Spiegelhalter DJ, Best NG, Carlin BP, van der Linde A (2002) Bayesian measures of model complexity and fit (with discussion). Journal of the Royal Statistical Society 64B, 583-639.
18. Spiegelhalter DJ, Thomas A, Best NG, Gilks W (2003) Bayesian inference using Gibbs sampling – manual of WinBUGS 1.4, Imperial College & MRC, UK
19. Wakefield JC, Best NG, Waller LA (2000) Bayesian approaches to disease mapping. Chapter 7 in Spatial Epidemiology: Methods and Applications. Elliot P, Wakefield JC, Best NG, Briggs DJ (eds.), Oxford University Press, Oxford.
20. Waller LA, Carlin BP, Xia H, Gelfand AE (1997) Hierarchical spatio-temporal mapping of disease rates. Journal of the American Statistical Association 92(438), 607-617.

UNDERSTANDING RISK CULTURE AND DEVELOPING A 'SOFT' APPROACH TO RISK ASSESSMENT METHODOLOGIES

S. RAO
SIES College of Commerce & Economics, Mumbai, INDIA

Abstract

One of the cornerstones of effective risk management and risk governance is effective risk assessment. Risk assessment methodologies today are rigorous and subject to a process of continuous improvement around the world. But they remain more concerned with the hard quantitative aspects of risks and do not devote much attention to the 'soft' human related aspects of culture and behavior. Apart from an attempt to model human errors, little else is included in our Risk assessment methodologies that reflect human risk culture and behavior. For e.g. should safety assessments of two technologically similar nuclear plants in two different geographical regions be adjusted only for the geographical conditions? It is quite logical that cultural differences of the respective population have also to be built in-in order to make the safety assessments more meaningful. This requires two things: One, a conceptual framework for risk culture and two, a mathematical framework for risk culture so that it is rendered conducive to risk modeling. The main focus of this paper is to suggest such a conceptual framework for risk culture. This framework is broadly evolved using three spheres of existence that have significant influence on the risk culture of humans–they are physical, psychological and ideal spheres. Family, Society and Geography predominantly influence the physical sphere of existence while Genetics, Upbringing and Innate abilities of individuals predominantly affect the psychological sphere. Religion, Education, and History, are the key influencing factors in the Ideal sphere. Subsequently, a mathematical perspective of risk culture is also briefly discussed.

1. Introduction

'The problems that exist in this world cannot be solved by the level of thinking that created them.'-Albert Einstein

How so true indeed! We need to definitely look at superior ways of thinking in order to come up with innovative and effective solutions to our problems, especially the social and environmental ones. A pertinent thought in this regard is the notion of Risk. 'Risks are possibilities that human activities or natural events lead to consequences that affect what humans value'[1]. On closer examination, at least some of our environmental/social problems could have been possibly avoided or their consequences mitigated had there been better risk management practices around at the time of their

B. Morel and I. Linkov (eds.),
Environmental Security and Environmental Management: The Role of Risk Assessment, 201–210.

creation. This implies that a continuous improvement of our risk management practices is absolutely essential for a better and safer future. One important way to achieve this objective is perhaps to make a beginning through better risk assessments of our risk situations. 'Traditional Risk assessment provides society with a narrow definition of undesirable effects and confines possibilities to numerical probabilities based on relative frequencies...the price society pays for this methodological rigor is the simplicity of an abstraction that does not take into account differences in social and cultural context ' Renn (1992) [2]. This means that a deeper understanding of risk culture and human risk-related behavior can complement and add value to the quantitative techniques of risk assessment and make them more effective.

Analyzing Risk related behavior could be quite challenging. We, as a global community seem to have more experience in dealing with the science of physical bombs than with the psychology of human bombs. Risk related behavior is complex and multi-dimensional. It can be quite impulsive/reflexive or it can occur after a lot of deliberations. It will certainly differ from person to person and will differ for the same person under varying circumstances/environment. Risk related behavior of individuals, groups of people and nations may be different if they are acting alone or in cooperation/collusion with other groups/nations. In short risk related behavior cannot be standardized or generalized easily. It is dynamic and depends upon a complex interplay of individual, collective and physical circumstances\environment as well. How then does one master the art of understanding risk behavior? The answer to this question perhaps lies in its genesis –the Risk Culture.

2. Risk Culture –Concept and significance

The term 'Risk Culture', as can be expected, has no standard definition. Its scope and characteristics are decidedly fuzzy and vary across the globe. The following definition is an attempt to provide a basic description of what 'Risk Culture' is – this description is only indicative and by no means exhaustive.

Risk Culture, is a combination of the patterns of risk related thinking, value systems and beliefs that humans hold (either consciously or sub-consciously) which culminate into their risk-related behaviour. Study of risk culture is therefore very crucial to detect patterns, analyze the causes and unwrap the mysteries of risk-related human behavior. 'Culture systems may, on the one hand be considered as products of action, on the other as conditioning elements of future action' Kroeber and Kluckhohn (1952) [3].This paper will be based on the latter characteristic - considering risk culture as a conditioning element for future action (i.e. future risk related human behavior).

A deeper understanding of the risk culture could help in the following ways:

1) **Improving Global safety and well being**: Insights into risk culture can be useful to tackle and prevent terrorism, crime, accidents etc. This could be achieved through better regulations and law enforcement, which would be more effective if they are adopted, based on the risk culture of the target population rather than a common 'one-size-fit-for-all' attitude. The benefits and importance of studying/analyzing risk culture and behavior in the context of environmental safety is very well summed up in the following observation at the UNEP World conference on Natural Disaster Reduction, Yokohama,

1994. "Community involvement and their active participation should be encouraged in order to gain greater insight into the individual and collective perception of development and risk, and to have a clear understanding of the cultural and organizational characteristics of each society as well as of its behavior and interactions with the physical and natural environment. This knowledge is of the utmost importance to determine those things which favor and hinder prevention and mitigation or encourage or limit the preservation of the environment for the development of future generations, and in order to find effective and efficient means to reduce the impact of disasters."[4]

2) **Improving quality of human life**: Better insights into the risk culture of a population, especially in a transnational context may reduce cognitive/perceptive biases and thereby provide the foundations for better transnational cooperation –especially in areas of cross-border security, trade and commerce etc. At a macroeconomic level, understanding risk culture of the population can pave the way for better governance. At a microeconomic level, it can open the path for newer business initiatives that are more responsive to stakeholder risk concerns and appetite. At all levels and under all circumstances, a better insight into risk culture builds a firm ground for better and effective risk communication and thus helps in improving the overall quality of human life.
3) **Towards Sustainable Development**: One of the key success factors in the path towards sustainable development is the ability to manage transitional risks. Managing transitional risks, especially in the context of adopting newer technologies and economic upheavals require keener insights into the risk culture and risk related behavior of the population. Trusting sections of the world population with newer emerging technologies needs lot of deliberations on the risk culture of the target population. Also this insight should be necessarily long term in the interests of world safety.

   Improper risk management/governance may lead to economic downturns and unemployment/underemployment that can cause sections of the population to resort to abnormal risk related behavior, which in turn may lead to crime, indiscriminate use of natural resources, industrial accidents and other such undesirable consequences in the long run. Thus understanding risk culture is very important from the viewpoint of progressing in the path towards sustainable development.

3. Developing a 'soft' approach to Risk Assessment Methodologies

Given the importance of risk culture as outlined above, it is only natural that we need to blend insights of risk culture into our risk assessment methods. For e.g. should safety assessments of two technologically similar nuclear plants in two different geographical regions be adjusted only for the geographical conditions? It is by now obvious that cultural differences of the respective population have also to be built in-in order to make the safety assessments more meaningful. This raises the question as to how this risk cultural aspect can be built in. This requires two things:

1. A conceptual framework for better understanding of Risk culture
2. A mathematical framework for quantifying risk-cultural concepts.

This paper would focus more on the conceptual framework for Risk culture and suggest a basis for the mathematical framework that is required to develop a 'soft' quantitative approach to risk assessments.

4. Developing a conceptual framework for Risk Culture

Hofstede (1980) states 'Culture is the interactive aggregate of characteristics that influence a human group's response to its environment.'[5] It is therefore important to understand such characteristics through a clear framework. The following is a suggested Risk culture framework for those characteristics [6] which influence a human's/human group's risk response to the environment.

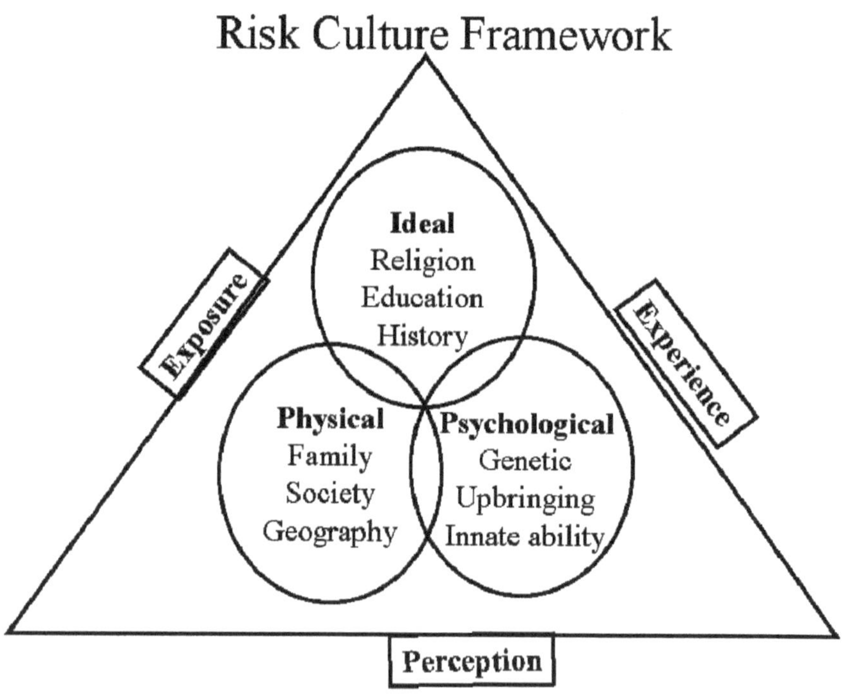

The risk culture is a combination of the exposure, experience and perception of individuals generated by three spheres of existence –The Physical, The Psychological and The Ideal. These three spheres never operate in isolation. A continuous interplay between these spheres determines the overall risk culture and risk related behavior of individuals/people.

4.1 THE PHYSICAL SPHERE OF EXISTENCE

The physical sphere of existence comprises of body, tangible objects of possession like property and the like which people can sense and use. This sphere is a product of family support and care, societal impact and geographical influences. The perceptions developed by individuals based on their exposure and experience in the physical sphere, shapes their risk culture and risk related behavior to a large extent. For e.g. a person from the West may find driving in a developing country very 'risky' because of road conditions, right-hand-drive etc. At the same time, in the physical sphere, there exist universal perceptions, which are common across the globe. For e.g. continuous quest for a sense of comfort and security for the body is almost a constant desire of mankind from his cave days to the present. In this he is generally desirous of safeguarding himself from harm. This in turn shapes his risk culture and risk related behavior.

4.2 THE PSYCHOLOGICAL SPHERE OF EXISTENCE

The Psychological sphere of existence refers to the thoughts and feelings of human beings that drive their motivation levels and cause their behavior. 'Culture is the collective programming of the mind which distinguishes the members of one group or category of people from those of another'- Hofstede (1994).[7] The psychological sphere supplies the logic and syntax for this programming of the mind. This programming is a function of the genetic factors of the individual, the upbringing and the innate capabilities enshrined in the individual. Whereas the physical sphere of existence is tangible and typically has distinguishable geographical boundaries, the psychological sphere is intangible and knows no geographical boundaries. Given the same exposure and experience conditions to individuals, their perceptions about similar risk situations may differ based on their ability to learn 'lessons' from past experiences and their ability to detect the variations of the current risk situation from those that they have faced in the past.

4.3 THE IDEAL SPHERE OF EXISTENCE

The Ideal sphere of existence is normative and refers to all that an individual/population hold as the best way to be. Apart from ethics and morals, this also includes what the individual perceives as legal and binding. Three most influencing factors in this sphere are Religion, Education and History. Depending on the exposure and experience levels of sections of population all the three factors can have a very cohesive effect on the target population and condition their minds to think alike even in difficult risk situations. This is also the trickiest of all the three spheres of the risk culture because it has the potential to cause severe discord in an individual – a gap between what he thinks the world around him should be and what it really is. This fundamental discord can then manifest itself as tendencies towards terrorism, serial killing, and a lot other social crimes. On the positive side, this sphere fills the individual with a 'sense of purpose' and keeps him focused on what goals he should have and how should he go about achieving them.

4.4 INTERWEAVING EFFECTS OF THE THREE SPHERES ON RISK CULTURE:

As noted earlier, the three spheres of existence are not independent and they closely influence each other. For e.g. in general, every human aspires for physical comfort and safety. An exception to this aspiration comes when there is a strong influence from the ideal sphere. He believes rightly (or wrongly at times as in the case of the 9/11 attacks) that he is driven by a worthy cause and prepares to 'sacrifice' all that he has including body and property for the cause. This triggers a risk related behavior, which is abnormal and very intense. Similarly psychological sphere factors of genetics, upbringing or innate abilities can influence the individual and change the normal risk related behavior of the individual in the physical sphere. Under these conditions, given the same physical circumstances as others (colleagues/siblings), he still demonstrates a different/abnormal risky behavior – this works for great sports personalities and rogue securities traders as well! The psychological sphere of existence is heavily dependent on the ideal sphere of existence as the ideals either consciously or sub-consciously permeate the human being in all thoughts and feelings. Wrong notions instilled for e.g., on grounds of misinterpretation of religious texts can cause the human being to demonstrate irregular risk related behavior towards people from certain religions/communities. Physical circumstances for e.g. forcible governance or control can lead to a feeling of insecurity in the individual and upset the equilibrium in his psychological sphere -the corresponding risk related behavior may well turn out to be abnormal. The ideal mode of existence is intricately linked to the physical sphere. For e.g., there can be disillusionment with religion, role models etc. if there are persisting problems of economic hardships/unemployment.

5. Influence of Religion, Education and History on Risk Culture:

The primary influence of the above three factors are in the Ideal sphere of existence but they can have interweaving impacts on the other spheres of existence also. Religion exists for man to find his spiritual equilibrium. In the process, almost all religions propose one or more role models, praise the virtues of that role model(s) and exhort the followers to live up to that standard. But this is easier said than done, because often centuries have passed after the role model ceased to exist and many time followers are left with outdated ideals in a modern era. To make matters worse, there are some times varied interpretations of the same holy books which lead to further confusion in the mind of the common man. All these lead to a certain lack of interest in affairs related to religion and this in turn has the consequence of upsetting or causing a loss of the spiritual equilibrium, which was once the objective of all religions. This loss of spiritual equilibrium manifests itself as dissatisfaction with the self and the environment around and this in turn leads to abnormal risk related behavior, which at times can be even rebellious or revolutionary. On the positive side, Religion can form the basis for oneness in thinking. For e.g. Thou shalt love thy neighbor as thyself (Judaism, Christianity) Hurt not others with that which pains yourself (Buddhism) Good people proceed considering that what is best for others is best for themselves (Hinduism) [8] is just one example of how different religions advocate the same virtue. The (mis) perceptions of followers of one religion vs. another have many a time triggered abnormal risk related human behavior and consequent social upheavals in history. The task of developing positive religious tolerance and openness is not just the duty of governments but also that of the religious leaders. One of the main challenges of the new century will be the reconciliation between the material opportunities and the psychological and spiritual needs of humans. [9]

Education shapes risk related thinking from a very early age and has perhaps one of the strongest influences on the risk related behavior of individuals and populations. Good quality Education encourages people to look at and be receptive to different perspectives /opinion about the same issues. This contributes to the maturity with which the individuals or population deal with risk situations. But the benefits of education can be easily distorted when religion intervenes in Education and dominates the minds of young people. This as we know, (from recent events in Afghanistan) is the reason for many youngsters taking on terrorism activities. In addition scientific education encourages people to think rationally and ask 'why?' which may not go down very well with religion where inevitably there are black boxes. Consequently, if people, especially the younger generation, have lots of unanswered 'whys' left in their minds, it again causes a loss to their spiritual equilibrium and triggers adverse risk related behavior. This means religion and science have to go together and not against each other. As Einstein said 'Science without Religion is lame and Religion without Science is blind'.

History has a slow but steady effect over the ages on particular races. The monuments, the heroic deeds of past rulers, the relics and archaeological evidence of ancient civilizations in their locality/nation leave their indelible mark on the population. The population identifies itself as originating from one history and that bond is very strong and introduces a sense of social togetherness and idealism. This works for the people from the armed forces and terrorists alike. Any challenge to this historic idealism triggers a culture shock and consequently abnormal risk related behavior.

6. Influence of Family, Society and Geography on Risk Culture

The above three factors primarily influence the physical sphere of existence of the individual/population. The family plays a pivotal role in ensuring that the individual remains satisfied with the physical sphere of existence. A broken family or a violent family background without doubt exerts undesirable influences on physical environment of children. This constant physical insecurity and dissatisfaction will eventually propel them towards a risk related culture and behavior that is different from children who hail from a more stable background.

Man is a social animal. He feels the need to belong to a society and 'conform' to its ways. This need to 'conform' holds even if the society to which he belongs is a 'cult' or a perverse terrorist group. Thus society ends up playing a significant role in shaping the risk culture of individuals and groups. This is a continuous circular iterative process-society influences every individual's risk culture and behavior and the risk culture of the society as a whole gets influenced by the collective risk related thinking and behavior of the same individuals. Society particularly contributes to the risk culture through amplifications of risk [11], socio-economic stratifications and through a multitude of other ways like methods of dressing, eating habits, living styles etc.

Geographic conditions influence the risk culture of individuals to a great extent-certain methods of living like for e.g. living in the desert countries makes the people more used to climatic changes and a hard lifestyle. Consequently their risk culture would be different from others living in more comfortable locations. Other geographic features like volcanic areas, earthquake prone areas, etc. make the respective sections of the population differently risk sensitive from the rest of the population.

7. Influence of Genetic factors, Upbringing and Innate Qualities on Risk Culture

Genetic factors, upbringing and innate qualities affect the psychological sphere of existence. History is replete with instances of tyrants who had no rational reason for persecuting innocent people and still they did – Many of these tyrants had parents and siblings who were very ordinary and simple people. Similarly many billionaire entrepreneurs and leading statesmen have also had very humble beginnings and a simple family background. This itself bears testimony to the fact that despite a similar family background/environment, some people embrace a different risk culture as compared to their own parents/siblings.

The field of behavioral genetics is evolving and at best controversial. It is also feared that 'behavioral genetic information could lead to a wide-range of risk-averse actions' [10] (by the regulators, policy makers and judiciary mechanisms).

A discussion on innate qualities and its influence on risk culture is incomplete without addressing the question- what exactly is the innate nature of man? Can it be generalized?

Over centuries, men have tried to understand their own nature. Around 300 BC, the Chinese philosopher Mencius proposed that human nature is originally good and moral virtues are innate. This was further echoed by the Japanese philosopher Fujiwara Seika (16th Century) who said that human beings are endowed with "illustrious virtue" .At around the same time as Mencius, the Chinese philosopher Gaozi argued that human nature is originally morally 'neutral'. Both these views are contested by their contemporary Chinese philosopher Xunzi, who held that human nature is innately evil and human beings are capable of becoming good through the civilizing activities of teachers and rulers.[12] In the Indian philosophy, it is believed that human nature is capable of manifesting itself in three ways- Asura (evil), Manushya (human) and Deva (Divine). Lot of methods including type of food to be consumed, association with scholars (satsang), namkirtan (singing hymns of God) and yoga etc. have been proposed to conquer the asuric nature and ascend towards the Deva nature. In the modern western social sciences, Theory X and Theory Y proposed by Douglas Mcgregor are widely known. Theory X proposes that people are inherently lazy and that people must be controlled, coerced etc. if common objectives are to be achieved. Theory Y proposes a positive view about people and says that people will exercise self-motivation, self-restraint and self-control. Other social scientists tried to generalize human nature or more particularly human personality but proposed no judgmental observations as to whether human nature is innately 'good' or 'bad'.

It is very obvious that if human beings are inherently good natured, their risk culture and behavior will manifest in a manner that is conducive to their personal and society's well being. The converse holds if we believe that human beings are innately evil. As can be seen from the above, neither old world philosophy nor new world social scientists have any consensus on this issue of whether human nature is inherently 'good' or 'bad'. This aspect of risk culture has profound implications as it means policy makers and regulators can have no firm clues as to what basis of human nature they should assume for their policy making and regulation –based on Theory X (Autocratic control-based style) or Theory Y (Empowering style)? If they do not like to take extreme positions and would like to take a more reasonable approach, then they need to know how much control-based and how much empowering the policy/regulation should be. Statistically speaking, this in turn depends on the proportion of Theory X and Theory Y people in the target population, which currently at least has no known methods of assessment. This situation compels supporting and opposing points of view on regulatory precepts like the precautionary principle.

8. A mathematical framework for quantifying risk cultural concepts

Quantitative techniques can be broadly classified under three heads –Deterministic, Probabilistic and Fuzzy. All three offer some possible methods of capturing risk culture numerically in a risk assessment. For e.g. deterministic approaches like goal programming, probabilistic approaches based on Bayesian techniques and Fuzzy approaches in describing cultural orientations towards risk can help in developing a 'soft' approach towards risk assessments. Additionally, MCDM techniques like AHP (Analytic Hierarchy Process) can also be utilized to establish the risk cultural preferences of populations. A detailed enumeration of the above methods is beyond the scope of this paper.

9. Conclusion:

Risk assessment methodologies are incomplete if they do not incorporate the risk cultural dispositions of the target population. A deeper understanding of risk culture will help make risk assessments more effective and risk management regulation /policy making more focused. This necessitates development of a 'soft' approach to risk assessment wherein the risk cultural aspects can be quantified and integrated into mainstream risk assessments. This in turn involves 1) Developing a conceptual framework for risk culture and 2) Developing a mathematical framework for risk culture. A conceptual Framework outlining the key components of Risk Culture and their interrelationships was discussed in the paper. A brief overview of possible quantitative techniques in this context was also discussed.

'That is true culture which helps us to work for the social betterment of all'
- Henry Ward Beecher

10. References

1. Christian Kirchsteiger, Draft Consensus Report Proposal Towards A Project On The Development Of A Compass For Risk Analysis, [2000]
2. Andreas Klinke and Ortwin Renn, Prometheus Unbound – Challenges of Risk Evaluation, Risk Management and Risk Classification, No.153, Working Paper, Akademie fuer Technikfolgenabshaetzung in Baden-Wuerttemberg, p 40, [1999]
3. Kamal Dean Parhizgar, Global Business Environments –Understanding Multicultural Behavior, Jaico Publishing, p61-62, [2004]
4. http://www.unep.org/unep/regoffs/roap/nettlap/Products/TrainingMaterials/Accidents/Part-III.html United Nations Selected notes on Accident Prevention and Preparedness, A safe world for the 21st century. World Conference on Natural Disaster Reduction, Draft Yokohama Message submitted by the Chairman of the Open-Ended Drafting Group of the Main Committee, Mr. Daniel D. C. Don Nanjira, [1994]
5. Kamal Dean Parhizgar, Global Business Environments –Understanding Multicultural Behavior, Jaico Publishing, p62, [2004]
6. S. Abid Husain, The National Culture of India, National Book Trust, p2, [1978]
7. Kamal Dean Parhizgar, Global Business Environments –Understanding Multicultural Behavior, Jaico Publishing, p62, [2004]
8. Kamal Dean Parhizgar, Global Business Environments –Understanding Multicultural Behavior, Jaico Publishing, p226, [2004]
9. Andreas Klinke and Ortwin Renn, Prometheus Unbound – Challenges of Risk Evaluation, Risk Management and Risk Classification, No.153, Working Paper, Akademie fuer Technikfolgenabshaetzung in Baden-Wuerttemberg, p 9, [1999]
10. http://www.ornl.gov/sci/techresources/Human_Genome/publicat/judicature/article5.html ,Mark A.Rothstein Judicature Genes and Justice The Growing Impact of the New Genetics on the Courts-The impact of behavioral genetics on law and courts, Vol 83(3), [1999].
11. Report of the OECD Workshop on Risk Assessment and Risk Communication in the Context of Chemical Accident Prevention, Preparedness and Response OECD Environmental Health and Safety Publications -Series on Chemical Accidents No. 1 Environment Directorate OECD, Paris, p66, 74, [1997]
12. Ian P.McGreal, Editor, Great Thinkers of the Eastern World, Harper Collins, p 27,35,36, [1995]

DIFFERENT PREVENTIVE STRATEGIES REQUIRE DIVERSIFIED RISK ASSESSMENT MODELS

Some examples taken from the European and the Italian experience.

S. MENONI
Dipartimento di Architettura e Pianificazione, Politecnico di Milano
Via Bonardi 3, 20133 Milano, ITALY

1. Introduction

The main point that will be discussed in this paper addresses the following question : does it make sense to refer to a universal and unique prototype of risk assessment while considering different goals, preventive measures and strategies ?

It can be anticipated that the position held here leads to a negative answer, and many examples will be provided to sustain it. The latter will pertain mainly to urbanised environments, for a number of reasons.

First, because there is a clear trend worldwide towards growing urban spaces and a preference for urban forms of life. If one examines the development history of large as well as smaller cities along the Mediterranean, a rapid and even dramatic increase can be clearly seen, particularly since the Fifties and the Sixties. The result being a spread of metropolitan areas and conurbations, with urban sprawl covering the space between old centres that were separated entities in the past.

The second point descends directly from the previous: urban areas host today almost half of the world population and 75% in developed countries [1], with very high densities not only in the Middle East or along the Maghreb shoreline, but also in some regions of Southern European states. This concentration translates into a high exposure to natural as well as technological hazards, especially in areas subject to several threats. The case of Naples being particularly evident, as the town hosts several dangerous chemical installations and is subject to earthquakes, landslides and especially to the major risk posed by and active explosive volcano, the Vesuvio.

The third reason for looking at cities, is the concentration and integration of connected systems, including public facilities, transportation networks and lifelines. In this kind of environment, the potential for systemic damage and for the development of chains of failures and damages is clearly very high.

The Nineties have been the decade in which severe events hit metropolitan areas, starting with the Northridge and Kobe earthquakes, respectively in 1994 and 1995, to the Ismit earthquake in Turkey in 1999. Not surprisingly, those events have been among the most costing ever experienced and raised several questions among governments and concerned parties about the creation of new and reform of already existing tools to mitigate and/or reduce expected losses.

B. Morel and I. Linkov (eds.),
Environmental Security and Environmental Management: The Role of Risk Assessment, 211–221.

2. Tools to be used for long term policies.

The first set of considerations discusses what kind of risk assessment tools are more pertinent for strategies seeking to achieve a significative reduction of environmental risks, particularly in urban contexts. In this case long term policies have to be considered: such an ambitious goal requires rather diversified means, sometimes alternative sometimes integrated, which may prove to be effective in terms of risk reduction only after a given period of time.

It has been already widely recognised that structural measures only, that is those acting mainly on the hazard, are insufficient to provide a satisfactory protection in the long run. To be effective they must be used in combination with non-structural measures, those affecting both exposure and vulnerability. Among the latter, land use, urban and regional plans have been granted extreme importance by the scientific community and public officials concerned about the consequences that wrong locations and poor urban management may have on shaping the potential disaster outcome.

Table 1: Tools to be used to control urbanisation in hazardous areas

Zoning	**Subdivision standards**	**Building codes**
A. Special seismic study zone	A. Performance standards for sensitive lands	A. Supplemental seismic standards
B. Open space/conservation zones	B. Standards and regulations for new development areas	B. Standards to be introduced in building codes
C. No-building zones	C. Regulations and codes for urban renewal	C. Seismic standards for retrofitting residential buildings
Hazardous buildings relocation	**Strategic public facilities**	**Lifelines**
A. Abatement ordinance for risky plants	A. Abatement ordinance for public facilities in dangerous areas	A. Substitution of old lifelines in dangerous areas
C. Increase the mutual distance between dangerous facilities and residential areas	B. Relocation of strategic facilities from dangerous areas	B. Retrofitting of lifelines in the most vulnerable situations
B. Incentives to relocate industrial plants in dangerous sites	C. New public facilities in safer areas	C. New infrastructures in safer areas
Real estate disclosure	**Property/develpment rights taking**	**Taxes, incentives**
A. Obliging contractors disclose risk to potential buyers	A. Property voluntary acquisition or expropriation	A. Tax benefits for those who retrofit their house
B. Particular conditions on selling and buying contracts	B. Purchase or expropriation of development rights	B. Incentives for those relocating from dangerous areas
Insurance policies	**Environmental impact studies**	**Mapping, research, studies…**
A. Insurance programs for goods exposed to risks	Including risk analyses in environmental impact assessment studies	A. Mapping programs in risky areas
B. Insurance programs not only for private citizens but also for local/provincial authorities		B. Suveys and research programs in risky areas
		C. Creation and updating of historic data on past disasters

The starting framework for discussing the actions that can be taken to reduce exposure and vulnerability, is provided by a table derived from an enlightening proposal made by Bolton and others [2].

In the first row the most frequent tools adopted until now have been put: zoning, subdivision standards, sensitive area ordinances, and building codes. In the second row decisions regarding special facilities (hazardous and not) are represented. In the last two rows non-regulatory tools are shown. Most of them consist of policies and programs designed to achieve prevention indirectly, by eliciting the risk factor in the real estate and insurance market.

With respect to the time when this summarising framework has been first presented, a number of experiences are available today to draw tentative, basic considerations regarding their advantages and limits. Experiences that will be illustrated have been developed in European countries and particularly in Italy. It is important to mention, in fact, that the actual implementation of the different set of tools has been often promoted by an action at the European community level, with many countries, like Italy (but Spain or Greece are not any different in this regard) deriving the entire body of their environmental laws and normative framework from European directives and programs.

This does not mean that the application of the latter is a plate translation of what has been set at a community level; many times, instead, national laws are innovative and put forward significant adaptations to local needs and specific features that are generally omitted while not neglected at the European level. Rather, this means there is a creative and innovative capacity at the local and national level, though, it cannot be forgotten that the initial input is provided by a more powerful external body represented by the European Commission with European institutions. Its power does not stems only from the fact that member states must adopt appropriate internal legislation according to what has been established by the EU, but also from the fact the Commission is able to provide cultural, technical and scientific guidance in fields, like the environment, where national governments often lack the necessary expertise, especially when front line and emerging themes are concerned. Furthermore, in many cases, problems cannot be solved, or only very partially, within national boundaries, while integrative approaches are the only solution for issues that are intrinsically global, transboundary or affect a common resource (the Mediterranean being the case).

2.1 RISK ASSESSMENT PROCEDURES FEEDING THE MOST "TRADITIONAL" REGULATORY TOOLS

Going back to the tools that have been presented, let's discuss each category in deeper detail.

As for ordinances, regulations or codes addressed to ordinary buildings, the following can be said. When rules for protecting structures from some kind of natural hazard (avalanches, floods or earthquakes) are set, it makes sense to ground them on a probabilistic risk assessment, reflecting in a rather standardised fashion the severity of the potential menacing event and its probability of occurrence. It is a sort of average protection that is looked for, extended to large areas with similar features. Those rules work well for threats the effects of which are well understood and widely known. They pertain to a rather long tradition of good construction practices, that were adopted also in the past as a prevention measure, especially in the aftermath of a disaster.

While talking about norms for special facilities or even about zoning, the situation changes completely, as those tools need much more than the previous to be shaped taking into account the specific characteristics of the place where they must be implemented. The concept of zoning, for example, first introduced in the urban plan of the German city of Frankfurt at the end of the XIX century by Franz Adickes, descends from the need to order the complexity of contemporary city functions, some of which are incompatible because they may provoke mutual nuisance when not actual risk. While simple in principle, zoning is one of the most difficult tool to apply sensibly, requiring a deep understanding of the relations that have matured in a specific urban environment between society, spaces and uses of land, buildings and facilities.

The main concept that has been adopted in zoning in relation to environmental threats is separation and safety distances, the idea being to keep away potentially vulnerable systems from natural or technological hazards. This can be done by creating a buffer along flooding rivers or between dangerous industrial plants and houses or infrastructures.

The buffer dimension is generally decided upon a probabilistic risk assessment in the first case, prescribing to avoid building in the floodplain corresponding to a return period of 20, 50 or 100 years flood; with respect to the second case, the European situation presents a rather composite and heterogeneous picture. Both probabilistic risk assessment and scenario approaches are adopted to set safety distances between hazardous installations and urbanised areas.

While the advantages of zoning criteria are rather obvious, limits deserve some extra words. First, it is clear that zoning can work only for future development or for a complete re-development project, while it can do very little in already built up areas. Second, the type of risk assessment tool that constitutes the basis for the future layout of the city is critical in determining the level of protection that will be attained.

Disregarding the problems that buffers along rivers may encounter in the future, considering the deep uncertainties climate change may impose on assessment carried out today on the basis of events recorded in the past, some observations will be made regarding the kind of assessments to be used for deciding the compatibility between urban functions and hazardous facilities. The Seveso II Directive, as amended in December 2003 [3], states that: «Member states shall ensure that their land-use and/or other relevant policies and procedures for implementing those policies take account of the need, in the long term to maintain appropriate distances between establishments covered by this Directive and residential areas, buildings and areas of public use, major transportation routes as far as possible, recreational areas and areas of particular natural sensitivity or interest, and, in the case of existing establishments, of the need for additional technical measures so as not to increase the risks to people» (art. 12 as amended).

Both probabilistic risk assessment and scenario approaches present drawbacks, taking into account that even the latter is rarely pure, as a likelihood is generally associated to each scenario and safety distances are decided considering both factors, the potential damage magnitude and its probability. The main defect is that a concrete factor, the distance, is defined upon a probabilistic estimate, which is at best an informed guess, that cannot be grounded on any historic series as in the case of earthquakes or floods. If in the case of the latter, doubts can be raised in the name of climate change, imagine the case of industrial so called "top events" that are luckily rare and concern a technology which is continuously evolving, thus falsifying comparisons between plants and installations built in different periods.

The acceptable level of risk is generally kept implicit, while any safety regulation or law sets one by its very nature. The consequences of a top event on the outside environment are generally underestimated, because the peak, most critical events are even not studied as their probability is considered too low to worry about. On other occasions, according to some technical procedures widely adopted in industrial risk assessment, the worst outcomes are compensated by the introduction of additional safety devices and countermeasures. Romerio [4] warns against the fact that many times it is not clear if the latter have been already adopted or are part of a project for the future. But even given for granted that by the time a new urban development will be created, safety devices and procedures will have been put in place, still the question whether or not those two factors, possible damage and countermeasures, can be balanced one against the other can be asked. This is particularly the case as prevention measures will be implemented mainly within the dangerous facility, while possible damages will reach people and goods far outside the plants' fences.

But who is the ultimate decision maker using the results of a risk assessment procedure to choose between the alternative options of development or non-development? The institution that will take this decision varies from one country to another; what can be still considered the rule is that residents and end users will have no role or very little role in deciding about the location of houses and urban facilities. In any case, decision makers will be dramatically influenced by the way the final results of the assessment procedure will be shown.

If in terms of numbers (i.e. expected radiation produced by an explosion or a fire, dimension of a toxic cloud), lay public including city managers, will not be able to grasp their meaning and visualize the amplitude of a negative outcome; if in terms of damage areas represented in a map, the situation does not change radically. People will tend to interpret the borders of the damage areas as real, unless the uncertainties and the many theoretical adjustments leading to those areas will not be stressed again and again. What if a possible accident scenario were represented in the form of a movie or even individual real-looking pictures taking as a background the area where the new development should take place?

As Schwartz [5] correctly puts it, a novel, a theatre play or a movie can be considered a scenario in the "scientific" meaning, and, by the way, the inventor of this tool, Khan working for the Rand Corporation in the Sixties, pursued as a goal to demotivate army people from continuing in the development of nuclear weapons by showing in a vivid way the potential day after.

2.2 RISK ASSESSMENT TOOLS TO BE USED FOR POLICIES ADDRESSING DEVELOPED URBAN AREAS

As zoning cannot really influence the situation in already consolidated urban context, other tools should be considered. The most obvious, though complicated and expensive, is relocation. In the case of natural hazards generally only buildings and facilities can be moved, while with technological hazards, at least in principle, also the origin of the threat can be moved away from the most congested parts of a city.

Apart from cases where relocation was attained for the completion of a public utility work, like a dike or an important infrastructure, relocation as a preventive strategy was rare in the past. Although safety can be considered a public good, relocation poses some equity questions, as taxpayers' money is used to resettle people who had put themselves in danger by choosing unsafe places to live in. However, in some instances, relocation may produce also a more general benefit to society, like in the case of flooding areas, where better drainage conditions in abandoned and renaturalized areas may increase the storage capacity of the basin, at least locally.

Modern societies have started considering it a viable option only lately. In the USA, some 10.000 dwellings and economic activities have been relocated after the 1993 Mississippi flood, while in Europe this measure is getting more acceptable in the public's eyes. In France, the government has decided to help financially citizens to leave high risk zones recurring partially to the Mitigation Fund against Natural Catastrophes created in 1995 [6].

In Italy some regions have passed laws to promote relocation after severe floods which occurred in the autumn of the years 2000 and 2002. Significantly, a non-negligible number of municipalities adhered to the relocation programs, thus showing that conditions are already mature also for this kind of rather unpopular measure.

What type of risk assessment is required to support relocation programs? Before answering this question, it must be underlined that relocation voluntary as well as non-voluntary, may succeed only if people participate to the entire process and are involved in the decision to leave the dangerous place and regarding how and where to resettle. What the World Bank [7] suggests for non-voluntary relocation perfectly fits also to voluntary relocation, which requires a close relationship between communities and their administrators, the continuity in the actions taken from the initial decisions up to the final reintegration in another site.

A recent report commissioned by the regional government of Lombardia, Italy, on this issue, showed that people are more likely to accept relocation if they feel constantly under threat, that is if the event is frequent or continuous. Therefore, risk assessments procedures should provide results in two forms: show where risk is higher in terms of hazard frequencies (not necessarily accompanied by the worst outcomes) and where, instead, potential losses can be dramatic but associated to very rare events. A government may decide that relocation should be promoted also in the second case, but it is clear that a lot more has to be done in terms of public information to obtain consensus.

Obviously, it would be preferable to achieve complete relocation from an area that has been recognised as exposed to high levels of risk. Nevertheless, sometimes also partial relocation can be useful, as it will decongest densely populated areas, easing the task for civil protection forces in case of need. This was the idea behind the regional government of Campania, Italy, when it decided to encourage the abandonment of the municipalities located in the most dangerous areas around the Vesuvio crater (the so called red zones), giving renters and owners money incentives to move away. This way authorities hope to lessen the time needed for a total evacuation from the present two weeks to one week, thus allowing for less uncertainties when a general public alarm will be decided.

2.3 PROBABILISTIC RISK ASSESSMENT OR SCENARIO MODELLING BACKING ECONOMIC TOOLS TO ACHIEVE PREVENTION?

Economic means like incentives or compensation can be used to obtain preventive and cautious behaviours, though this is not an automatic outcome of this kind of tools. Rather, the latter must be carefully designed to achieve prevention.

The most popular economic tool used in this regard in the realm of risks is insurance. It should be reminded, though widely recognised, that insurance is not at all a preventive measure per se, it is only a mean for mitigating the consequence of a disaster, providing in a quicker and more structured way with respect to other options, funds to rebuild and restore the damage. In case of non-renewable goods, like for example historic towns or monuments, or non-renewable natural resources, both in quantitative and qualitative sense, insurance is a misused instrument, as the good or the resource – once lost – cannot be compensated, but only, and to a very limited extent, substituted.

Insurers have traditionally based the premiums to be paid by insured on probabilistic risk assessment, that is considering almost uniquely the hazard level in a given area, as only this term can be more or less represented if not in terms of expected frequencies at least as probabilities.

In the latest years, however, insurers have expressed increasing concern for the potential destructive effects of an unforeseen event in congested metropolitan areas. Two types of uncertainties seem to be particularly critical, one regarding the hazard, the second concerning the vulnerability of exposed systems.

As for the hazard, insurers worry about events that can be forecast only to a very limited extent, like in the case of climate changes or the consequences of a catastrophic accident in a dangerous facility. Furthermore, they are increasingly interested in getting more knowledge regarding vulnerability and exposure, which may consistently determine the ultimate severity of damages (see the interesting report by Linnerooth-Bayer et al., [8]).

Two main consequences are of interest here. The first is thee entering of a new tool, the scenario modelling, better fit that the probabilistic risk assessment to combine hazards and different degrees of vulnerabilities of exposed systems, for those events involving higher uncertainties. The finer the detail with which parameters are able to grasp vulnerabilities arising from specific urban patterns, the higher the likelihood of damage scenarios drawn from the model. A result exacerbating the costs of premiums to be established but also reducing the potential for dramatic expost surprises.

On the other end, the discovery of such diversified outcomes related to the same kind of event in different settings and the need to spread growing risks associated with high levels of uncertainties, thwarts any attempt to regionalize the use of such tools as insurance, that can work only on a global market. It is not by chance if the European Commission has mandated to the CEA (Comité Européen des Assurances) the study of an insurance system to be adopted at the European level. One of the possible reference model for such a system is the French Cat-Nat, that has been recently extended to cover the damage provoked by a technological accident.

3. Risk assessment for emergency planning

While the various strategies discussed in the previous section seek for solutions to reduce the present level of risk, emergency planning is aimed at assuring the best possible response to mitigate and alleviate the consequences of a negative event, once it has occurred. While contingency plans used to be a rather generic set of actions to be taken in case of need and a list of responsible people and organisations to be contacted first, in recent years those plans have been increasingly designed to respond actual needs and urgent requirements that may arise in a disaster. A new "generation" of contingency plans have been developed in several European countries, in which the logistics and the timing of the developing response are considered with stronger emphasis than it used to be in the past.

For emergency preparedness purposes, probabilistic risk assessment is probably misplaced: specific actions can be designed, especially in a complex urban environment, only following a rather detailed analysis of the chain of damages and systemic failures that may be triggered by an initial input (be it a natural or a technological hazard).

According to a decision of the regional government of Lombardia (cfr. Delibera Regione Lombardia n. 46001, 28/10/1999) a scenario is the «synthetic description provided also through maps of the possible effects a disaster may have on people and infrastructures».

Emergency plans should address mainly the systemic and organisational failures that contribute, together with the physical damage, to create the damage and failure chain. But what analytical input should be provided for contingency plans? Only one scenario or a number of?

The most common answer that has been given is "the worst possible scenario", sustaining that if there is the capability to face this one, all the others will pose less problems. This approach, however, does not seem too satisfactory for several reasons: it promotes the false feeling that if a civil protection organisation is not able to tackle the worst possible scenario, then it is helpless and it is not worthy to worry about emergency preparedness anymore. This is, by the way, the attitude of unprepared or inadequately prepared organisations, referring to a useful classification proposed by Lagadec [9] and though it seems rather unacceptable, it is not infrequent to find it even in official documents.

But also from a theoretical point of view such a premise may conduce to faulty conclusions: it cannot be taken for granted, for example, that the worst scenario includes all the features of any other possible scenario. It may be the worst for one or more parameters, but not necessarily comprise the entire set of (infinite) eventualities. Why not being prepared then for less catastrophic scenarios, but more likely to occur and in which a successful action may safeguard against the degradation towards much more undesirable situations?

As Guilhou and Lagadec put it in their work [10], scenarios should be first considered as a mental tool to prepare for surprises, so the more imaginative the better, no matter how likely or probable the scenario is. In this regard, even though the organisation may not be able to face exceptionally disastrous events, it may still be useful to mentally prepare for them, to enlighten what may be done and what cannot according to the present level of preparation, how far the latter is from a positive solution of the stress posed by the event under examination. By not asking those questions, again, the organisation may find itself far less prepared than it thinks to be even in facing more trivial situations.

In this regard, the Seveso II Directive shows an important pitfall, as it prescribes external emergency plans to be developed on the basis of safety reports, thus for the most dangerous category of hazardous installations. The resulting interpretation in Italy, for example, is that those plans are mandatory only for the most dangerous category of plants, while they are left to public administrators will for all the others. Does this make sense in a highly congested urban environment, where an accident in even a small chemical factory may have a very high destructive outcome? Clearly no. Preparation provides far larger benefits than the costs it involves, and should be strongly promoted looking not only at the hazard severity but also at the risk situation as a whole, in which many times it is the dimension of vulnerability and exposure that creates the conditions for a disaster.

An interesting example comprising the observations that have been carried out in the previous lines is provided by the Guidelines prepared by a working group of firemen and railways safety responsible officials in Italy in 1997 [11]. In this document a number of generic scenarios have been evaluated with respect to potential accidents that may occur in long train tunnels. Regarding the most critical scenario that has been identified, that is an accident in a train carrying dangerous substances developing an initial fire facing a passengers train running in the opposite side, the following conclusions have been drawn: «The probability of such an event is rather small, especially if confronted to all the others that have been taken into account. [...] Although it may be considered a "top event", that is the worst possible event that can be imagined, as its frequency is rather low it has not been considered by the working group as a reference scenario. This scenario is not considered for two reasons: because in any case there would be very few chances to manage it successfully, knowing the structural features of the tunnels and the available emergency forces, secondly because the structural measures that would be required to improve tunnels' safety are incompatible both in economic and feasibility terms». The first puzzling point that is striking is the idea that a scenario is considered only if there is a preconceived idea that there is the capability to face it, while the latter should be the result of re-organisation and further preparation and not a pre-condition to act; the second misconception derives from mixing considerations regarding prevention measures in the long run (to make safer the structure) and measures to be taken in order to improve emergency management.

4. Areas of concern for the future

There are two fields in which the normative and legislative development has not reached yet the desirable results: the adequate consideration of the potential damage to the environment resulting from a major accident on one side and the transportation of dangerous goods on the other.

With respect to the first, the new amendment to the Seveso II Directive, by quoting the Baia Mare accident causing a massive cyanide contamination in the Danube river, could have been the occasion for stressing with stronger emphasis the need for preserving environmental resources like air, water, groundwater, soil from the acute consequences of such an accident. Unfortunately, this has occurred in the new amendment only to a very limited extent. It is a missed opportunity, because a strong input in this regard from the European Commission could have provided an incentive for developing research in this rather neglected field.

The damage to vital environmental systems following a severe accident has been insufficiently recorded and analysed, with few exceptions. The latter provide an interesting and encouraging starting point, which, however, requires more investment in studies and resources.

In a very similar situation lays the question of the transportation of dangerous substances, with a rather heavy potential of accidents with transboundary effects and consequences affecting common natural resources.

Despite the fact that it represents a very serious risk, concerning both urban areas (when trucks and train means are considered) and marine and fluvial environments, it has been neglected at the European as well as national level.

The lack of a stringent legislation leads to a situation where the risk itself is relatively underestimated and only imprecisely known. Information is spread among a variety of different institutions and administrations, and also the advanced technical solutions that would permit to trace dangerous trucks and ships do not provide any significative result without a directive requiring such an information to be stored and without designating a specific institution to keep and update it.

On the basis of the available information it is very hard to draw a convincing risk assessment regarding the potential threat associated with the transport of dangerous goods and to select likely scenarios or those deserving higher priorities.

5. Preliminary conclusions

In this paper it has been argued that risk assessment cannot be considered a monolithic tool, to be developed by experts and then delivered to decision-makers who should make their decisions on its ground. Different types of policies and preventive strategies require different inputs, the ideal being experts working with and not for decision-makers. The need to use a variety of tools, such as probabilistic risk assessment, scenario modelling, simulations differentiated not only according to the policy for which they are designed but also to the geographical scale and to the final users, requires a rethinking of the entire matter, in order to better meet new and emerging demands. This seems to be the case not only for the tool itself, but also for the manner in which it is applied to a specific urban, regional and cultural context.

Furthermore, it has been shown how the development of risk analyses tools and procedures is strongly linked to the existence of a legislation requiring them. The Seveso directives provided a strong incentive to develop and improve tools for estimating accident scenarios the consequences of which may be worse outside than inside the plant. The lack of legislative and political pressure to produce risk assessments in some fields, such as technological accidents affecting natural vital resources or the transportation of dangerous substances, leads to a worrying situation of underestimation and relative ignorance. Risk analysis is a field which cannot be entirely developed in laboratories: in order to produce credible results it requires the association to administrative procedures and the systematic collection of data and information.

6. References

1. United Nations, *World urbanization prospects: the 2003 Revision*, 2004
2. Bolton P., Heikkala, S., Green, M.M., and May, P(1986) *Land use planning for earthquake hazard mitigation: a handbook for planners,* Special Publication n.14. Boulder: Institute of Behavioral Sciences, University of Colorado.
3. Directive 2003/105/EC of the European Parliament and of the Council of 16 December 2003, amending Council Directive 96/82/EC on the control of major accident hazards involving dangerous substances.
4. Romerio, R. (1998) The risks of the nuclear policies, *Energy Policy* 26: 3, 239-246.
5. Schwarz, P. (1991) *The art of the long view. Planning for the future in an uncertain world*, Doubleday Publisher, New York.
6. Michel-Kerjan, E. (2001) *Insurance against natural disasters: do the French have the answer? Strengths and limitations,* Cahier n. 2001-007, Laboratoire d'Econometrie, Ecole Polytechinque de Paris.
7. The World Bank, Environment Department (1993) Public involvement in environmental assessement: requirements, opportunities and issues, in *Environmental Assessment Sourcebook Update*, October, n. 5.
8. Linnerooth-Bayer J., Mace, M., Verheyen, R. (2003) *Insurance related actions and risk assessment in the context of the UNFCCC*, Background paper for UNFCCC Workshops, May.
9. Lagadec, P. (1995) *Cellules de crise. Les conditions d'une conduite efficace*, Les Èditions d'Organisations, Paris.
10. Guilhou, X and Lagadec, P. (2002) *La fin du risque zéro*, Les Èchos, Eyrolles Editeur, Paris.
11. Gruppo di lavoro misto FFSS-CNVVF, (1997) *Linee guida per le gallerie ferroviarie*, D.M. 03/96, Fasc : 4101, 25 July, unpublished document.

COUPLING PUBLIC PARTICIPATION AND EXPERT JUDGMENT FOR ASSESSMENT OF INNOVATIVE CONTAMINATED SEDIMENT TECHNOLOGIES

T. P. SEAGER
School of Civil Engineering, Purdue University, West Lafayette IN, USA

S. H. ROGERS
Industrial Economics, Inc., Cambridge MA, USA

K.H. GARDNER
Center for Contaminated Sediments Research
University of New Hampshire, Durham NH, USA

I. LINKOV
Cambridge Environmental, Inc., Cambridge MA, USA

R. HOWARTH
Environmental Studies Program, Dartmouth College, Hanover NH, USA

Abstract

This project investigates the utility of Multicriteria Decision Analysis (MCDA) as a tool for testing stakeholder responses to and improving expert assessment of innovative contaminated sediments technologies. Within the broader context of environmental decision-making theory, this case study focuses on a planned dredging project in Dover, New Hampshire where sediments containing PAHs and heavy metals will be removed from 2.7 miles of the Cocheco River (a navigable estuary). Faced with limited alternatives for dredged material disposal, local officials decided to place the contaminated materials in a sealed and lined disposal cell in a riparian area. However, the decision process employed (process of elimination) may have been severely taxed by innovative technological alternatives. To assess the feasibility of innovative technologies in this case, a group of stakeholders with a vested interest in the materials management decision were queried about the basic criteria they would apply to assessing decision alternatives, experts at the Center for Contaminated Sediments Research (CCSR) at the University of New Hampshire provided performance estimates related to those criteria, and an MCDA outranking study identified those stakeholder groups likely be in conflict or willing to reach consensus. Of the three innovative technologies tested, one was found to be unsuitable for this site while two others were likely to have support from different stakeholder groups. Those groups with strongly held views were modeled with the greatest confidence while groups with less strongly

D. Morel and I. Linkov (eds.),
Environmental Security and Environmental Management: The Role of Risk Assessment, 223–244.

expressed preferences may be satisfied by more than one alternative and have a greater willingness to compromise.

1. Introduction

Environmental resources typically have multiple public uses. Consequently, competing socio-political interests are often brought to bear on environmental problems. Although different alternatives are likely to be preferred by different stakeholder groups, incorporation of public and stakeholder participation in environmental management, policy and decision-making may lead to less expensive, less contentious, and more satisfying decisions. On the other hand, environmental systems are especially complex, involve high degrees of uncertainty and require specialized knowledge to properly understand. Therefore, expert judgement is essential to determine cause-effect relationships and assess which strategies or technologies are likely to lead to the desired outcomes. Whenever expert and public models of decision-making are in conflict, the two groups are likely to become estranged. This may especially be the case in management of contaminated materials such as sediments, in which multiple sources of pollution and responsible parties may be difficult to identify, contaminants are long-lived, and source-control strategies (often effective in air and watershed management) are not likely to result in acceptable rates of attenuation or recovery.

Because many environmental issues are multifaceted and involve scientific, socio-political, and economic dimensions, decision-making processes can be fragmented and contentious (USEPA 2000). Both politically driven (in which powerful interests may exert disproportionate influence on the outcome) and expert-driven (in which intensively bureaucratic processes operate outside public influence) can undermine the democratic ideal of managing shared resources for the public good. Alternatively, significant, early, and meaningful public input including written surveys, focus groups, public meetings, hearings, written comment, workshops, referenda, town meetings and interviews, can aid decision-makers as well as make a less contentious and more transparent process.

However, disparate public groups are unlikely to reach consensus on which among many technological alternatives is best. Each group is likely to prefer the alternative that performs best in the areas they consider most important. Consequently, environmental problems are group decision problems from which no single best solution is likely to emerge, and multiple criteria and perspectives must be brought to bear. Multi-criteria decision analysis (MCDA) can provide a framework for structuring complex environmental decisions. However, MCDA is rarely employed in the context of public participation for management of contaminated materials (Linkov et. al 2004a). There is an acute need for research that is capable of synthesizing expert assessments (especially for innovative technologies) with public (specifically, stakeholder) participation methods (Seager et al 2005). In this study, stakeholders involved in a real dredging operation are queried to test their response to hypothetical innovative alternatives. The results could help create more satisfying decision-processes and serve to prioritize scarce research and development resources so that technologies both responsive to public concerns and viewed positively by experts are pursued first.

2. Problem Statement

While many federal agencies have been charged with incorporating public participation into environmental decision-making, numerous implementation problems remain. Current decision processes typically used in federal agencies emphasize physio-chemical modeling and engineering optimization. Although federal agencies are required to consider social and political factors, the typical decision process does not provide for explicit consideration (Kiker et al. 2005). Many regulatory agencies currently employ an optimization approach such as cost-benefit analysis (CBA) that reduces all facets of the problem to a single measure. In general, mono-criterion approaches often exclude the diversity of public opinions. For example, monetization of values pertaining to species extinction, global warming, & energy independence are deeply controversial, while concepts such as fairness (including the distribution of costs and benefits) are conceptually outside the scope of CBA. Consequently, federal decision making processes have been criticized for failing to sufficiently include public input (c.f., Stahl 2002, 2003).

The current decision-making paradigm at expert-driven government agencies may have deep roots in a progressive view of government that sought to minimize ideological conflicts by separating governing processes into scientific (i.e., objective) and value oriented (i.e., subjective) issues (Nelson 1987). Given the complexity of environmental systems and the highly specialized knowledge required to assess decision outcomes reliably, environmental decision making has historically been perceived as a scientific, administrative function carried out by public trustee organizations with little input from the public. For example, the United States Army Corps of Engineers (USACE) has previously employed a "decide-announce-defend" approach to decision making that often left affected public groups feeling alienated (Beierle & Cayford 2002). Given the presumption that analysis can be objective, the optimization (i.e., mono-criterion) approach seems to offer the promise of maximizing public benefits. However, when this approach has failed to satisfy public groups, experts often responded to criticism by calling for greater education efforts (perhaps to teach the public to think more like experts), rather than diversifying decision criteria. A competing view counters that all analysis is actually motivated by argument. Whereas the maximization view originates in microeconomic theory and is predicated on the hypothesis that alternatives, preferences, criteria, and tradeoffs can be judged as objective truths, the analysis-as-argument view is taken from the concept of jurisprudence in which emphasis in decision-making is placed on history, ethics, and rhetorical persuasion including evidence, conclusion, context, metaphor, analogy, values, and audience (Sagoff 1988). In fact, all real decision-making processes inherently rest on both value judgments and objectives facts. Consequently, they are not pure distillations of either view.

Since the 1970's, the trend has been towards more open and transparent decision processes (Diduck & Sinclair 2002, Wood 1995), eventually culminating in the National Research Council recommendations for an analytic-deliberative process that is responsive to the interests of all the parties involved and addresses uncertainties in an understandable manner (NRC 1996). The emphasis on deliberation in this recommendation brings the public into the decision process and has been implemented in an increasing variety or ways including advisory committees, citizen juries, town meetings, negotiations and focus groups (c.f., Beierle & Cayford 2002, Borsuk, et. al 2002, McDaniels & Roessler 1998, Gregory 2000, Ananda & Herath 2003, Gregory & Wellman 2001, Gregory & Keeney 1994, Wilson & Howarth 2002). In general, public participation can mean any of several mechanisms for involving the lay public or their representatives in an administrative decision-making. However, the interaction of both analysis (by experts) and deliberation (among the public) calls for a framework that can simultaneously facilitate analysis of facts and present several different value perspectives.

3. Introduction to the Case Study

All environmental decisions regarding contaminated materials involve risk and require expert knowledge. However, the issues surrounding environmental decision-making are perhaps felt most acutely in the management of contaminated sediments and soils, as exemplified by the public controversy surrounding the recent USEPA decision to dredge the PCB-contaminated upper Hudson River. In general, contaminated sediments and soils present long-lasting problems that can not be addressed solely by source control. The pollutants dissipate slowly, and local effects can be acute. Although the complexity of issues publicized in more notorious case studies may make testing experimental decision approaches difficult, millions of cubic yards of contaminated sediments are dredged from navigable waterways every year and management of these materials can be a difficult and contentious task at any scale.

The focus of this study is on stakeholders, who are defined as individuals or groups of people with close interest or direct interaction with a problem. Stakeholders are a subset of the public, and generally thought of has having greater knowledge about the problem and more to gain (or lose) by influencing the decision process. Consequently, stakeholders are likely to be more motivated than the general population to participate and influence a decision process. This study hypothesizes that understanding stakeholder values may allow technical experts to characterize individual preferences for new management alternatives, enabling mangers and engineers to prioritize research and development of the most promising new alternatives. The principal objective of this study is to investigate methods of synthesizing MCDA and expert knowledge with time and cost-effective stakeholder participation methods that could speed the introduction of beneficial reuse technologies and stakeholder-based processes into smaller, heavily resource constrained communities exemplified by the City of Dover, NH.

3. Research Methods

3.1 INTRODUCTION

All MCDA methods have several common elements including generating and assessing performance of alternatives, quantifying or characterizing values, and ranking or contrasting the performance of alternatives with respect to the values expressed (Belton & Steward 2002). In many cases, the most valuable aspect of an environmental decision process is in helping participants refine or clarify values that are difficult to express – especially within the constraints of a public decision-making process. Some aspects of a typical MCDA process most intensively involve experts (such as assessing the performance of alternatives), whereas others most intensively involve decision-makers (such as deciding which performance criteria are most important). In this case, alternatives were hypothetical and developed exclusively by experts (with the exception of the actual alternative selected by local officials as a basis for comparison). While this exact approach may not be appropriate or successful in an actual decision process, the results of this study are expected to help test which technologies are most responsive to stakeholder concerns at this site. In practical terms, this method could be employed to prioritize different alternatives for pilot-scale study, or further research and development, and speed the introduction of new technologies into the marketplace. However, the results are site specific. That is, technological options view favorably at one site may be inappropriate for others, depending upon the views of the stakeholders involved and the specifics of the site, such as type and level of contamination, availability of beneficial end-use opportunities, or cost.

In this case, approximately 75,000 cubic yards of sediment, some of which are contaminated with polyaromatic hydrocarbons (PAHs) and heavy metals, are scheduled to be removed from the Cocheco River in and around Dover NH beginning in the Fall of 2004 (Rogers et al. 2004). The Cocheco River is located in the southeastern part of New Hampshire and flows toward the Gulf of Maine and the Atlantic Ocean. The proposed section of the river to be dredged is that just below the dam in the center of the City of Dover to its confluence with the Piscataqua River. There are many motivations for the dredging project, including maintenance of a navigable channel, which is considered essential to the long-term economic development plans to return the City to its former status in the 19th Century as in inland port. Because it is a navigable waterway and thus under federal jurisdiction, the USACE has been helping the city of Dover coordinate the process and will be performing the dredging.

There has been and still is much debate in the community over the dredging decision. However, the focus of this case study is solely on the disposal of the contaminated sediments, rather than on dredging operations themselves. In this regard the actual decision process employed by City officials was unstructured in comparison to MCDA approaches. Regulatory constraints required secure disposal of contaminated materials (i.e. prohibiting ocean dumping). Although a privately owned landfill is only 12 miles from Dover, the operators refused to accept the dredge spoils. The next largest landfill (in Maine) was prohibitively expensive due to the transportation costs and tipping fees, so secure landfill disposal was judged to be infeasible.

The next most secure option appeared to be a constructed upland disposal site. The City looked at approximately a dozen sites in close proximity to the dredging operation, but ruled out those that were presently undisturbed or not suitable for de-watering of dredged spoils (i.e. too steeply sloped). Officials felt that priority consideration should be given to sites that have already been compromised, such as the former landfill and abandoned recreation area that officials eventually chose.

Nonetheless, this site required application for a waiver from NH Department of Environmental Services (NHDES).

During the process of completing dredge permit and disposal cell waiver applications; there was opportunity for public involvement in two ways: public comment on written application materials and in public hearings. Three separate applications could have initiated public comment or hearings: the USACE dredge proposal, the waiver application, and wetlands permit application. The first two generated insufficient public comment to warrant a hearing, even though the City was required to notify abutters and plans were publicly available in several locations. However, the wetlands permit application generated significant interest. A number of questions regarding the long-term viability of the disposal site were raised, but ultimately the site was presented as the only feasible disposable alternative, and the waiver was granted. Table 1 summarizes the alternatives that were considered by the City of Dover and USACE along with the various factors that were taken into account in this determination.

Table 1: Feasibility of Disposal Alternatives

Alternative	Consideration
Turnkey Landfill	Refused to accept because of volume and characteristics of sediments
Ocean Dumping	Unacceptable because of level of contaminants
Upland Disposal Sites Along the River	Land was undisturbed (in natural state) or unsuitable (e.g. grades too steep)
Secure landfill site in Maine	Transportation costs were too high
Local Landfill (Superfund Site)	Contaminants were not suitable.
Former Landfill Site/ Dover Public Works	Costs of upkeep and monitoring were acceptable. Proximity to river minimizes transportation risks and costs. Officials say there will be minimal environmental effects. Others are skeptical. Waiver needed to build disposal cell.

As part of the waiver approval agreement, the City is obliged to pay for and tend to the monitoring and upkeep of the disposal cell. USACE is currently putting together an Operations and Maintenance Plan for the cell, which is part of the final approval process. The cell will be left uncapped for one year while the sediments dewater. A fence will be placed around the cell during this time to reduce the risk of public harm. The de-watering liquid will travel to the Dover wastewater treatment system, which according to hearing materials in the file, "confines the contaminants to a manageable location" (NHDES 2001). The construction of the cell will be inspected by the USACE with the city of Dover becoming the final owner of the site. Some of the disposal area is part of Maglaras Park where PAHs from the disposal of the 1985 dredging were discovered on the soccer and baseball fields. There are plans to use the new spoil site as fields again, once the cell has been capped and covered and if safety standards allow.

While the decision-making process employed by local authorities in the Cocheco project is practical and may have resulted in a wise choice, from the stakeholder and MCDA perspectives, it is clear that the decision process was not explicitly structured to elicit or be responsive to stakeholder concerns. Additionally, the introduction of new alternatives—such as beneficial reuse technologies—would have increased the complexity of the choices so significantly that it could have overwhelmed the simplified, heuristic decision process employed. Therefore, the Cocheco River case was selected as an appropriate opportunity to test new methods of stakeholder participation that could guide researchers towards a better understanding of human concerns without the high stakes often associated with actual decision processes.

There are a number of methods of stakeholder value elicitation and public participation available to choose from. The actual methods employed in this case study rely upon a combination of processes mentioned in the literature cited above, considering the resource limitations of the research group. The study was designed to have three primary points of contact with stakeholders:

1. A semi-structured interview to establish the primary concerns.
2. A written survey.
3. A more structured verification interview.

Following the initial interviews, the information regarding stakeholder values was forwarded to experts for additional characterization of the decision criteria (by identifying attributes of each, and if possible, appropriate metrics for these attributes) and assessment of the performance of new technologies in the areas identified. Figure 1 shows a schematic of the methods employed while the details are described in the paragraphs below.

Key Stakeholder Groups

Citizen/Environmental Advocacy Groups
Conservation Law Foundation, SaveDover, Cocheco River Watershed Coalition

Business Interests
Greater Dover Chamber of Commerce, George's Marina

State & Local Government/State Agencies
City of Dover Environmental Projects Office, New Hampshire Dept. of Environmental Services, Dover City Council, Dover Conservation Commission

Local Citizens/Abutters to the Project
Individuals identified in public comment documents, or solicited from knowledge of proximity to affected areas

3.2 CASE STUDY RESEARCH AND IDENTIFYING KEY STAKEHOLDERS

Research began by investigating the background information available on the local case study. Business interests, government officials, local citizens and environmental advocacy group leaders and members were identified in public documents generated during the permitting and public hearing process (sidebar). Additionally, abutters to a proposed disposal site were identified, as were employees at the local wastewater treatment plan, individuals mentioned in newspaper articles, and anyone else recommended to researchers. The stakeholders fell into four general categories as detailed in the box above.

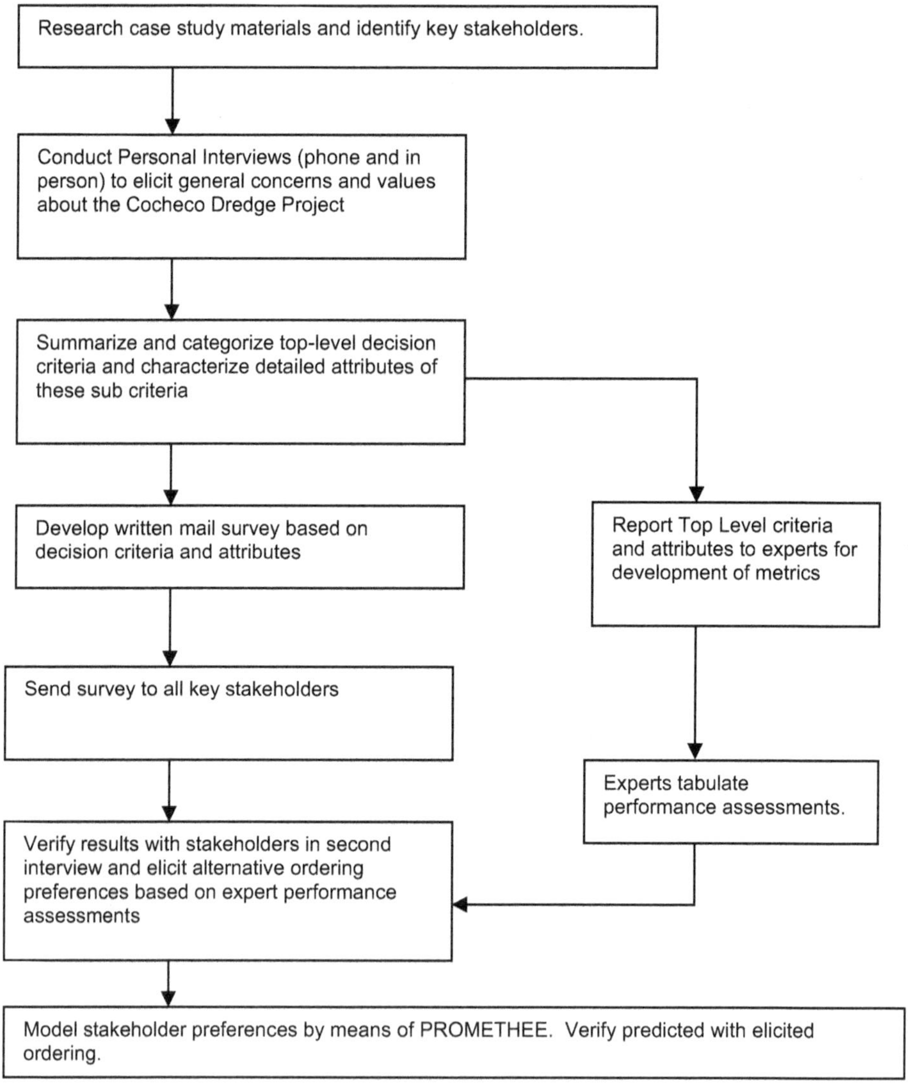

Figure 1: Schematic Representation of Stakeholder and Expert Engagement

3.3 INITIAL INTERVIEWS

Using a semi-structured, reflective interview, representatives from each stakeholder group were interviewed personally or on the phone to identify key decision criteria and project objectives. In general, "semi-structured interviews have some degree of predetermined order but still ensure flexibility in the ways issues are addressed by the informant" (Dunn 2000). A sample interview centered on some of the following questions:

- What has been your level of involvement with the Cocheco River Dredge and Disposal Project?
- What concerns do you have with the disposal/management options for the contaminated sediment?
- How did you participate in the decision making process? What were your perceptions of the process?
- When it comes to the management of contaminated sediments in general, what are your biggest concerns, most important values and/or guiding principles for evaluating the situation?

Among the salient concerns voiced during the initial interviews were four recurring themes: financial costs and benefits, environmental quality, human habitat, and ecological habitat. Although stakeholders differed in emphasis, each of these qualities was mentioned during most of the interviews. At this stage, stakeholders helped to characterize the major decision criteria by discussing how they could be measured or manifested in specific attributes. For example, economics was identified as an important decision criterion, but economic considerations may have facets differing in importance to different stakeholders. Project costs (80% of which are slated to paid from Federal sources), maintenance costs, and community economic development (e.g., jobs) all were identified as driving the overall economic assessment. Sample responses are listed in the box below.

3.4 THE WRITTEN SURVEY

The four top-level criteria identified in the interviews were reported to the sediment management technology expert group to define specific measurable attributes that could be incorporated into a written stakeholder survey. In many instances, the attributes identified could be interpreted as relating to more than one of the top-level decision criteria, suggesting that the criteria are not completely independent – a quality that is typical of real world environmental decision problems (Lahdelma et al. 2000). Also, several of the attributes identified proved difficult to translate into measurable quantities. For example, although experts agreed that air & water quality measures could be an important aspect of how stakeholders interpret overall environmental quality, devising comprehensive metrics for air and water that could be accurately and economically estimated and easily communicated to stakeholders proved to be challenging.

The survey was sent to 15 key stakeholders and served as a tool to qualitatively and quantitatively measure their values about the four criteria in general.

The case study served as the reason for conducting the survey, which the respondents knew, but they were also encouraged to answer the questions based on their values in general. Stakeholders were asked to assign percentage weights for each of the four major decision criteria, and to rank all attributes in order of importance in several different groupings of four to eight. Attributes were compared pair-wise and ranked to determine the dominance of some over others using PROMETHEE as embedded in *Decision Lab 2000* software (Visual Decision Inc. 2000, Brans & Mareschal 1994, Brans & Vincke 1985). When one attribute was consistently preferred to another attribute in all groupings, the preferred attribute was awarded a 'win point.' Intransitivities (in which one attribute may be preferred over another on two questions but the order reversed in a third) were handled by awarding a partial win point (such as two-thirds for the first alternative and one-third for the second in the parenthetical example). A final ranking from 1 (most important) to 11 (least important) was established on the basis of the win points. In each case, respondent profiles emerged from the attribute rankings that were consistent with the directly elicited percentage weightings. In the end, the direct percentage weightings were sufficient to facilitate the MCDA.

Sample Interview Responses to the Questions Detailed Above & Concerns Extrapolated from Public Comment

"I would accept increased costs for a better solution, especially because the whole country is bearing the cost of the project..." *This respondent was also concerned with the materials that had previously been buried in the proposed disposal site and said* "there is awful stuff buried there."

"Why would Dover want this recreational field (proposed after disposal site is covered) if there is a possibility that children might be exposed to highly carcinogenic compounds?"

"How will hotspots in the river be dealt with? Will more contaminated sediments be disposed of differently?"

"Will there be a public sign off that the cell has been constructed to approved specifications?"

"The proposed disposal site seems like the best option (even though we live 800 feet from the site), however, disposal must be done in a proper manner."

"The proposed disposal site raises serious concerns about the potential for future impacts to the aquatic environment."

"...because the dredge spoil de-watering area cover will eliminate direct contact with both the proposed and existing dredge spoil sediments that are present at the site, the city feels this outcome is more protective of public health and the environment than the existing conditions."

"Has there been groundwater testing near this site? Should children be allowed to play in areas where waste is exposed and what should the city do to make the area safer?"

"I would have liked to see a full analysis of the project consisting of a larger scope (one that would be required by an EIS) instead of just an E.A. Cost benefit analysis and other questions that would be answered with a more searching review."

"During the decision process, I felt like there were times when dissenting voices were shot down."

In general, "Environmental regulation is important but advisory needs to be tempered with the realities of life (e.g. individual property rights) and done in the most environmentally sound manner."

The reuse and cap of the disposal site and cell is a "three way win for the Department of Environmental Services and the city of Dover because it will provide: 1. A double cover for the old landfill. 2. A new cap for the old contaminated cell. 3. A new state of the art, monitored containment cell for the newly dredged material."

In response to the survey, stakeholders took pains (e.g., by writing in the comments sections or margins) to point out that these attributes were interrelated rather than independent. For example, a single attribute such as preservation of floodplain and shoreline was perceived as being a meaningful attribute of ecological and human habitat by most stakeholders, but also perceived by some as relating to environmental quality and economics. In particular, several attributes identified with "environmental quality" were also identified as relating to other decision criteria. Figure 2 is representative of how some stakeholders mapped the attributes relative to the four top-level criteria. However, not all stakeholder (or experts) would relate the attributes to top-level criteria is exactly the same way.

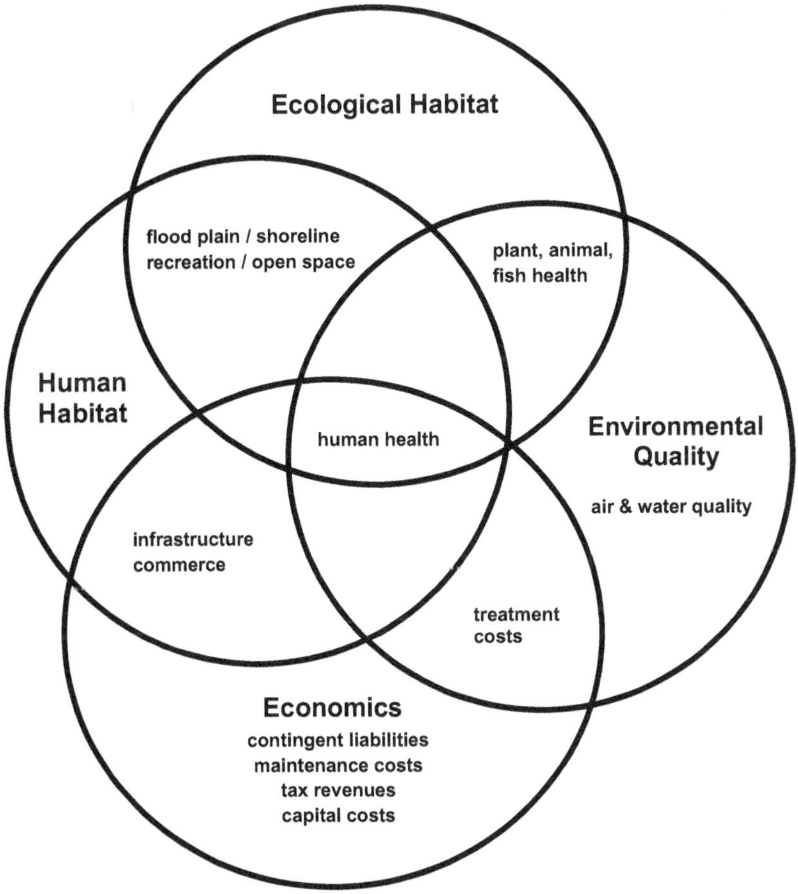

Figure 2: Typical assignment of attributes to top-level decision criteria.

3.5 EXPERTS TABULATE ALTERNATIVES PERFORMANCE TABLE

Graduate students and faculty at the University of New Hampshire's Center for Contaminated Sediments Research (CCSR) were interviewed about their research on innovative beneficial reuse alternatives. This expert group chose wetlands restoration, cement manufacture, and flowable concrete fill as potentially viable technologies for the Cocheco site. Interviews served as a forum to elicit assessments of these alternatives, as well as the capped cell alternative chosen by local officials, to create a performance table that compares the attributes of each alternative to the others based on the four criteria identified by stakeholders in initial interviews. The results are summarized in Table 2.

Table 2: Expert Performance Assessment of Alternatives

Alternative	Cost ($/cy)	Environmental Quality	Ecological Habitat (acres)	Human Habitat (acres)
Cement Manufacture	$30 *+3.0*	High *+2.0*	0 *-1.0*	0 *-1.0*
Flowable Fill	$55 *+1.0, -2.0*	Medium *-2.0*	0 *-1.0*	0 *-1.0*
Wetlands Restoration	$75 *-1.0*	High *+2.0*	+10 *+3.0*	0 *-1.0*
Upland Disposal Cell	$40 *+2.0, -1.0*	Medium *-2.0*	0 *-1.0*	+4 *+3.0*

Notes: Expert assessment determined the performance of each of the four salient criteria that stakeholders' identified as important. The actual alternative planned for use in the Cocheco River Project is the Upland Disposal Cell. Dominance rankings are given in Italics according to the number of clearly inferior (positive) or superior (negative) alternatives.

Experts were challenged to develop metrics that could faithfully represent the diversity of stakeholder views regarding each criterion, could be assessed reliably (in the sense that different expert assessments would result in similar results), and convey meaning. For example, improvements in human habitat could be measured in terms of acres, property values, or specific functional units (such as the number of building lots or youth soccer fields) just as ecological habitat might be measured in terms of acres, or in terms of increased wildlife populations. Ideally, both experts and stakeholders would participate in devising the proper metrics to assess performance. It is even possible for analysis to continue in the event different groups use different criteria, or judge the performance of the alternatives differently against the same criteria. However, it became clear during the process that simple measures would be easier to assess, and that in most cases the differences between the technologies were so striking that the patterns of dominance would not be changed by refined metrics. In this particular case, the performance assessments must be interpreted as hypothetical--both because the technologies themselves are not fully developed and because the details of the case study are not sufficiently developed to allow for detailed assessments

In outranking MCDA methods, alternatives may be judged against one another by comparing their performance on each criterion. (In contrast, utility MCDA approaches assess each alternative singularly, without regard for the performance of other alternatives). For example, experts expected cement manufacture to be the least expensive dredged material management option—consequently outranking all three other alternatives with respect to cost. Moreover, cement manufacture is tied with wetlands restoration for the highest environmental quality assessment, with both outranking the other alternatives. Wetlands restoration dominates all the others in the creation of ecological habitat, whereas the upland capped cell dominates in human habitat-chiefly due to the plan to return use of the recreation areas following capping of the cell. It is clear from Table 2 that flowable concrete fill is inferior to one or more alternatives in all respects for this site. Based solely on this table, there would be no reason to choose this option in the case study. However, another situation might be better suited for this option.

To guard against bias among the stakeholders (in the absence of full vetting of the details of each alternative) the title of each alternative was presented to stakeholders as Option #1, Option #2, etc. and the respondents were asked to evaluate each alternative upon the merits presented in the table. Also, the performance *rankings* (+3.0, -1.0, etc.) were not included in the performance table made available to stakeholders.

3.6 VERIFICATION INTERVIEWS

Interpreted survey results were verified in a personal one-on-one interview with each stakeholder who completed the survey and was willing to participate in the interview (12 total). During this interview, stakeholders were presented with the 'win points' attribute rankings based upon analysis of the written surveys and were asked to confirm and/or revise their interpreted values as well as elaborate on their concerns pertaining to the disposal of contaminated dredged sediment in general and in particular with the Cocheco Case. They were also asked to comment on the survey method as a tool to increase public participation in the decision process. The following are sample questions asked of stakeholders during the verification process:

- How does our interpretation of your values match with what they actually are?
- What are your thoughts on the design of the survey? What were its strengths and weaknesses?
- Do you have suggestions for improvement of the survey?
- Is there anything you would like to tell us that was not captured in your survey?
- Can you please elaborate on the written comments you made on the survey?

Follow up questions were asked based on the stakeholders' responses to the above questions. Many of the stakeholders were critical of the actual survey tool, stating that it was hard to differentiate among the interrelated criteria and that they were not used to expressing their values about this type of decision. However, the majority agreed that researchers had faithfully captured their values in general and were willing to elaborate on them during the interviews. All participants were agreeable to sharing their views and appreciated being incorporated into the research process. Additionally, the interviews served as an opportunity to begin eliciting stakeholder input on the beneficial reuse technologies. At the end of each interview, stakeholders were presented with the alternative performance table, asked to make blind rankings of the four technological options, and send it back to the Center at their convenience.

3.7 CHARACTERIZING STAKEHOLDER PREFERENCES USING MCDA.

Few MCDA approaches have been adapted to group decision-making problems associated with the environment. For example, mathematically complex methods such as multi-attribute utility theory (MAUT) are so demanding of stakeholders, who must define inter- and intra-criteria weighting functions, that time and effort required to elicit extensive information from a dozen (or more) decision makers can be cost or time prohibitive. Moreover, a utility-maximizing approach essentially reduces the decision process to a mono-criterion objective function. These approaches are least appropriate for problems that involve criteria or alternatives that are incommensurable (Seager 2004). For assessment of innovative technologies in particular, use of simplified MCDA methods may be required. For these reasons, Salminen et al. (1998) recommend use of multiple decision-analysis approaches, recognizing that different recommendations may result from use of different MCDA tools. However, management of dredged contaminated materials is often undertaken by resource-constrained local officials on a modest scale. Particularly when decisions are spaced decades apart in time, as in the Cocheco case, officials may be unwilling to invest in multiple decision analysis methods. In this case, a simplified outranking method may be sufficient.

In environmental problems, outranking methods have several advantages over other methods:

- Outranking methods are typically only partially compensatory. That is, massive overperformance on one criterion (such as cost) may not make up for underperformance on others (such as environmental protection).
- Stakeholder value elicitation can sometimes be simplified to direct elicitation of intercriterion weights. More sophisticated analyses can incorporate psuedo-criteria, such as indifference thresholds and partial preferences, for handling uncertainty in performance assessment or various degrees of stakeholder sensitivity (Lahdelma & Salminen 2002). However, initial analyses can proceed with basic information and results are usually easy to communicate to stakeholders (Geldermann & Zhang 2001).
- Outranking methods are tolerant of partially-quantitative information, such as ordinal criteria. This can be a significant advantage in environmental problems where fully-quantitative performance assessments are difficult or impossible to devise.
- Alternatives need not be fully ordered. In fact, outranking methods may identify alternatives that are incomparable due to qualitative differences.
- Outranking methods are capable of representing multiple views, or exploring acceptability weight spaces. That is, in the absence of intercriteria weights or

reliable performance information, outranking methods are able to calculate a range of weights or performance specifications that would result in any one alternative dominating others (Lahdelma et al. 1998, Lahdelma & Salminen 2001).

On the other hand, outranking approaches suffer from at least one significant disadvantage. Because the performance of each alternative is assessed relative to all others, it is possible for ranking of highly competitive alternatives to change depending upon which lower ranked alternatives are included in the analysis. That is, discarding what seems like an uninteresting or non-competitive alternative that performs poorly on many criteria (but performs well in one or two) may change the ordering of alternatives near the top, which is characteristic of the somewhat *ad hoc* approach of outranking from the perspective of rational choice theory (i.e., utility theory and its applications in economics, statistics, decision sciences, and political science).

There are several standardized outranking approaches available. PROMETHEE, which stands for Preference Ranking Organisation METHod of Enrichment Evaluation, is one of the simplest, easiest to implement (partly because commercial software is available) and communicate to non-expert audiences. Dominated alternatives (i.e. inferior in every respect) can be identified easily and tradeoffs identified among the remaining alternatives highlighted. Individual stakeholder orderings are established by weighting decision criteria as a percentage of the overall decision in a manner consistent with expressed stakeholder values. Dissimilar individual stakeholder preferences are never summed or averaged. Each stakeholder may have a different ordering of the preferred alternatives. Consequently, PROMETHEE is especially useful for calling attention to potential conflicts or alliances between different stakeholder groups.

In PROMETHEE, rankings are based upon calculation of positive and negative 'flows,' which are measures of the weighted average ranking of each alternative according the performance table. For example, in an equal-weighting (or balanced scenario), the positive flow for cement manufacture is calculated as the sum of positive rankings +3.0 (from economics), +2.0 (from environmental quality), zero, and zero (from both human and ecological habitat), divided by the total number of spaces in the matrix made up of competing alternatives (in rows) and criteria (columns), which is 12. The result is 5 divided by 12, or .42. Negative flows are computed on the basis of negative rankings. Lastly, overall comparison of positive flows, negative flows, or the sum of these may determine alternative orderings. Often, the alternative orderings provided by the positive and negative flows are identical. When they are not, PROMETHEE may have identified alternatives that are incomparable. In this case, one alternative may exist that has both outstanding strengths and serious shortcomings. Selecting this alternative may reflect a strongly held preference for the criteria assessed as strengths -- a position that may generate controversy.

An example ordering is shown in Figure 3, with positive flows reported in a small box above negative flows. Of the seven stakeholders that participated in the ordering of preferred alternatives, the decision analysis correctly predicted the elicited ordering of all four alternatives for three of the stakeholders. In the other four cases, the stakeholders' first and second choices matched exactly. These results suggest that the researchers can rely upon the stakeholder value elicitation instruments to communicate a reasonably well quantified expression of values that can be employed to prioritize development of the current alternatives, or screen new alternatives that may be introduced into the decision process later. Moreover, while the decision matrix in this case was fairly simple for stakeholders to analyze heuristically, the consistency between predicted and elicited results suggests that the decision analysis may be a valuable tool to assist decision-makers in evaluating more complex situations in a manner consistent with stakeholder values.

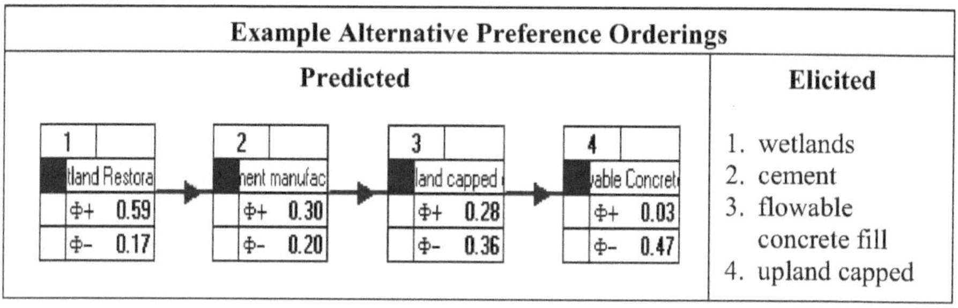

Figure 3. Based on individual preference functions, Decision Lab can predict the order in which any stakeholder would prefer available alternatives using PROMETHEE. Predicted results for all stakeholders were compared to the actual ordering of alternatives elicited from stakeholder inspection of the performance table given to stakeholders during the verification process.

Considerable uncertainties are built into both the value elicitation and performance assessment instruments. Therefore, it is important to investigate the stability of the alternative orderings—i.e., the sensitivity of the ordering to the criteria weightings. When small changes in criteria weights result in a change in preference ordering, decision makers may surmise that the preferences are weakly held, and that opportunities for compromise may exist. Alternatively, when the preference orderings are quite stable, they may be the result of strongly held views. Two or more such groups with different preferences may be in conflict.

To simplify the analysis, like-minded stakeholders were grouped into four general types according to the criteria they held to be most important. Groups most concerned with human health, eco-environment, balance, and costs emerged. The criteria weights used to represent each group are presented in Table 3, and the first two preference alternatives predicted for the group. Note that in some cases the ranking of alternatives would change subject to whether the most positive, least negative, or best combination of flows is used as the basis of the ordering. In these cases, the alternatives may not be directly comparable, and a strict preference may not be completely expressed. Also, the balanced group presents an interesting case in which two alternatives are equally preferred (in both positive and negative flows) as second best.

Table 3: Criteria Weightings of Typical Stakeholder Groups

	Human Habitat	Ecological Habitat	Env Quality	Cost	1st Choice	2nd Choice
Human Health (3)	0.5	0.1	0.25	0.15	Upland Cap +0.6, -0.25	Cement +0.32, -.20
Eco/Env (6)	0.2	0.3	0.4	0.1	Wetland +0.57, -0.17	Cement +0.37, -0.17
Balanced (2)	0.25	0.25	0.25	0.25	Cement +0.42, -0.17	Upland Cap +0.42, -0.33 Wetland +0.42, -0.33
Cost Group (1)	0.25	0.05	0.1	0.6	Cement +0.67, -0.10	Upland Cap +0.65, -0.28

Note: The numbers in parentheses indicate the number of respondents in each group. Positive and negative flows are separated by commas below the name of the preferred alternatives.

Using principal components analysis, *Decision Lab* is capable of graphical analysis of the preferences of each different group relative to one another. Figure 4 depicts each group on separate axes. In general, the axes point towards the preferred alternatives and away from those alternatives that are least preferred. (Small distinctions between closely ranked alternatives may not be corrected interpreted by this rule, as the GAIA plane represents a 2D projection of a 3D volume). The length of the axes is indicative of the conviction with which the each view is held. The GAIA plane is most useful at identifying potential conflicts between different groups. Although all parties can agree that flowable concrete fill is not appropriate for this site, the Cost Group and the Eco/Environmental group are plotted at obtuse angles to one another, and likely to be at odds. In this case, agreement between these groups may be difficult or impossible to achieve. However, a consensus satisfying to the majority of the groups may be reached by a compromise solution. The Balanced and Human Health groups are plotted between the two groups with extreme views, and may be leaning either way.

To investigate the possibility that one of the alternatives may emerge as a consensus choice, the strength of conviction of each group must be investigated. One approach is to estimate the minimum change in expressed criteria weighting required to effect a change in the preference ordering. This approach obviates the need to reliably and precisely establish exact criteria weights. Instead, the rankings may be interpreted more as one of many likely (or unlikely) outcomes. For example, a slight overweighting of any one criterion in the Balanced group would break the tie for second place between wetlands restoration and upland capped cell. The likelihood that, upon further reflection, the Balanced group might reconsider their views must be considered. Figure 5 presents the weightings considered representative of each group as a column chart. The stability intervals over which the first two preference orderings are stable are represented as error bars.

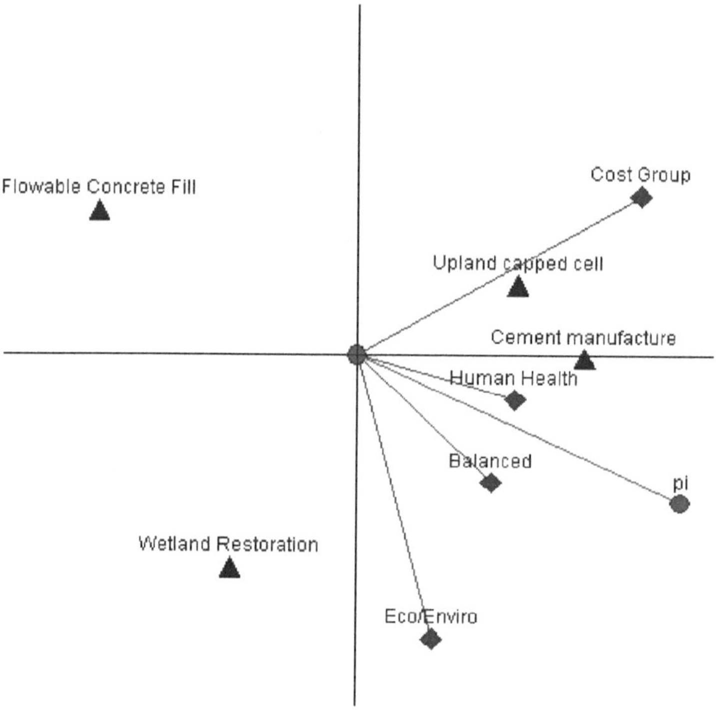

Figure 4: GAIA Plane analysis graphically depicts the relation between different stakeholder groups (diamonds) and the alternatives they are expected to prefer (triangles). In general, the groups that have the greatest potential for disagreement are represented by axes that are pointing away from one another. The "pi" axis is an average of all groups, representing the consensus if all groups are counted equally.

The stability intervals show the strength of conviction in the Cost Group and the Eco/Env. Each may be considered highly unlikely to revise their views so drastically that their primary concern (cost or ecological habitat and environmental quality, respectively) actually becomes a secondary concern—although it may be possible. Similarly, the Human Health group seems fairly well placed within the middle range of the stability interval on each criterion. However, the preference ordering predicted for Human Health may be ambivalent about whether upland capped cell or cement manufacture is preferred. The latter is the first choice of the Cost Group, suggesting that Human Health may be persuaded to agree with Cost Group on the advantages of cement manufacture. In fact, cement manufacture is the only alternative that exists as first or second choice among all groups, perhaps suggesting a compromise that could be accepted by all groups. This is because, despite the drastically different priorities of the Cost Group and Eco/Environmental, according to the expert assessments, cement manufacture partially satisfies each.

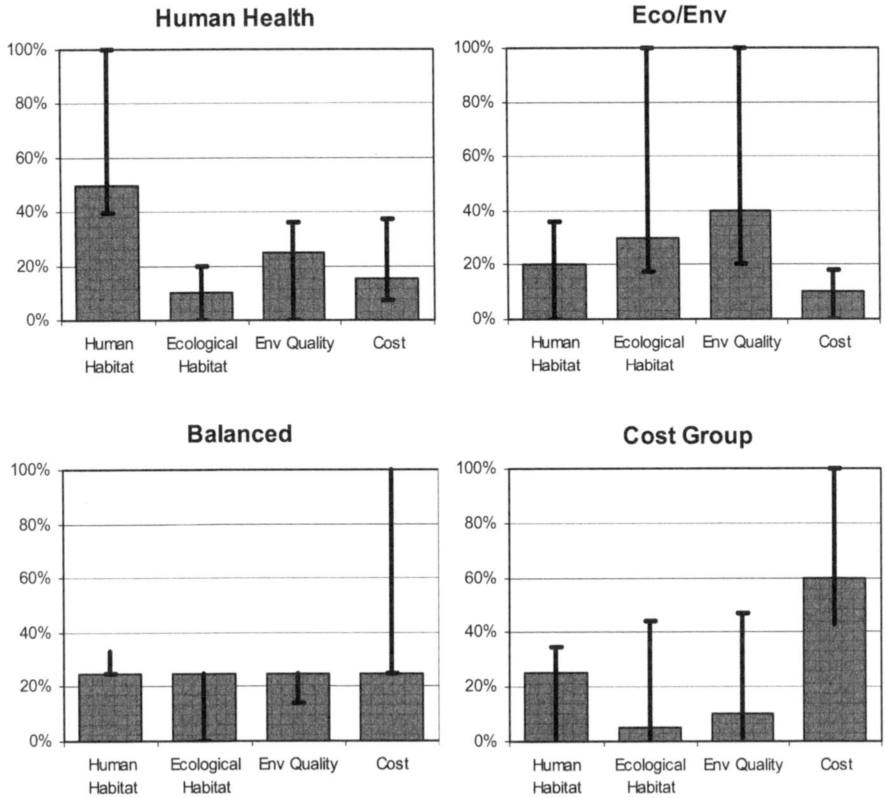

Figure 5: Stability intervals (represented as error bars) indicate the range of criteria weights over which the first two predicted preference orderings are unchanged. Upper bounds are indicative of the extent to which a criterion can be overweighted (at the equal expense of other criteria) without changing the top two predicted preferences. Lower bounds represent the potential extent of underweighting. Note that any change in the Balanced group would result in a reordering, due to the exact tie for second place.

4. Case Study Conclusions

The principal purpose of the MCDA approach employed is not necessarily to find the 'best' decision, but to improve the understanding of different stakeholder values and provide a framework for experts to assess the potential of innovative technologies with regard to those values. The approach of eliciting these values in parallel to development and assessment of the new alternatives may allow for smoother introduction of new technological alternatives (such as beneficial reuse of contaminated sediments) at a more fully developed point in the decision process. So long as expert assessments of the new technologies are consistent with the criteria and metrics established in conjunction with stakeholders, the outranking methods presented may provide an effective tool for assessment of which stakeholder groups may be most likely to support the new alternative, or where potential compromises (or opposition) may be discovered. In progressing this research, the following general observations may be made:

1. The stakeholders involved were eager to have their values heard and incorporated into the management decision process, but critical of written survey methods (although they did confirm the effectiveness of the survey at conveying a simplified, basic message.)
2. The research experts recognized the importance of stakeholder values to management of environmental problems, but were especially challenged by the process of devising measurable, quantitative metrics that would faithfully reflect the decision criteria expressed.
3. The systematic outranking analysis is more effective at identifying dominated alternatives (such as flowable fill in this case), discovering the sensitivity of second-best alternatives to preference weightings, and in general sorting out complex trade-offs than are stakeholder or expert heuristic processes.

The stakeholder value elicitation/public participation and decision analysis process studied may have potential for other environmental managers as a guideline on how to cost effectively incorporate the public and affected parties into the decision process in a meaningful way.

5. Acknowledgements

This work was funded in part by the Cooperative Institute for Coastal and Estuarine Environmental Technologies (CICEET), a partnership between the National Oceanic and Atmospheric Administration (NOAA) and the University of New Hampshire.

6. References

1. USEPA. 2000. *Toward Integrated Environmental Decision-making.* USEPA Science Advisory Board: Washington DC. EPA-SAB-EC-00-011.
2. Linkov I, Varghese A, Jamil S, Seager TP, Kiker G, Bridges T. 2004. Multi-criteria decision analysis: Framework for applications in remedial planning for contaminated sites. In *Comparative Risk Assessment and Environmental Decision Making.* Linkov I, Ramadan A, eds. Kluwer Academic Publishers: Boston MA.
3. Seager TP, Lambert JH, Gardner KH. 2005. Fostering innovation in contaminated sediments management through multi-criteria technology assessment and public participation. *Risk Analysis.* Under review.
4. Kiker GA, Bridges TS, Varghese A, Seager TP, Linkov I. 2004. Application of multi-criteria decision analysis to environmental decision-making. *Integrated Environmental Assessment and Management.* 1(2):95-108.
5. Stahl CH, Cimorelli AJ, Chow AH. 2002. A new approach to environmental decision analysis: Multi-criteria integrated resource assessment (MIRA). *Bulletin of Science, Technology and Society.* 22(6):443-459.

6. Stahl CH. 2003. *Multi-Criteria Integrated Resource Assessment (MIRA): A New Decision Analytic Approach To Inform Environmental Policy Analysis. Vol 1.* Dissertation submitted to Faculty of the University of Delaware in partial fulfillment of the requirements for the degree of Doctor of Philosophy in Urban Affairs and Public Policy.
7. Nelson R. 1987. The economics profession and the making of public policy. *J Economic Literature.* 25(1):49-91.
8. Beierle T, Cayford J. 2002. *Democracy in Practice: Public Participation in Environmental Decisions.* Resources for the Future: Washington, DC
9. Sagoff M. 1988. *The Economy of the Earth: Philosophy, Law, and the Environment (Cambridge Studies in Philosophy and Public Policy).* Cambridge U Press: Cambridge UK.
10. Diduck A, Sinclair J. 2002. Public involvement in environmental assessment: The case of the nonparticipant. *Environmental Management.* 29(4):578-588.
11. Wood C. 1995. *Environmental Impact Assessment A Comparative Review.* Longman Scientific & Technical: Essex England.
12. NRC 1996. *Understanding Risk: Informing Decisions in a Democratic Society.* Stern P, Fineberg H, eds. National Academy Press: Washington DC.
13. Borsuk M, Clemen R, Maguire L, Reckhow K. 2001. Stakeholder values and scientific modeling in the Neuse river watershed. *Group Decision and Negotiation.* 10(4):355-373.
14. McDaniels T, Roessler C. 1998. Multiattribute elicitation of wilderness preservation benefits: A constructive approach. *Ecological Economics.* 27(3):299-312.
15. Gregory R. 2000. Valuing environmental policy options: A case study comparison of multiattribute and contingent valuation survey methods. *Land Economics.* 76(2):151-173.
16. Ananda J, Herath G. 2003. Incorporating stakeholder values into regional forest planning: a value function approach. *Ecological Economics.* 45(1):75-90.
17. Gregory R, Wellman K. 2001. Bringing stakeholder values into environmental policy choices: A community-based estuary case study. *Ecological Economics.* 39:37-52.
18. Gregory R, Keeney R. 1994. Creating policy alternatives using stakeholder values. *Management Science.* 40(8):1035-1048.
19. Wilson M, Howarth R. 2002. Discourse-based valuation of ecosystem services: Establishing fair outcomes through group deliberation. *Ecological Economics.* 41:431-443.
20. Belton V, Steward T. 2002. *Multiple Criteria Decision Analysis: An Integrated Approach.* Kluwer Academic Publishers: Dordrecht NL.
21. Rogers SH, Seager TP, Gardner KH. 2004. Combining expert judgment and stakeholder values with PROMETHEE: A case study in contaminated sediments management. In *Comparative Risk Assessment and Environmental Decision Making.* Linkov I, Ramadan A, eds. Kluwer Academic Publishers: Boston MA.
22. NHDES. 2001. Wetland Permit Application. File #2001-932
23. Dunn K. 2000. Interviewing. In *Qualitative Research Methods in Human Geography.* Hay I, ed. Oxford University Press: Oxford UK.
24. Lahdelma R, Salminen P, Hokkanen J. 2000. Using multicriteria methods in environmental planning and management. *Environmental Management.* 26(6):595-605.
25. Visual Decision, Inc. 2000. See documentation for *Decision Lab 2000* on-line http://www.visualdecision.com/
26. Brans J, Mareschal B. 1994. How to decide with PROMETHEE. http://www.visualdecision.com/Pdf/How%20to%20use%20PROMETHEE.pdf
27. Brans JP, Vincke PH. 1985. A preference ranking organisation method: The PROMETHEE method for multiple criteria decision- making. *Mgt. Sci.* 31(6):647-656.
28. Seager TP. 2004. Understanding industrial ecology and the multiple dimensions of sustainability. In *Strategic Environmental Management* by O'Brien and Gere Engineers. John Wiley & Sons: New York NY. ISBN: 0-471-09221-5.
29. Salminen P, Hokkanen J, Lahdelma R. 1998. Comparing multicriteria methods in the context of environmental problems. *European J Operational Res.* 104(3):485-496.
30. Lahdelma R, Salminen P. 2002. Pseudo-criteria vs. linear utility function in stochastic multi-criteria acceptability analysis. *European J Operational Res.* 141(2):454-469.
31. Geldermann J, Zhang K. 2001. Software review: 'Decision Lab 2000.' *J Multi-criteria Decision Analysis.* 10(6):17-323.

32. Lahdelma R, Hokkanen J, Salminen P. 1998. SMAA – stochastic multiobjective acceptability analysis. *European J. Operational Research.* 106(1):137-143.
33. Lahdelma R, Salminen P. 2001. SMAA-2: Stochastic multicriteria acceptability analysis for group decision making. *Operations Research.* 49(3):444-454.

ENVIRONMENTAL RISK ASSESSMENT AND MANAGEMENT: PROMOTING SECURITY IN THE MIDDLE EAST AND THE MEDITERRANEAN REGION

Report of the Working Group on Environmental Risk Assessment and Management [1]

J. GANOULIS
UNESCO Chair for Water,
Laboratory of Hydraulics, Department of Civil Engineering,
Aristotle University of Thessaloniki, 54124 Thessaloniki, GREECE

L. SIMPSON
National Science Foundation Fellow
Master's degree candidate, Technology and Policy Program
Massachusetts Institute of Technology, USA

1. Introduction

Risk Assessment and Risk Management are two parts of a process known as Risk Analysis (RA) [1], [2]. This is a complex scientific domain involving different disciplines (e.g. physics, chemistry, toxicology, engineering, law, economics, sociology and political sciences) and inter-disciplines (e.g. ecology and environmental sciences) and addresses difficult technical, environmental, economic and social problems, where hazards may threaten populations or ecological systems. Environmental Risk Analysis (ERA) deals mainly with the evaluation of uncertainties in order to ensure reliability in a broad range of environmental issues, including utilization of natural resources (both in terms of quantity and quality), ecological preservation and public health considerations. In the USA and more recently in Europe, Japan and other developed countries, ERA has rapidly become not only a scientific framework for analysing problems of environmental protection and remediation but also a tool for setting standards and formulating guidelines in modern environmental policies [3], [4].

The main element in ERA is the definition and characterization of the notion of risk. Risk has different connotations and interpretations depending on the historical developments of the specific corresponding scientific disciplines. Engineering risk [5], [6], environmental, ecological or health risks [7] [8] may have different verbal or mathematical expressions. It is generally accepted that risk is a measure or a function of two variables: (1) the frequency or the probability of failure and (2) the consequences of failure. Most of the time, events of a very rare degree of appearance (small

[1] *Members included*: Ganoulis, J. (chair, Greece), Haruvy, N. (co-chair, Israel), Goucharova, N. (Ukraine), Zaidi, M. (USA), Lukashevich, A. (Belarus), Grebenkov, S. (Belarus), Green, M. (Israel), Tsiikerman, A. (Israel), Levner, E. (Israel), Girgin, S. (Turkey), Lipchin, O. (Israel), Simpson, L. (USA).

B. Morel and I. Linkov (eds.),
Environmental Security and Environmental Management: The Role of Risk Assessment, 245–253.

probability) lead to severe damage (like the devastating tsunami of 26th December 2004 in Indonesia). Frequent occurrences of a random event, like heavy precipitation, may generate only limited damage.

ERA is a general framework providing tools and methods for problem formulation, load-resistance or exposure-response characterization and risk quantification [6]. More important for practical applications is the risk management process, where decision-making may be analysed by evaluating incremental benefits against different degrees of risk [9].

The task of this discussion group was two-fold:

(1) Review recent development in ERA and its importance for promoting environmental security, stability and cooperation in the Middle East and the Mediterranean, and
(2) Investigate effective ways for implementing ERA in the Mediterranean region and especially in the Middle East.

2. Recent advances and importance of ERA for stability and peace

Risk-based assessments have been rapidly developed during the last five decades mostly in the fields of engineering, environmental sciences, chemistry, toxicology and human health. Quantitative Risk Assessments (QRA) and Probabilistic Risk Assessments (PRA) have evolved in civil, nuclear and chemical engineering under heavy regulatory and public pressure [10]. More traditionally, risk of extremes (e.g. floods and droughts) [6] and structural safety under extreme loads (e.g. earthquakes and high speed turbulent flows) were investigated using probabilistic approaches in water, civil and aeronautical engineering.

During the past few decades ERA has benefited greatly from the rapid progress in the science of RA, mainly in the areas of environmental, public health and ecological risks. New scientific tools and methodologies have been developed and well defined regulatory frameworks and directives have been adopted by professional institutions and national agencies, such as the American Society of Civil Engineers (ASCE), the USA-Corps of Engineers (USACE), the US-Environmental Protection Agency (US-EPA) and the European Environmental Agency (EEA). Recently the concepts of *Comparative Risk Analysis (CRA)* [11], *Cumulative Risk Assessment* [12], [13] and *Integrated Risk Management* [14] have been developed, and may be applied to environmental and public health issues in order to reduce risks by implementing new legislation regarding drinking water and food safety [3], [4], [15].

Environmental security is a core element for promoting peace and stability between peoples. This was emphasized at the World Summit on Sustainable Development (WSSD), held in Johannesburg, August - September 2002 [16]. At this summit the international community identified the mutually beneficial management of the world's natural resources and environmental preservation as key factors for peaceful coexistence, cooperation between countries and sustainable development.

WSSD recognized that environmental crises related to non-sustainable utilization of natural resources, water shortages and climate variation could affect every aspect of life from ecosystems to human health, food security, human rights and cultural heritage. Such crises could significantly affect various levels of economic development, increase the risk of conflicts between countries at various degrees and increase regional political instability. The WSSD plan of implementation recommends integrated environmental management and effective strategies for water use as important tools to promote environmental security, political stability and peace. ERA may be used as a framework for reducing ecological and human health risks and implementing cost-effective measures for sustainable use of natural resources.

In the Mediterranean region and particularly in the Middle East, the semi-arid climate, ethnic conflicts and unstable socio-economic conditions have put natural environmental resources like water, air and soil under severe natural and anthropogenic pressures. Tal and Linkov [17] reviewed the main environmental issues of a transboundary nature in the Middle East and described how CRA may be used as a general framework in order to reduce possible environmental conflicts between countries in the region.

Management of transboundary environmental resources and especially water involves different preventive and curative actions, use of economic instruments, integrated plans for water conservation and demand management, watershed and ecosystem management and wetland conservation [18]. However, in order to effectively address the complexity of real environmental and ecological problems, the use of technical and technological instruments, like monitoring, modelling, risk management and decision analysis, is not sufficient. Effective cooperation between different institutions, involvement of local stakeholders and public mobilization are also very important components for implementing specific action plans [19].

UNESCO has developed two major international programmes for transboundary natural resources management at a regional level. The first is called UNESCO-ISARM (Internationally Shared Aquifer Resources Management) and deals with technical, institutional, economic and social considerations of groundwater management in transboundary regions. The second is the PCCP programme (from Potential Conflict to Cooperation Potential). PCCP focuses on various methodologies, including training and capacity building, in order to prevent and alleviate potential conflicts in sharing transboundary water resources. UNESCO Chair and Network INWEB (International Network of Water–Environment Centres for the Balkans), which is an active regional network in South Eastern Europe (SEE), participated in both programmes. It has recently completed an inventory of internationally shared water resources in the Balkans [20]. The scope and methods for developing such inventories for transboundary groundwater are given as an example in Appendix 1.

The working group recognized the need to examine how the development of an open and broad based partnership of groups and individuals coming from different disciplines could help to implement the ERA framework at a regional level. The main idea was that by combining the expertise and state-of-the-art knowledge of different scientific communities and disciplines, such as engineering, economics, environmental and social sciences, regional partnerships may contribute to the development of new methods and models in order to more efficiently resolve conflicts and controversial issues in the management of transboundary environmental resources.

3. Developing a regional cooperative ERA network for the Middle East

The main objective of the working group was to set up a network of scientists and institutions, and using ERA as a framework, to address environmental problems in the region. The group discussed a methodology for establishing such a network, which will be called *MEERA-Net (Middle East Environmental Risk Analysis Network).* The following issues were discussed:

3.1 JUSTIFICATION

Traditionally, scientists concerned with risk assessment have worked mainly individually or cooperatively with scientists and institutions within their own country. This is satisfactory for small-scale environmental projects. However, to deal with larger scale projects, it is essential that scientists from various countries within a region work together. Such cooperation may be facilitated and enhanced by a structured regional cooperative network.

Cooperation is particularly difficult in areas such as the Middle East, where countries not enjoying good political relations need to work together to solve environmental problems. The situation is particularly volatile due to the possibility of conflicts, even war between countries, the consequences of which would exacerbate environmental problems. Cooperation is also difficult due to economic and social disparities between countries in the region. Mediation is needed to deal with collecting and sharing data, applying ERA methodologies and alleviating potential conflicts in managing internationally shared natural resources, in particular water resources.

The establishment of MEERA-Net was seen as the best way to provide a sustainable cooperation in order to develop an objective regional database, to effectively implement ERA and to facilitate funding from international organisations.

3.2 SCALE

The MEERA-Net aims to involve scientists from all core countries in the region i.e.: Egypt, Israel, Palestine, Jordan, Syria, Lebanon, Cyprus and Turkey. Main partners should be those of the discussion group willing to join on a voluntary basis. Such a group would be able to tackle regional environmental problems (first level) and exchange experience on ERA. At some point however, a wider cooperation may be necessary, connecting MEERA-Net to other existing regional networks. This would involve a second level of inter-regional cooperation incorporating SEE (Balkans) and Northern Africa. A third level (extra-regional or international) would include networks from the European Union and North America.

3.3 BENEFITS

MEERA-Net aims to improve transboundary cooperation at a regional level. It intends to propose institutional arrangements for effective environmental management and formulate ideas and plans for sustainable use of natural resources. The network aims to further the development of local economies and to promote peace through a harmonisation of environmental legislation, guidelines, and ERA methodologies. With these goals in mind, it should attract international funding and, by addressing existing national and transboundary environmental problems, benefits will result to all member countries and the region as a whole.

3.4 ISSUES

Firstly inventories indicating the existing situation from environmental and institutional points of view, and mainly in transboundary areas, should be compiled (see the example in Appendix 1). Environmental issues in the region should be evaluated by the network and weighted so as to determine priorities and pilot projects. A topic that will clearly receive priority is *sustainable water management* including issues of supply, demand management, sharing international waters, recycling and re-use, waste-water treatment and disposal, desalination, coastal waters, ecology, biodiversity, and climate change. Another important topic is that of *air pollution* (including issues of transportation and industrial activities). All topics should be considered from multiple perspectives: economical, ecological, social, public health, risk perception, and climate change.

3.5 TOOLS

The tools and methods to achieve the network's objectives were then discussed. It was agreed that MEERA-Net should be an open network that promotes a multi-disciplinary approach to all environmental issues, and involves ERA experts from different fields including science, engineering, economics, law and sociology. It should use new information and communication technologies to reach as wide an audience as possible and to facilitate ease and speed of communication. The network's approach should be flexible so that it can adapt to different areas', countries' and regions' varying cultural, social and legal practices. MEERA-Net will not be a research network in itself, but a means to address existing environmental problems.

MEERA-Net will also be concerned with informing the general public about environmental issues, and sensitising them to the ideas of prevention, protection and preservation. A special initiative to promote environmental awareness at all levels of public education, including primary schools, was seen as an important tool that will promote sound environmental practices.

MEERA-Net will promote joint training projects and the sharing of expertise, and aim to enhance the ease of exchange of knowledge, state of the art information, experience and advice.

3.6 PROCEDURE

Suitable experts from a wide range of disciplines should be identified and invited to join the network. Participants of this working group agreed to be founder members of the network. It was suggested that one representative from each country be chosen as the contact person for that country. Dr. Nava Haruvy was chosen as the contact person for Israel. Contact people from the other countries in the region still need to be nominated.

Once the network has been established, it will use the following three-step procedure to tackle any environmental problem selected from the list of priority issues previously drawn up.

The first step is to set-up an *information community*. This means acquiring all relevant information and data. It is important to acquire an exhaustive list of information so that no part of a project will be overlooked and to understand the attitudes of scientists from different countries. The means of acquiring data could be broken down as follows:

1.1 Mapping existing data and information
1.2 Data validation through analysis
1.3 Development of inventories
1.4 Monitoring and establishing common monitoring networks
1.5 Reviewing guidelines and standards in affected countries
1.6 Sharing data and information
1.7 Electronic cooperative networks and use of internet for data communication
1.8 Education, training, and capacity building
1.9 Exchange publications and papers

The second step is to convert the information community to a *knowledge community* by analysing and evaluating the data. To structure the information and determine its importance the following list was developed:

2.1 Problem formulation and risk assessment
2.2 Modelling of the environmental situation
2.3 Analysis of past experiences and suggestions for new pilot-studies
2.4 Develop tool boxes for problem resolution
2.5 Organise workshops, exchanges of experiences, and seminars
2.6 Learn from local traditional techniques and craftspeople
2.7 Comparative analysis and "best practices" guidelines

The third step is to convert the knowledge community to a *community of practice*. This step aims to develop useful recommendations and create a plan to implement them. In order to reach a final consensus, mediation is needed between many different groups of people, including:

3.1 Financing and sponsoring organisations
3.2 Decision makers at all levels
3.3 Private and public administrative organisations
3.4 End users
3.5 The general public

4. Conclusion

In order to effectively solve regional environmental problems, countries need to work together. This is best accomplished by creating regional cooperative networks, such as MEERA-Net, to facilitate communication, data sharing and common and joint actions for the mutual benefit of the whole region.

5. References

1. Duffus, J.H. (2001) Risk Assessment Terminology, *Chemistry International 23(2),* <http://www.iupac.org/publications/ci/2001/march/risk_assessment.html>.
2. U.S. Environmental Protection Agency (USEPA) (1992) *IRIS. Glossary of Risk Assessment Related Terms,* Washington, DC: U.S. IRIS online <http://www.epa.gov/ngispgm3/iris/>.
3. U.S. Environmental Protection Agency (USEPA) (1984) *Risk Assessment and Management: Framework for Decision Making*, EPA 600/9-85-002, Washington, DC: U.S. December 1984.
4. U.S. Environmental Protection Agency (USEPA) (1998) *Guidelines for Ecological Risk Assessment.* Federal Register 63: 26846-26924, 14 May 1998.
5. Duckstein, L. and E.Plate (eds.) (1987) *Engineering Reliability and Risk in Water Resources,* E.M. Nijhoff, Dordrecht, The Netherlands, 565 pp.
6. Ganoulis, J. (1994) *Engineering Risk Analysis of Water Pollution: Probabilities and Fuzzy Sets.* WILEY-VCH, Weinheim, Oxford, NY, 306 pp.
7. Council on Environmental Quality (1989) *Risk Analysis: A Guide to Principles and Methods for Analyzing Health and Environmental Risks.* PB89-137772. National Technical Information Service, Washington, DC.
8. Ganoulis, J., Mpimpas, H., Duckstein, L. and Bogardi, I. (1996) Fuzzy Arithmetic for Ecological Risk Management. In: Y. Haimes, D. Moser and E. Stakhin (Eds.), *Risk Based Decision Making in Water Resources VII,* ASCE, NY, pp. 401-415.
9. Linkov, I., et al. (2004) Multicriteria Decision Analysis: A framework for structuring remedial measures at contaminated sites. In: *Comparative Risk Assessment and Environmental Decision Making,* 15-54, NATO Science Series, Kluwer Academic, The Netherlands.
10. Apostolakis, G. (1990) The concept of probability in safety assessments of technological systems. *Science* **250**: 1359-1364.
11. Linkov, I. and A. Bakr Ramadan (eds.) (2004) *Comparative Risk Assessment and Environmental Decision Making,* NATO Science Series, Kluwer Academic, The Netherlands, 436pp.
12. U. S. EPA (2003) *Framework for Cumulative Risk Assessment, Washington,* DC EPA/630/P-02/001F.
13. U.S. Environmental Protection Agency (USEPA) (1997), *Guidance on Cumulative Risk Assessment. Part 1. Planning and Scoping*, Science Policy Council, Washington, DC.
14. Ganoulis, J. (2004) Integrated Risk Analysis for Sustainable Water Resources Management In: *Comparative Risk Assessment and Environmental Decision Making,* 275-285, NATO Science Series, Kluwer Academic, The Netherlands.
15. U. S. EPA (2002) *Lessons Learned on Planning and Scoping for Environmental Risk Assessments,* Washington, DC.
16. The World Summit on Sustainable Development, Johannesburg, South Africa (2002) On line: <http://www.johannesburgsummit.org/>.
17. Tal A. and I. Linkov (2004) The Role of Comparative Risk Assessment in Addressing Environmental Security in the Middle East, *Risk Analysis*, in press
18. Ganoulis, J., L. Duckstein, P. Literathy and I. Bogardi (eds.) (1996) *Transboundary Water Resources Management: Institutional and Engineering Approaches.* NATO ASI SERIES, Partnership Sub-Series 2. Environment, Vol.7, Springer-Verlag, Heidelberg, Germany, 478 pp.
19. Ganoulis, J., I. L. Murphy and M. Brilly (eds.) (2000) *Transboundary Water Resources in the Balkans: Initiating a Sustainable Co-operative Network,* NATO ASI SERIES, Partnership Sub-Series 2: Environmental Security, Vol. 47, Kluwer Academic, Dodrecht, Boston, London, 254 pp.
20. Ganoulis J. et al. (eds.) (2004) *Inventory of Internationally Shared Aquifers in the Balkans: Characteristics and State of Policies,* UNESCO-IHP, INWEB, UNESCO-ISARM Consultative Workshop, Thessaloniki, 21-23 Oct. 2004 (draft) on line <http://www.inweb.gr>.

6. APPENDIX

6.1 SCOPE, METHODS AND TOOLS FOR DEVELOPING AN INVENTORY OF TRANSBOUNDARY AQUIFERS IN THE BALKANS

by

- UNESCO Chair / INWEB (International Network of Water Environment Centres for the Balkans)
- AUTh (Aristotle University of Thessaloniki) (Greece)
- Hellenic National Commission for UNESCO (Athens)
- IAH / TARM Commission (International Association of Hydrogeologists / Transboundary Aquifer Resource Management)

In cooperation with

- UNESCO-IHP (International Hydrological Programme)
- UNECE (Economic Commission for Europe: Working Group on Monitoring & Assessment) (Switzerland)
- ESCWA (The Economic and Social Commission for Western Asia) (Lebanon)
- OSS (Observatoire du Sahara et du Sahel) (Tunisia)

6.2 OBJECTIVES

In order to help resolve problems of water use, environmental protection and economic development in SEE and the Mediterranean, there is a need to improve cooperation between countries sharing transboundary aquifers.
The objectives of the programme are:

- to review the available data collected during the first phase of the ISARM Inventory of the Transboundary Aquifers of OSS Mediterranean Countries.
- to learn from the experience gained in the ESCWA region and to coordinate with ESCWA on the preparation of a Mediterranean inventory.
- to compile an inventory of the Balkans region following the ISARM framework document guidelines.
- to identify the key issues for sustainable management of transboundary aquifers in SEE and the Mediterranean area.
- To create an appropriate database at the UNESCO-WMO IGRAC Centre and develop a GIS-based inventory of shared aquifers in the Mediterranean region and SEE. This inventory may serve as a basis for formulating specific local projects for cooperative actions.

6.3 BACKGROUND

The need for international cooperation on the sustainable management of transboundary groundwater resources in the Balkans is particularly acute, and there are many examples where potential conflicts in the use of international groundwaters could arise. Prior to 1992 there were only six international river basins in the Balkans, whereas after the collapse of the Yugoslav Federation, the number of internationally shared river basins in the area more than doubled. Today in fact there are 13 international shared river basins as well as 4 transboundary lake basins. Institutions dealing with water problems in the region need support to use modern information and communications technologies for monitoring, modelling and water management studies.

The water resources in the semi-arid northern part of the UN-ESCWA region are dominated by groundwater characterized by significant and large shared aquifers located in the Eastern and Southern sections of the Mediterranean basin. UN-ESCWA gives high priority to sustainable development and joint management of transboundary groundwater resources and has carried out important aquifer studies in the region. UN-ESCWA is a partner in the ISARM initiative and is currently concluding two important ISARM case studies on the regional Basalt Aquifer system in Jordan and Syria and on Paleogenic Carbonate Aquifers, which are both represented within the Mediterranean basin.

The work for this project is based on previous programmes and initiatives such as:

- ISARM, UNESCO, IAH, UNECE, FAO (Food and Agriculture Organisation)
- The European Water Initiative
- The Athens Declaration
- The UNECE Guidelines on Monitoring and Assessment of Transboundary Groundwaters
- The OSS programme on arid zones
- The ESCWA findings on legal frameworks.

6.4 METHODOLOGY

ESCWA, UNECE, UNESCO and OSS will compile a draft document presenting the state of knowledge on shared aquifers in the region based on the ISARM questionnaire. Using this information, experts from OSS and SEE will be invited to identify key issues and problems for the sustainable management of aquifers in the region at a three-day workshop in Thessaloniki, Greece. At the workshop they will work together to initiate the preparation of an Atlas of International Aquifers in the Mediterranean Region, to define support mechanisms to improve cooperation for the better use and protection of shared groundwater resources in the region, and to identify tools to help countries better evaluate their groundwater resources.

Part 3.

Case Studies in Environmental Security and Environmental Management

REUSE OF WASTEWATER IN AGRICULTURE- ECONOMIC ASSESSMENT OF TREATMENT AND SUPPLY ALTERNATIVES AS AFFECTING AQUIFER POLLUTION

N. HARUVY
Netanya Academic College
1 University Rd , Netanya 42100, ISRAEL

Abstract

Treated wastewater in Israel is the unlimited and a reasonable alternative for water supply to agriculture as the high-quality fresh water supply is gradually transferred to urban uses. Still, the domestic and industrial effluents carry pollutants including micro and macro organic and inorganic matter which potentially pose hazards to health, the environment, crops and soils, and may deteriorate aquifer quality. Nevertheless, for irrigation in Israel, using treated wastewater is the best means to facilitate agricultural production under conditions of water scarcity, since it uses a water resource that is available in large quantities and that already requires treatment, in order to prevent environmental damage. Salinity level is higher in effluents than in influents, while regular treatment processes do not get rid of salinity, unless combining relatively expensive desalination processes.

In this paper the focus is on salinity/chlorides as representing groundwater pollution. Hydrological model is used to predict the flow of chlorides through the unsaturated zone of the subsoil and into the groundwater below. We assumed that there is a threshold value for chloride concentration in the water supply for domestic consumption, and considered that when the concentration of chlorides in the supplied water reaches this threshold, desalination of groundwater should be initiated.

When irrigation is with treated wastewater, desalination processes are initiated earlier than under conditions of irrigation without effluent, and this increases the water supply costs. The damage to groundwater by effluent irrigation is assessed in terms of increased costs that arise from processes including water pumping and transporting, wastewater treatment and earlier initiation of desalinization. We compared several scenarios regarding salinity threshold levels and irrigation combinations with and without wastewater. This was applied to a specific case study in the Coastal aquifer of Israel composed of 8 hydraulic cells. The resulting salinity levels decrease with stricter restrictions and increase with wastewater irrigation. Water supply costs increase with stricter restrictions while with wastewater irrigation, desalination costs increase but total costs decrease.

B. Morel and I. Linkov (eds.),
Environmental Security and Environmental Management: The Role of Risk Assessment, 257–262.

1. Background

Annual renewable water resources in Israel are limited and amount to 1,900 MCM (Million Cubic Meter) per annum. Agriculture is the main consumer of fresh water and uses 65% of the total annual freshwater withdrawals. With the future increase of population, the domestic water consumption is increasing rapidly, reducing the amount of fresh water available for agriculture, and at the same time increasing the output of urban wastewater. As the high-quality fresh water supply is gradually transferred to urban uses, treated wastewater is the unlimited and reasonably alternative for water supply to agriculture. Since urban wastes should be treated and discarded anyway, wastewater irrigation serves also as environmental quality agent usually being the cheapest option for wastewater disposal (Haruvy, 1997a; Haruvy, 1997b; Haruvy, 1998; Haruvy, 2000).

Domestic and industrial effluents carry pollutants including micro and macro organic and inorganic matter, which potentially pose hazards to health, the environment, crops and soils, and may deteriorate aquifer quality (Graber et al, 1995; Haruvy et. al, 1997); Nevertheless, for irrigation in Israel, using treated wastewater is the best means to facilitate agricultural production under conditions of water scarcity by using water source that is available in large quantities and that already requires treatment, in order to prevent environmental damage.

While most constituents may be reduced to satisfactory levels through advanced treatment processes (Feigin et al., 1990), salinity level is higher in effluents than in influents. Regular treatment processes do not get rid of salinity, unless combining relatively expensive desalination processes (EPA, 1992; Cecen and Gonenc, 1995). Treated wastewater in Israel is divided according to treatment processes as following: tertiary-30%, secondary- 50% and primary - 20%. Its expected rate of reuse out of total water will increase from 25% in 2000, to 37% in 2010 and to 46% in 2020.

Crop yields are affected by water salinity levels (Maas and Hoffman, 1977) as by other wastewater constituents (Paranichianakis et al, 1999; Reboll et al., 2000), and so is soil (Page and Chang, 1985; Tarchitzky et al, 1999). Environmental impacts of salinity and nutrient constituents in wastewater in hypothetical region were described by Haruvy (2004). In this paper I focus on salinity/chlorides as representing groundwater pollution, and apply it to a region in Coastal aquifer composed of 8 hydrological cells combining wastewater treatment processes and agricultural reuse.

2. Methodology

Irrigation with effluent may accelerate the contamination of groundwater, mainly by chlorides, nitrogen, heavy metals and organic compounds. We used a hydrological model to predict the flow of chlorides through the unsaturated zone of the subsoil and into the groundwater below. Pumping capability is influenced by amount of leaching to groundwater. We assumed that there is a threshold value for chloride concentration in the water supply for domestic consumption, and considered that when the concentration of chlorides in the supplied water reaches this threshold, desalination of groundwater should be initiated. Part of the treated water source is then desalinated to a given salinity level, and diluted with other water sources until reaching the permitted level. When irrigation is with treated wastewater, it is necessary to initiate desalination earlier than under conditions of irrigation without effluent, and this increases the water supply costs. The damage to groundwater by effluent irrigation is computed in terms of increased costs that arise from water pumping and transporting, wastewater treatment and earlier initiation of desalinization.

We compared several scenarios regarding threshold levels and water sources, and assessed the water supply costs. Water supply alternatives include: local groundwater, national carrier's water, and wastewater. The hydrological model predicts the flow of chlorides and estimates the resulting groundwater pollution and water treatment and supply costs. Damage of groundwater pollution is assessed by increase of costs. Some results were presented by Haruvy (2000) and Haruvy et al. (2000), and the scenarios were presented in detail by Haruvy et al. (2002).

This was applied to a specific case study in Israel composed of 8 hydraulic cells in the Coastal aquifer: Region A including 4 hydraulic cells and aquifer salinity levels higher than 230 mgl Cl including agricultural area of 8,000 ha irrigated mainly by wastewater; Region B including 4 hydraulic cells and lower aquifer salinity levels of 75-164 mgl Cl. It is populated by a big city with 150,000 citizens and including wastewater treatment plant, which effluents are conveyed to Region A; The total area includes urban area of 7,988 ha and agricultural area of 16,530 ha (Table 1).

Table 1: Land use in ha

	Region A	Region B	Total
Agriculture			
Built	1,020	1,140	2,160
Citrus	1,215	2,897	4,112
Other orchards	700	1,050	1,750
Field crops	5,308	3,200	8,508
Total agriculture	**8,243**	**8,287**	**16,530**
Urban			
Built	25	2,853	2,878
Open	82	4,655	4,737
Industry	4	369	373
Total urban	**111**	**7,877**	**7,988**
Grand total	**8,354**	**16,164**	**24,518**

We estimated water demand and respective water supply sources. Agricultural water demand is 59.34 MCM, and urban water demand is 27.29 MCM. Total recharge of aquifer 49.53 MCM and the amount that can be pumped is 34.31 MCM (Table 2)

Table 2: Water use in MCM

	Region A	Region B	Total
Agriculture			
National carrier	0.22	11.96	12.18
Local aquifer	9.41	19.39	28.81
Wastewater	14.82	3.53	18.35
Total Agriculture	**24.46**	**34.88**	**59.34**
Urban	2.57	24.71	27.29
National carrier	0.02	21.76	21.78
Local aquifer	2.55	2.95	5.50
Total urban	**2.57**	**24.71**	**27.29**
Grand total	**27.03**	**59.59**	**86.63**
Initial salinity- mgl cl.	239.82	179.64	192.17

We compared several scenarios where scenarios 1-4 differ by more sever restrictions (lower threshold levels) while scenarios 5-8 differ by irrigating without and with wastewater irrigation (350 mg cl) (Table 3).

Table 3: Comparison of scenarios

Scenario	Threshold level for town Mgl cl	Threshold level for agriculture Mgl cl	Wastewater salinity level	Including wastewater	Priority of pumped water
1	250	---	250	+	Town
2	250	250	250	+	Town
3	150	150	250	+	Town
4	50	50	250	+	Town
5	250	---	---	-	Town
6	250	---	---	-	Agriculture
7	250	---	350	+	Agriculture
8	250	---	350	+	town

Water allocation under scenarios 1-4 is described above in Table 2, Water allocation under scenarios 5-8 is described in Table 4.

Table 4: Water allocation total area (MCM)

Scenarios	**5**	**6**	**7**	**8**
Agriculture				
National carrier	46.41	27.02	12.18	28.06
Local aquifer	12.93	32.32	28.81	12.93
Wastewater	0.00	0.00	18.35	18.35
Total agriculture	**59.34**	**59.34**	**59.34**	**59.34**
Urban				
National carrier	5.90	25.30	21.79	5.90
Local aquifer	21.39	1.99	5.50	21.39
Total urban	**27.29**	**27.29**	**27.29**	**27.29**
Grand total	86.63	86.63	86.63	86.63

The derived salinity levels as assessed by the hydrological model increase with time (except cells near sea without pumping). Salinity levels foe Region A are presented in Table 5, and salinity levels for Region B- in Table 6.

Table 5: Derived salinity levels for the various scenarios- Region A (mgl cl)

Years	1	2	3	4	5	6	7	8
20	174	174	171	166	175	174	174	176
40	186	185	168	148	200	185	195	209
60	208	201	171	129	231	208	224	246
80	230	217	173	117	260	232	253	281
100	250	228	175	109	285	256	279	311

Table 6: Derived salinity levels for the various scenarios- Region B (mgl cl)

Years	1	2	3	4	5	6	7	8
20	337	337	329	318	337	337	337	344
40	415	409	347	284	415	416	460	459
60	514	468	342	226	515	533	605	607
80	607	500	349	196	610	659	740	742
100	689	522	351	182	693	783	854	857

One can see that salinity levels decrease with stricter restrictions (scenarios 1-4) and increase with wastewater irrigation (scenarios 5-8).

We computed relevant costs affected by water sources and needed treatments to arrive the determined salinity threshold levels. One can see that costs increase with the stricter restrictions (scenarios 1-4). Also, desalination costs increase but total costs decrease with wastewater irrigation (scenarios 5-8) where scenario 5 is compared to 6 and scenario 8 is compare to 7 (Table 7). Although treatment costs that include desalinization are higher when irrigation is with treated wastewater, the total water supply costs are lower, because of the lower costs of wastewater to farmers. Nevertheless, the salinity levels in the groundwater are relatively high, and this should also be included in the account, in the form of derived environmental damage.

Table 7: Computed water supply and treatment costs for the various scenarios (million NIS)

Costs	2	3	4	5	6	7	8
Region A							
Total costs	2,226.7	2581.2	2,932.6	2,572.3	27,56.9	2,028.3	2,028.5
Desalination costs	625.2	1,036.2	1,513.5	284.3	0	300.7	301.1
Region B							
Total costs	5,562.4	5,596.2	8,021.8	6,612.8	54,26.9	5,286.0	5,674.6
Desalination costs	785.9	872.2	6,758.6	764.6	0	0	480.3
Grand Total	**7,789.1**	**8,177.4**	**10,954.5**	**9,185.1**	**8,183.8**	**7,314.3**	**7,703.1**

3. Summary and conclusion

Wastewater irrigation is an important water source in Israel, but it may enhance aquifer pollution unless treated appropriately. We estimated environmental effects by assessing water supply and needed treatment costs by focusing on salinity. Economic assessment methods are important for environmental security and water supply alternatives, and respective treatment processes should be compared to decrease costs and prevent aquifer pollution.

We assess the water supply costs for several scenarios varying by resulting salinity levels and/or water supply sources. We see that costs increase with stricter restrictions. Also, desalination costs increase but total costs decrease with wastewater irrigation. Although treatment costs that include desalination are higher when irrigation is with treated wastewater, the total water supply costs are lower, because of the lower costs of wastewater to farmers. The higher salinity levels in the groundwater should also be included in the account in the form of derived environmental damage

4. References

1. Cecen, F., Gonenc, I.E., 1995, Criteria for nitrification and denitrification of high-strength wastes in two upflow submerged filters. Water Environ. Res. 67, 132-142.
2. EPA, 1992, Manual, Guidelines for Water Reuse, US, Environmental Protection Agency, Washington, DC
3. Feigin, A., Ravina, I., Shalhevet, J., 1990, Irrigation with Treated Sewage Effluent, Ecological Series, Springer Verlag, New York
4. Graber, E.R., Gerstl, Z., Fischer, E., Mingelgrin, U., 1995, Enhanced transport of atrazine under irrigation with effluent. Soil Sci. Soc. Am. J. 59, 1513-1519.
5. Haruvy, N., 1997a, Agricultural reuse of wastewater: nation-wide cost-benefit analysis, Agric., Ecosyst. Environ., 66: 113-119.
6. Haruvy, N., 1997b, Wastewater irrigation – economic considerations as affecting decision-making. J. Financ. Manage. Anal., 10: 69-79.
7. Haruvy, N., Hadas, A. and Hadas, A., 1997, Cost assessment of various means of averting environmental damage and groundwater contamination from nitrate seepage. Agric. Water Manage., 32: 307-320.
8. Haruvy, N., 1998. Wastewater reuse – regional and economic considerations. Resour. Conserv. Recy., 23: 57-66.
9. Haruvy, N., 2000. Agro-economic aspects of irrigation with treated effluents. Conference of the World Bank on "Irrigation with Marginal Water", Washington, USA.
10. Haruvy, N., Hadas, A., Ravina, I. and Shalhevet, S., 2000, Cost assessment of averting groundwater pollution, Water Sci. Technol., 42: 135-140.
11. Haruvy, N. (2004), Irrigation with treated wastewater in Israel – assessment of environmental aspects: 375-384 In: Linkov, I. and Ramadan, A. (eds.) "Comparative Risk Assessment and Environmental Decision Making" Kluwer
12. Maas, E.V., Hoffman, G.J., 1977,. Crop salt tolerance - current assessment,. J. Irrig. Drain Div., Am. Soc. Civ. Eng. 103, 115-134.
13. Page, A.L., Chang, A.C., 1985, Fate of wastewater constituents in soil and groundwater trace elements. In: Pettygrove, G.S., Asano, T. (Eds), Irrigation with reclaimed municipal wastewater – a guidance manual. Lewis, Chelsea, 1985, pp. 13.1-13.16.
14. Paranichianakis, N.V., Angelakis, A.N., Chartzoulakis, K.S., Jones, H.G.., 1999, Growth and mineral uptake of Sultanina grapevines irrigated with reclaimed wastewater and the potential for using such water in Crete, Greece. In: Ferreira, M.I. (Ed), Third International Symposium on Irrigation of Horticultural Crops, Estoril (Lisbon), Portugal, 28 June-2 July, 1999. Acta Hort. 2000, 537, 725-730.
15. Reboll, V., Cerezo, M., Roig, A., Flors, V., Lapena, L., Garcia-Agustin, P., 2000, Influence of wastewater vs groundwater on young citrus trees. J. Sci. Food Agric., 80, 1441-1446.
16. Tarchitzky, J., Golobati, Y., Keren, R., Chen, Y., 1999, Wastewater effects on montmorillonite suspensions and hydraulic properties of sandy soils. Soil Sci. Soc. Am. J. 63, 554-560.

MINIMIZING THE RISKS ASSOCIATED WITH WATER SCARCITY

S. ARLOSOROFF
Israel Water Engineers Association
2 Menorah Street, Tel Aviv 69416, ISRAEL

Abstract

Environmental and socio-economic risks are strongly linked to water scarcity, especially in semi arid regions like the Middle East. In many of these countries, Water Demand management (WDM) is the cheapest available 'source' of water. This paper will discuss a potential strategy to combat associated risks resulting from water scarcities, a condition which might accompany the Middle East socio-economic growth for many years to come.

That strategy is defined as "Water Demand Management", "Water Conservation", or the "Increase of Water Use Efficiency". These 3 definitions mark a major paradigm shift from the conventional supply management of water. To conserve water successfully costs and risks for demand-side management must be considered. Effective management produces additional quantities of water for the immediate needs of the society by creating virtual quantities of water through conservation strategies or by increasing agricultural and industrial production per unit of water.

The experience of Israel in Water Demand Management (WDM) is presented as a potential and powerful instrument to enhance socio-economic prosperity and growth with limited water quantities. Israel has achieved significant positive as well as negative results. This paper will deal with Israel's Water Resources Management strategy as a case study of adequate water demand and supply management in a highly water scarce conditions. It will try to introduce the concept, possibly emerging out of the Israel case study, that the largest and cheapest un-tapped water resource in the world, may be improved "Water Demand Management".

1. Background:

The global situation can be summarized in the following paragraphs, while the situation in many developing countries is much more critical. Recent studies claim that more than 40% of the world food and agricultural needs are produced on irrigated lands. As the developing countries and especially the urban populations in these countries continue to grow at a rapid rate, the forecasted food and agricultural demand will increase the pressures on dwindling water resources. As most of the feasible water resources in river basins as well as in the aquifers, have already been tapped and are

B. Morel and I. Linkov (eds.),
Environmental Security and Environmental Management: The Role of Risk Assessment, 263–267.

being consumed by various sectors, particularly agriculture, one can not avoid asking the question from where and how will the demand for more food and water be met?

Israel was established in 1948, a semi arid country, having a population of 650,000, a GDP of $300/capita, and water allocation of approximately 300 cubic meters of water per person for all uses. In 2003 Israel has reached a population of 6.5 million, a GDP of $16,000/capita, and maintained the water distribution of 300 cubic meters of water per capita.

The country balances its agricultural production and imports following the total development of its water resources. Its national Water Demand Management Strategy (WDMS) involves an intensive national campaign of water conservation, improved efficiencies of water use, as well as the initiation of a comprehensive waste water treatment and re-use trading treated effluents with farming fresh water allocations.

2. Water Demand Management/Water Conservation

The levels of water use and related scarcity, at present within the three Middle East entities being considered in this study (Israel, Jordan and Palestine), are at approximate rates of 300, 140, and 80 cubic meters/capita/year (CMCY) respectively. These levels are for all uses (municipal, industrial, and agricultural) and represent critical levels (World Bank definition for semi-arid countries, water use is critical and acute for levels below 500 CMCY).

It is assumed that already by 2010, unless significant changes in supplies and demands will be implemented, the water scarcity and water quality deterioration will cause severe socio-economic damages and risks to the social life of regional societies. Considering the great majority of the regional conventional water resources have already been developed by the three stakeholders and over-pumping from ground water storage already prevails during dry periods, it seems that only two realistic options remain in order to minimize the regional social economic risks and potential damages:

1. management of water demand
2. augmentation of supplies from inter boundary sources and/or desalination of brackish and sea water resources.

Addressing these strategies 'Ten Commandments' of Water Demand Management have been developed:

1. A national assessment of water resources must be done in order to determine the total national groundwater development and control.
2. A legal basis and/or water code must be developed, institutional responsibility to encourage efficient water use.
3. Comprehensive water metering of all producers and consumers.
4. Research and development for optimal water use, allocations, and technological developments
5. Water pricing rates with progressive block rates
6. Wide scale introduction of pressure irrigation systems, drip irrigation systems, and automation of horticulture.
7. Completion of secondary and tertiary treatment of sewage

8. Establish water trading/exchange policy whereby farmers receive treated effluents for fresh water allocations; developing a 'virtual water policy'.
9. Establish special funds in order to retrofit most irrigation systems and industrial processes.
10. Even after improving efficiency of water use, augmentation of supply will have to come from implementation of brackish and Sea Water Reverse Osmosis Desalination plants (SWROD).

The above list summarizes Israel's unique and comprehensive strategy condensed in the "10 Commandments" policy and complex implementation program. It was designed and aimed to achieve a successful national water demand management program. It took Israel almost 40 years to support research, develop the technological and agronomic means, create the financial and economic incentives and sanctions, and harvest the results. The results are evident mainly in the agricultural sector as well as in the urban and industrial sub-sectors.

Population and economic growth synergize to put real demands on water management in the region.

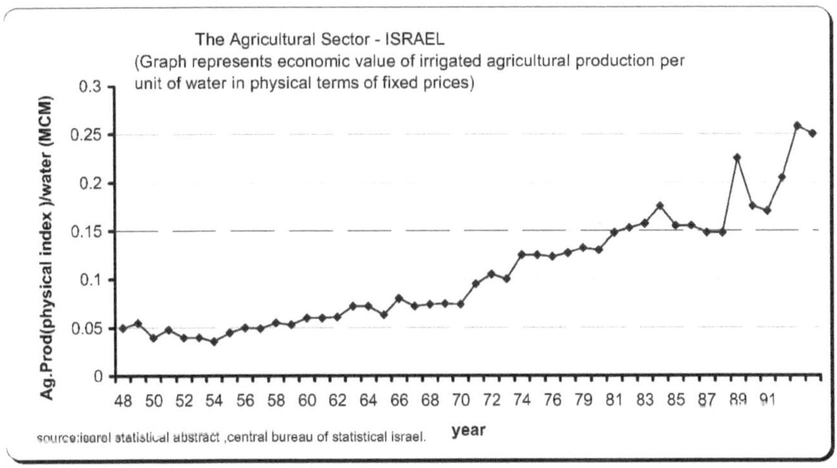

One can see from the figure that the real agriculture production per unit of water has risen by more than 300% enabling the country as well as the farmers to maintain adequate production and income despite reduction of water allocations, increase in water pricing, incentives and sanctions on total water metering, and progressives water rates averaging above $.20/m^3. Strides in the efficiency of agricultural water use is due in large part to the development of the famous locally researched practice of *drip irrigation*, the automation of field levels, and changing crops being cultivated based on their marginal production per unit of water. Similar developments were implemented in industry and selected urban centers, for instance retrofitting old fixtures into modern toilet flushing, adopting flow regulators as well as drip irrigation and automation in parks and gardens, etc.

The impact of these measures is seen from the relative stability of industrial and urban water use per capita that have remained almost constant despite a 300% in increase of GDP per capita. If Israel had not embarked on that nation wide program, it would be using an additional 150-200% of its current water availability. This would be 'accomplished' at prohibitively high costs through desalinating huge quantities of sea water and/or importing all its grains vegetables, fruits, dairy, beef etc., causing significant economic hardship.

Following the above Water Demand Management campaign the Israeli government has given priority to the national waste water treatment and the provision of this water for farming areas, thus enabling the trading of treated effluents with fresh water allocated to the farmers. In 2003, 50% of the total water allocated for irrigation was already treated effluent.

The two previously discussed programs have increased substantially available direct and indirect water resources for Israel. If both of these strategies were to be adopted by Israel's neighbors they would decrease the risk of water scarcity for the region.

3. Potential Regional Solutions:

The potential resources for inter boundary transfers –are marked by the arrows on the figure above. The needed steps to be taken are discussed later that are an integrated part of the regional strategy to minimize the risks associated with water scarcity. However, the set of options and strategies to supplement these essential local programs should be seen as integrated with the inter boundary options. It is important to keep in mind though that these strategies are irrelevant unless a peaceful situation prevails.

Looking at the map above, by moving from North to South on (the map lies horizontally, therefore the right side is north) the first two connections are linked to the Litani river in Southern Lebanon discharging approx. 500 MCMY (million cubic meters per year). This is currently the cheapest source of water. Other connections involve the discharge of water in the east straight into the Jordan basin with links to the Palestinians and Jordanians. Or from the Awali River, the western discharge (after producing power for Lebanon) links along the shore until it meets the Israeli National Water Carrier. This is a costlier option because of the longer connection and the need to fully treat the water before it enters the national water system of Israel that is of potable quality.

In the South the arrow marks water from the Nile basin, before it is discharged into the sea and after it is used by Egyptian farmers. Its gravity drainage flows back into the Nile delta streams by a canal to the Northern Sinai planned by Egypt. Some water will reach the Gaza region and South Israel, while Israel will supply an equal quantity to the Palestinians in the West Bank and/or to H.K. Jordon.

The arrows from the Mediterranean Sea, in the middle section of the map, refer to large desalination plants. The first one is to be operational in 2005 at a cost of $0.5/cubic meter. Water to the top of the Palestinian mountains and H.K. Jordan would cost approximately $0.5/c.m. and $0.8/c.m. respectively. There are cheaper ways to deliver water to the Palestinians and to Jordan, however they call for very sensitive political and economic agreements. In order to accomplish this participation of the three sides and the donor community will have to take place regarding the marginal costs of these options.

The average cost to supply unlimited amounts of water to Palestine or H.K. Jordan would be approximately $1-1.2/c.m. However the present economic conditions of the two areas is such that donors incentives will need to be put in place to decrease the prices to consumers and encourage the willingness to pay. There is a proposal that the desalination plants as well as the pumps and piping system will be totally owned, operated, and maintained by the United States or other donors. In order to remove the suspicion and sensitivity of the Palestinians and Jordanians, to" Israeli hands manipulating our water supply" control of this supply will have to be by a foreign entity. Instead of being in that virtuous circle, the West Bank and Gaza are in the opposite vicious circle.

4.Conclusion

The pioneering works and the related R&D efforts of Israel have influenced many other countries and regions as water conservation is becoming a global need. Other countries in the Middle East and Eastern Europe suffer from similar levels of water scarcities making efforts to spread these concepts extremely important. The risk assessment policies are not a product of academic methodology but demand a change in the political reality and the proper use of the civil society forces.

In the case of the Middle-East, cooperation between the neighboring countries, transfer of knowledge and experience, as well as donor support in the form of capital and operational and maintenance costs will allow water cooperation to create a framework through which future occurrence of disputes or war could be minimized. As water scarcity poses serious risks to socio-economic development and regional prosperity, one should see the cooperation in this sector as a potential spearhead strategy for other sectors. Removal of unnecessary hostile relations is a budgetary solution and not a matter of life and death. With the financial levels of these countries augmented by foreign capital, this inter-boundary cooperation can occur well within their capacity.

INTEGRATED RISK-BASED MANAGEMENT OF WATER RESOURCES IN THE JORDAN RIVER BASIN

E. LEVNER
Department of Computer Science
Holon Academic Institute of Technology 58102 Holon
ISRAEL

Abstract

The aim of this paper is to develop a multi-criteria tool for optimal allocation of water resources in the Jordan River Basin between riparian countries. The model addresses multiple interests of each country in the region in accordance with international laws and allocates water amounts so as to mitigate the environmental risks for all the countries.

1. Introduction

Five riparians share the water resources of the Jordan River Basin – Israel, Jordan, Lebanon, the Palestinian Authority and Syria. The dispute over fair allocation of water resources between Israel and Arab countries has always been a major reason for international conflicts. Since the day of its foundation and up to date, Israel is concerned with water problems, and, in particular, with the problem of fair reallocation of the waters on the Jordan River Basin (see, for example, [2, 3, 6, 10-14, and 17] for a more detailed description of the water problem in the Jordan River Basin and related political issues). According to Prof. Dan Zaslavsky, former Israel Water Commissioner, the latter problem is one of the following major water problems Israel has to face today [18]:

1. Over-pumping of underground aquifers;
2. Water pollution by industrial agricultural and municipal wastes;
3. Water problems with our neighbors and their management;
4. Management, decision making and responsibility -- to whom to address these above problems.

Today, the waters of the Jordan River and its inflows are extraordinarily polluted by industrial, agricultural, municipal, and other anthropogenic sources. The demand for new sophisticated methods for ensuring sustainable development of the exploited Jordan River Basin ecosystem is increasing on the nation-wide and international levels. Without careful global management and international coordination of anti-pollution efforts the waters of the Jordan Basin are threatened with catastrophic changes; under

B. Morel and I. Linkov (eds.),
Environmental Security and Environmental Management: The Role of Risk Assessment, 269–279.

an unprecedented high rate of natural resource degradation due to pollution and overexploitation, the whole water ecosystem may lose its integrity and collapse.

Israel's water claims for the Jordan River waters have been always based on its rights as a legitimate riparian for the equitable utilization and a fair share of this trans-boundary water resource, which flows through its territory and works for satisfying its economic and social needs. Within any multiple-agent system, and, particularly, in large trans-boundary water ecosystems consisting of industrial, agricultural, municipal, recreational and other participating agents, there are political, organizational, social, psychological and other barriers between the participants [8]. They may be caused by contradicting criteria and demands preventing sustainable development and environmental integrity in a region. In recent years, it has been well recognized in the management science literature that many success stories of the multi-criteria mathematical programming (MCMP) can be explained by breaking down or smoothing out the barriers in large multi-agent systems - this is a general view pursued by the MCMP methodology, and this view is being developed in the present paper.

The main purpose of this paper is to develop a multi-criteria decision support tool for rightful allocation of water resources in the Jordan River Basin that addresses interests of each country in the region in accordance with international laws. Our model allocates water amounts so as to minimize the environmental risks, or, in other words, to reduce the force of the total discomfort for all countries in the region.

2. Historical Review of Plans for the Jordan River Basin Management

There have been a plethora of water-distribution plans for the Jordan River Basin proposed in the past, that finally culminated in the renowned Johnston plan. Among them, the best known, and those related to the current state-of-the-art are the following: Leonides Plan (1939), Lowdermilk Plan (1944), Hays Plan (1948), MacDonald Report (1951), Bunger Plan (1952), C.T. Main's Plan (1953), Israel's Seven Year Plan (1953), Cotton Plan (1954), Arab Plan (1954), Baker-Harza Plan (1955), and finally, the Johnston Plan (1955). The Appendix 1 brings the list up to date. Recommendations for water allocation in the Jordan Basin according to major early plans are presented in Table 1.

The present paper suggests a re-examination of the water allocation shares based on the Johnston Plan. The plan is modernized to take into account new social, economic, and ecological realities in the Middle East, and the requirements of international laws.

The equitable allocation of trans-boundary waters is defined under international laws by the Convention on the Law of Non-Navigational Uses of International Water-courses adopted by UN in 1997 [15]. "Equitable" does not mean "equal", but, rather, a variety of factors, including population, hydrology, geography, climate, economic and social needs, risk of appreciable harm, etc., should be taken into account in the allocation of water rights. According to [15], the sharing of international waters must be agreed upon basing on the following main factors:

Table 1. Water allocation in the Jordan Basin according to major early plans

	Allocation (MCM/yr)				
Plan	Lebanon	Syria	Jordan	Israel	Total
Main's Plan (1953)	Nil	45	774	394 (32.5%)	1213
Arab Plan (1954)	35	132	698	182 (17.4%)	1047
The Cotton Plan (with the Lithani River integrated) (1954)	450.7	30	575	1290 (55%)	2345.7
Unified (Johnston) Plan (1955)	35	132	720	400 (31%)	1287
New Arab Plan [9] (2001)	9%	35%	20% (Jordan) + 16% (Palestine)	19%	1400

A. The geography of the basin: the extent of the drainage area in each basin state.
B. The hydrology of the basin: the contribution of water by each basin.
C. The climate affecting the basin (in particular, the total rainfall).
D. The past and existing utilization of the waters of the basin.
E. Economic and social needs of each basin state (the projected water consumption).
F. The population dependent on the waters of the basin.
G. The costs of alternative means of satisfying the water needs of basin states.
H. The availability of other water resources in the basin.
I. The risk of causing appreciable harm and injury to co-basin states.

The main difficulty is that each factor above is not to be considered independently, but, rather, to be estimated in integration with all other factors with priorities (weights) of the factors being taken into account. The mathematical model below translates this principle into standard multi-criteria optimization formulation, which, in turn, is reduced to a single-criteria quadratic programming problem. The model is a modification of similar optimization (mathematical programming) models developed recently in [2, 3, 6].

3. Integrated Water Resource Management in the Jordan River Basin and Risks

There are many definitions of environmental risk. Following the approaches in [7, 9, 16] we adopt the following, perhaps somewhat utopian, vision of the objectives of integrated water resource management:

- To balance competing uses of water and to efficiently allocate water resources through thorough coordination of social values, and environmental costs, factors and benefits.

- To coordinate and resolve conflicts by including all units of government, agencies and water stakeholders in the decision-making process.
- To promote water conservation, reuse, source protection, and enhance good water quality.
- To foster public health and safety.
- To mitigate the environmental risks caused by water misbalance and pollution.

Speaking informally, the environmental risk is a quantitative measurement of ecological hazards with their economic, social and related consequences being taken into account. Following the U.S. EPA definition [16], we define the *environmental risk assessment* as a quantitative appraisal of the actual or potential impact on humans, animals, plants and technological infrastructures of contaminants from a hazard.

Before the *risk assessment* is explained formally and in more detail, let us define the terms *hazard* and *risk* – how we understand them in the present paper. *Hazard* is the potential for harm. Assume that the amount of water allocated to a country is x Mcm/year, constituting y% of the total amount of the water resource available, whereas a corresponding proportion of the population in the country with respect to the total population in the basin is z%, and suppose that $y < z$. Clearly, the hazard then is the inconsistency in the water supply, measured by the difference $z - y$. In what follows, in a similar way we define the potential for harms caused by all factors A to H indicated in the previous section, that is, the differences between the allocated share y, on the one hand, and corresponding proportions of geographical, economic and social needs, on the other hand. Another type of the hazard is defined by the fact that the waters of the Jordan River are heavily polluted; a hazard here is the danger for human health.

Ecological risk is the likelihood of harm occurring in an ecosystem and the severity of its outcome [16, 9]. There are many ways in which the evaluation of risks can be carried out. These range from the numerically complicated systems to a qualitative expert judgment of risk, that is, {low, medium or high}. Most of the formal ways define risk R as the product of the *weight* w of a hazard, also called a *risk factor*, and amount of damage, Q caused by the hazard, in a monetary, or material, or grade form: $R = wQ$.

A *risk factor* (also called a *risk weight,* or *a risk factor number*) is the product of the likelihood (probability) and severity of harm arising from a hazard. A *likelihood* rating is based on the qualitative scale shown below.

1. *Not likely.* There is really no likelihood of an accident or pollution occurring. Only under freak conditions could there be a possibility of an accident or illness. All reasonable precautions have been taken so far as is reasonably practicable. This should be the normal state of the water source.
2. *Possible.* If other factors were present, a pollution or illness might occur, but the probability is low and the risk is minimal.
3. *Quite possible.* The accident or pollution may happen if additional factors precipitate it, but it unlikely to happen without them.
4. *Likely.*
5. *Very likely.* If the situation continues as it is, there is almost a 100% certainty that an accident or pollution will happen at least once.

Now let us establish a *severity* rating for the identified hazards using the following scale:

1. *Nil.* No risk of injury, or contamination, or disease
2. *Slight.* Causing minor injury or harm.
3. *Moderate.* Causing moderate injury, or harm, or disease.
4. *High.* Causing death or serious injury to an individual.
5. *Very high.* Causing multiple deaths and/or widespread illnesses to population.

A *risk factor number* is obtained by multiplying the likelihood rating by the severity rating. A number between 1 and 25 would result. Such a rating enables the most serious risks to be considered first, i.e. the higher the number the higher the risk.

We will classify the risk factor number as follows:

- 16 – 25 *Extreme* Risk level unacceptable.
- 10 – 16 *High* Undesirable.
- 7 – 10 *Medium* May be acceptable.
- 1 – 6 *Low* May be acceptable.

In our first example considered above, risk R is measured as $w_i(z_i - y_i)^2$, in the case of a single factor i, and $\Sigma_{i=1,\ldots,n} w_i(z_i - y_i)^2$, in the case of n factors, where w_i is a weight (risk factor number) of factor i. In our second type example, the risk to human health caused by several toxic stressors, may be presented as

$$R = \boldsymbol{w} \bullet \boldsymbol{W} = \Sigma_{j=1,\ldots,L} w_j(W_j)W_j,$$

where:

L : the number of stressors, *fons et origo* of water pollution;

$w_j = w_j(W_j)$: the risk factor number for the jth stressor;

W_j : the amount of the water pollution caused by the jth stressor.

4. Mathematical Model

This section outlines the mathematical form of the HAIT model in terms of multi-criteria mathematical programming problem. (For the sake of simplicity of the presentation, the costs of water supply are not regarded and constraints related to recycled water are omitted).

Notation

Fresh water sources: the Jordan River and its tributaries (Hasbani River, Dan Spring, Banias River, Hermon Springs, the Yarmouk River, the Zarqa River, and others). This fresh water is used by all riparian countries. The total fresh water flow of the Jordan River is accepted to be from 1250 to 1650 MCM/year.

Wastewater: represents the total wastewater generated from sectors.

Sectors: the industrial, agricultural, municipal and tourist sectors (integrated water users) in each country.

Indices:

i = water source, $i = 1,\dots, I$,
j = country, $j = 1,\dots, 5$,
d = sector, d = 1, …, D.

Variables

QF_{ij} Quantity of fresh water allocated from fresh water source i to country j,
QF_{ijd} : Quantity of fresh water allocated from water source i to sector d in country j,
QF_{dj} : Quantity of fresh water allocated from all water sources to sector d in country j,
QF_j Total quantity of fresh water allocated from all fresh water sources to country j,
QW_{dj}: Quantity of wastewater generated from sector d in country j,
All variables are to be non-negative.

Parameters

AD_{dj} : Amount of water demanded by sector d in country j,
AR_{dj} : Accepted ratio of wastewater generated from sector d in country j, in %% from the total fresh water allocated in sector d of country j,
MW_{dj} : Maximum allowed amount of wastewater occurred in sector d in country j,
B_i: Maximum capacity of the fresh water source i.
Q: Total quantity of fresh water allocated from all fresh water sources to all countries in the basin.
Z_{fj} : proportions of geographical, hydrological, climatic, economic and social needs, defined by eight factors A to H in Section 2 (f=1, …,8). The values Z_{fj} are found from the statistical surveys and are presented below in Table 2.

Constraints

The first group of constraints state that fresh water consumed in a sector must satisfy the corresponding demand:

(1) $$QF_{dj} = \Sigma_{i=1,\dots,I}\, QF_{ijd} = AD_{dj}, \text{ for all } d \text{ and } j.$$

The second set of constraints state that the wastewater must come originally from fresh water and should not exceed the maximum allowed norm:

(2) $$QW_{dj} = AR_{dj} \times OF_{dj} \leq MW_{dj}, \text{ for all } d \text{ and } j.$$

The third set of constraints state the bounds on natural capacities of the fresh water sources:

(3) $$\Sigma_{j=1,\dots,5}\Sigma_{d=1,\dots,D}\, QF_{ijd} \leq B_i, \text{ for all } i\,.$$

The fourth set establishes the total quality of fresh water allocated to country j:

(4) $$\Sigma_{i=1,\dots,I}\,\Sigma_{d=1,\dots,D}\, QF_{ijd} \leq QF_j, \text{ for all } j\,.$$

Multiple Criteria

$R1_j = \Sigma_{f=1,\ldots,8}\, w_f(QF_j/Q - Z_{fj})^2$ (j=1,... 5): Ecological risk defined in Section 3, that is, the weighted sum of squared differences between the allocated share QF_j/Q, for country j and proportions Z_{fj} of geographical, hydrological, economic and social needs, defined by eight factors A to H in Section 2.

$R2_j = \Sigma_{d=1,\ldots,D}\, w_d\, QW_{dj}$ (j=1,..., 5): Ecological risk of the second type defined in Section 3, that is, the risk to human health caused by toxic wastewaters from all the sectors.

Problem Formulation

$$\text{MINIMIZE } R1_j = \Sigma_{f=1,\ldots,8}\, w_f(QF_j/Q - Z_{fj})^2 \ (j = 1,\ldots, 5)$$

$$\text{MINIMIZE } R2_j = \Sigma_{d=1,\ldots,D}\, w_d QW_{dj} \ (j = 1,\ldots, 5):$$

subject to constraints (1)-(4) above; all the variables are to be non-negative.

For solving that multiple-criteria optimization problem, we proceed with a standard min-max approach, according to which we find a solution providing the minimum value to MAX ($R1_j$, $R2_j$, | j =1,..., 5) subject to constraints (1)-(4). Thus, we reduced a multi-criteria problem to a standard single-criterion quadratic programming problem.

Input Data

Main input data used in computational experiments are presented in Table 2. For simplicity, and as a first approximation, all the factor weights w_f are taken equal to 1. The values presented in Table 2 are Z_{fj}, basic proportions (in %) of geographical, hydrological, climatic, economic and social needs, defined by eight factors A to H in Section 2. These values are found from the statistical surveys, most of them being adapted from [1, 5, 12].

Table 2. Input data: Z_{fj}, basic proportions (in %)

	Factor A: Geography (Catchment Area)	Factor B: Hydrology- Average Annual Discharge	Factor C: Climate	Factor D: Past Utilization	Factor E: Economic Needs	Factor F. Population	Factor G: Alternative Options to Water	Factor H: Other Water Resources
Israel	11	12	32.5	60	35	16.4	54.5	28
Jordan	39	38	27	25	16	14.5	11	27
Lebanon	4	8	12	1	11	11.2	8	15
Syria	40	31	12	12	35	47.1	26	5
Palestine	6	11	16.	2	3	10.8	0.5	35
Total	100	100	100	100	100	100	100	100

For convenience, the proportion values Z_{fj} in Table 2 are given in %. They are computed as follows: If the factor is proportional to the water need that its value is obtained as

(5) $$Z_{fj} = F_j/\Sigma_{j=1,\ldots,5} F_j.$$

where F_j is the factor j value measured in natural or physical units.

In the case when the factor is reciprocal to the water need then its value is obtained as

(6) $$Z_{fj} = (1/F_j)/\Sigma_{j=1,\dots,5}(1/F_j)$$

In order to clarify the contents of Table 2, consider, for example, column *Factor D: Past Utilization* (see Table 3). Israel is currently the dominant user of the waters of the Jordan Basin. Table 3 presents the existing utilization of the Jordan and Yarmouk Rivers (the data being cited from [1, 5, 10, 12]).

Table 3. Factor D: Existing utilization of the Jordan and Yarmouk Rivers (Mcm/yr)

Country	Israel	Jordan	Lebanon	Syria	Palestine	Total
Quantity (MCm/yr)	810	340	5	165	20	1340
Percent	60	25	1	12	2	100

Then, according to the proportionality rule (5), Z_{41} = (810/1340)100 = 60.4 (%).

Consider now column *Factor C: Climate* (see Table 4).

Table 4. Factor C: Climate. Average Annual Rainfall (mm), according to [1, 5, 12]

Country	Israel	Jordan	Lebanon	Syria	Palestine	Total
Quantity (mm)	184	222	508	508	361	1783
Percent	32.5	27	12	12	16.5	100

Then, according to the reciprocal proportionality rule (6), Z_{31} = [(1/184)/0.01672]100 = 32.5 (%). All other entries in Tables 2-5 are computed analogously.

In order to process factor G: *The costs of alternative options of satisfying the water needs of basin states,* we assume here that this factor is measured proportionally to Gross Domestic Product (GDP). Notice, however, that other assumptions may take place (see, e.g. [12], where this factor is measured reciprocally to the GDP per capita); obviously, different assumptions may lead to different final solutions.

Table 5. Factor G: The costs of alternative options. GDP (2003)

Country	Israel	Jordan	Lebanon	Syria	Palestine	Total
Amount (Billions $)	120.9	23.64	17.82	58.01	1.6	222
Percent	54.5	11	8	26	0.5	100

Numerical Results. Omitting the intermediate computational details, the obtained solution to the quadratic programming problem above is presented in Table 6.

Table 6. Optimal solution to the quadratic programming problem

	Allocation $QF_j/Q \times 100$ (%)				
Plan	Lebanon	Syria	Jordan	Palestine	Israel
HAIT/ECOST	4	10	20	16	50

5. Conclusions and Directions for Future Research

This paper extends the research currently being done in the field of water allocation and water resource management in the Jordan River Basin. We show that a multi-criteria mathematical programming (MCMP) approach, though not being totally new from the mathematical point of view, opens up fresh opportunities for coordinating technological, social, economic and ecological contradictory demands of different participating parties and creates a mechanism for a fair water allocation between neighboring countries. The proposed risk-based multi-criteria approach may serve as a framework for further research aimed at obtaining fair and rightful water allocation in the region.

Several extensions of this work seem to be plausible. In the context of the Jordan River basin, the allocation of water is more than a technological problem – economic and social considerations are of the utmost importance as well, and, thus, some factor weights must be larger than others. Thus, it seems appealing to evaluate the weights of the considered factors and amplify their role in the model. A further extension could be to expand a group of risk factors, in particular, to incorporate the risk of poor water quality and water deficit, together with their social consequences. In our future research, we intend to integrate the proposed techniques with the computer-based sensitivity analysis so that to take into account input data uncertainty and various scenarios for future economic growth. Concluding our study, we would like to cite Frederick Frey and Thomas Naff [4]: "*Precisely because it is essential to life and so highly charged, water can — perhaps even tends to — produce cooperation even in the absence of trust between concerned actors.*"

6. Acknowledgement

The author thanks anonymous referees for their useful comments and suggestions.

7. Appendix. Development Plans For The Jordan River Water System

Plan	Agency/Organization/Sponsor
1913 Franhia Plan	Ottoman Empire
1922 Mavromatis Plan	Great Britain
1928 Henriques Report	Great Britain
1935 Palestine Land Development Company	World Zionist Organization
1939 Ionides Survey	Transjordan
1944 Lowdermilk Plan	USA
1946 Survey of Palestine	Anglo-American Committee of Inquiry
1948 Hays-Savage Plan	World Zionist Organization
1950 MacDonald Report	Jordan
1951 All Israel Plan	Israel
1952 Bunger Plan	Jordan/USA
1953 Main Plan	UNRWA/United Nations
1953 Israeli Seven-Year Plan	Israel
1954 Cotton Plan	Israel
1954 Arab Plan	Arab League Technical Committee
1955 Baker-Harza Plan	Jordan
1955 Unified (Johnston) Plan	USA
1956 Israeli Ten-Year Plan	Israel
1956 Israeli National Water Plan	Israel
1957 Great Yarmouk Project	Jordan
1964 Jordan Headwaters Diversion	Arab League
1991 Integrated Joint Development Plan	Japan (University of Tokyo)
1993 Declaration of Principles: PLO/Israel	Israel and PLO (Annex III, IV)
1994 Treaty of Peace: Jordan/Israel	Jordan and Israel (Article 6, Annex II)
2000 JOWA Model [6]	INCO-DC and ARIJ, Bethlehem, Palestine
2001 New Arab Plan [12]	Birzeit University, Palestine
2001 WWS401c Project [2]	Princeton University, USA
2002 WAS Model [3]	MIT, USA - Tel-Aviv University, Israel

Source: Adapted from [10, 11], the recent works of 2000-2002 being appended.

8. References

1. CLS (Commission on Life Scinces), *Water for the Future, The West Bank and Gaza Strip, Israel, and Jordan,* The National Academy of Science, Washington, USA, 1999.
2. Feiveson, Hal (ed.), 2000, *Water Rights in the Jordan Valley,* Project WWS401c, Woodrow Wilson School at Princeton University, 2000. Available at [www.wws.princeton.edu/~wws401c/rational.pdf]
3. Fisher, Franklin M. , Shaul Arlosoroff, Zvi Eckstein, Munther Haddadin, Salem G. Hamati, Annette Huber-Lee, Ammar Jarrar, Anan Jayvousi, Uri Shamir, Hans Wesseling, 2002, Optimal Water Management and Conflict Resolution: The Middle East Water Project, *Water Resources Research*, 10, November 2002.
4. Frey, Frederick W., and Thomas Naff, Water: An Emerging Issue in the Middle East, *Annals of the American Academy of Political and Social Science,* 1985, 482, 65-84.
5. GTZ (Association for Technical Cooperation, Germany), *Middle East Regional Study on Water Supply and Demand Development,* Phase I, Regional overview. Multilateral Working Group on Water Resources. Eschborn, Germany, 1996.
6. Isaac, Jad, Hans Kupfersberger, Rudolf Orthoffer, and Hillel Shuval, *et al.,* 2000, *Developing Sustainable Water Management in the Jordan Valley,* INCO-DC Project JOWA, Applied Research Institute, Bethlehem-Jerusalem, Dec. 2000 Available at [systemforschung. arcs.ac.at/jowapubl].
7. Levner, Eugene and Jean-Marie Proth, 2004. Strategic Management of Ecological Systems: A Supply Chain Perspective, in E.Levner, I. Linkov and J.-M. Proth (eds), *Strategic Management of Marine Ecosystems*, Springer, Berlin, 2005.
8. Levner, Eugene, Dror Zuckerman, and Gabriel Meirovich, 1998. Total quality management of a production-maintenance system: A network approach, *International Journal of Production Economics,* pp. 407-421.
9. Linkov, Igor and Ramadan, Abu Bakr (eds), *Comparative Risk Assessment and Environmental Decision Making*, Springer, Berlin, 2004.
10. Murakami, Masahiro, 1995. *Managing Water for Peace in the Middle East: Alternative Strategies,* United Nations University Press, Tokyo, 1995, 319 p.
11. Naff, Thomas, and Ruth C. Matson. 1984. *Water in the Middle East: Conflict or Cooperation?* Boulder, Colo., USA, and London: Westview Press. P. 1-62.
12. Sawalhi, Bassam I., Mimi, Ziad A., and Aliewi, Amjad S., 2001. Multi-criteria decision tool for allocating the waters of the Jordan Basin between all riparians, *Proceedings of the International Conference "Globalization and Water Resources Management : The Changing Value of Water",* AWRA/IWLRI-University of Dundee, August 6-8, 2001.
13. Shamir, Uri, 1998. Water agreements between Israel and its neighbors, in J. Albert *et al.* (eds), Transformations of Middle Eastern Natural Environments: Legacies and Lessons, Yale School of Forestry and Environmental Studies, Yale University, Bulletin 103, 1998, 274-296.
14. Shuval, Hillel, 2003. An equitable approach to resolving the water conflicts on the Jordan River Basin under condition of scarcity, Proceedings of the World Water Congress, June 2003, Philadelphia, USA.
15. *United Nations Convention on the Law of the Non-Navigated Uses of International Watercourses,* 1997, World Bank, Washington. Available at http://www. thewaterpage.com/UN Convention 97.html.
16. *USA EPA, Ecological Risk Assessment Guidance for Superfund,* , USA Environmental Protection Agency, PA, 1991.
17. Wolf, Aaron T., 1995, *Hydropolitics along the Jordan River • Scarce Water and its Impact on the Arab-Israeli Conflict,* UN University Press, New York, 283 pp.
18. Zaslavsky, Dan, 2000, *Definition of Israel's Water Problems*, The lecture presented in the Begin-Sadat Center for Strategic Studies, August 2000. Available at www.biu.ac.il/SOC/besa/water/zaslavsky.html

OUTCOME OF NATO WORKSHOP ON RADIOLOGICAL RISK IN CENTRAL ASIA

M.K. ZAIDI
Radiological and Environmental Sciences Laboratory (RESL)
1955 Fremont Avenue, Idaho Falls, Idaho, USA.

F. STEINHAUSLER
Division of Physics and Biophysics, University of Salzburg
Hellbrunnerstr. 34, A 5020 Salzburg, AUSTRIA.

I. MUSTAFAEV
Institute of Radiation Problems, National Academy of Sciences
Baku, AZERBAIJAN

R.E. DUNKER
Environmental Monitoring Laboratory, Department of Physics/Health Physics,
Idaho State University, Pocatello, Idaho, USA.

Abstract

A North Atlantic Treaty Organization Advanced Research Workshop (NATO-ARW), "Radiation Safety Problems in the Caspian Region", was held in Baku, Azerbaijan in September 11-14, 2003. The objectives of this workshop were the exchange of information in the field of radiation safety and radioecology. Fifty-eight participants from twelve countries investigated hazards to regional populations from ionizing radiation. A major topic was the discussion of the investigations carried out in the Caspian Region. The main source of radioactivity in the Caspian Sea is due to global fallout, subsequent river run-off from catchments areas and the long-term oil-gas extraction.

1. Introduction

An Advanced Research Workshop (ARW), entitled "Radiation Safety Problems in the Caspian Region," funded by the North Atlantic Treaty Organization (NATO), was held in Baku, Azerbaijan, from September 11-14, 2003 (Directors: Mohammed K. Zaidi, Radiological and Environmental Sciences Laboratory (RESL), Idaho Falls, Idaho, USA and Islam Mustafaev, Institute of Radiation Problems, National Academy of Sciences, Baku, AZERBAIJAN). The ARW had two main objectives:

B. Morel and I. Linkov (eds.),
Environmental Security and Environmental Management: The Role of Risk Assessment, 281–286.

1. Exchange of information between researchers and regulators working in the topic area radiation safety in the biosphere
2. Identification of potential areas of international cooperation, aimed at assessing and mitigating regional radiological problems.

Altogether fifty-eight (58) participants, originating from scientific, public and Non-Governmental Organizations (NGO) representing the following twelve (12) countries: Austria, Azerbaijan, Croatia, Georgia, Kazakhstan, Kyrgyzstan, Russia, Turkey, Turkmenistan, Ukraine, USA, and Uzbekistan, contributed to meeting these objectives.

2. Radiation safety as a key scientific and social problem in the Caspian region

Radiation safety is considered a main issue in achieving a sustainable development in the Caspian region. Over the past 50 years, the area shared by several Central Asian countries, Russia and the Caucasus, has experienced multiple activities related to the nuclear fuel cycle, which resulted in radioecological, as well as radiation safety and radiation security related problems. In addition, the impact from non-ionizing radiation is of significant importance to the region [1,2]. The participants discussed the following main topic areas:

1. Natural and technologically enhanced natural radiation: Leaching of radionuclides and wind erosion of tailings from extensive uranium mining; natural radionuclide exposure due to fertilizer production and the use of coal as fuel in electrical power production; occupational exposure for employees engaged in the exploration of oil and gas.
2. Man-made radiation: Radioecological consequences due to the deposition of fall-out from nuclear weapons tests; environmental contamination and its impact on man resulting from the accident at the Chernobyl nuclear power plant in April 1986.
3. Nuclear material security: Illicit trafficking of nuclear and other radioactive materials and its potential consequences in terms of terrorism
4. Non-ionizing radiation: Regional electromagnetic pollution, in particular the environmental impact of military radar installations.

There is a considerable level of concern about these topics among members of the public, leading to multiple NGO-activities, expressing the anxiety felt by the ordinary citizen. At the scientific level, there is an impressive amount of know-how at various academic and research institutions. However, the lack of nationally available funding, as well as inadequately developed national and international cooperative framework agreements result in a noticeable deficit of exchange of information and experience in addressing these problems.

This NATO ARW was able to contribute to the improved understanding of science and technology underlying these pressing issues. The most important data and their interpretation presented at the workshop are summarized below.

3. Natural and technologically enhanced natural radiation

The legacy of the uranium-mining district Koshkar-ata, close to the city of Aktau (Kazakhstan), represents a typical example of a radioecological issue in the region in need to be addressed [3]. The initial lake (surface area: 64 km^2), created in order to serve as a basin for sedimentation, is drying up. The exposed area totals about 10 km^2, resulting in dried up areas with elevated gamma dose rates up to 0.89 μS/h. The Semipalatinsk test site was one of the major sites used by Russia for testing nuclear weapons for 40 y [4]. The village Dolan has been identified for many years as the most highly exposed location in the vicinity of the test site [5]. Also in Kyrgyzstan uranium mining and milling has left a large-scale environmental problem behind: after the termination industrial activities in this area more than 80 tailing piles are awaiting remediation [6,7].

Uranium production is also of high importance in Turkey. Uranium reserves at four different locations account for 57.8% of the total reserves, with an average U_3O_8 concentration of about 0.04% to 0.05%. It is noteworthy from a radiation protection point of view that Turkey also has thorium reserve amounting to 380 000 t containing ThO_2 at 0.21%. Another facet of the Turkish approach to uranium production is the case of the recovery of uranium as a by-product from the phosphoric acid production at the Yarmica fertilizer plant. The yellow cake produced contains U_3O_8 ranging from 13% to 18.4% from acidic, and 30% to 46.4% from basic stripping solutions, respectively [8].

Turkey is also experiencing the environmental impact due to the use of coal for the production of electric power. The mean uranium concentration of Yatagan lignite samples is 68 Bq/kg, increasing to 266 Bq/kg in the resulting fly ash [8].

The oil and gas industry, a major industry in the Caspian region, can have a significant impact on the occupational exposure of its workers. As the production water is brought to the surface, some of the dissolved radium precipitates in solid form. Most commonly, the radium co-precipitates with barium sulphate, forming a hard and relatively insoluble scale. The radiological impact ranges from increased external radiation in the vicinity of vessels and tanks filled with brines, to elevated inhalation doses resulting from sand blasting of scales during their removal from tanks. Furthermore, radon levels have been found to range from 20 Bq/m^3 to 1 178 Bq/m^3. Also the recycling of contaminated metal originating within the oil and gas industry can pose a contamination problem for the operator of a steel recycling plant.

4. Man-made radiation

The region has extensive experience in monitoring of man-made environmental radioactivity.

In Azerbaijan detailed oceanographic and radionuclide distribution studies have been carried out in the Caspian Sea, e.g., by the indigenous development of a dedicated radiation detector system. This system is capable of carrying out maritime gamma surveys down to depths of 1000 m. This allowed the study of the behaviour of anthropogenic radionuclides in the Caspian Sea, using ^{90}Sr, ^{137}Cs, ^{239}Pu and ^{240}Pu as tracers. It has been established that global fallout and subsequent run-off by the rivers from the catchments areas are the main radiation source terms. Increased ^{90}Sr levels have been interpreted as remobilisation of ^{90}Sr from soil and subsequent transport by rivers into the Caspian Sea [7,8]. In this context it is noteworthy that a new project foresees the automated remote monitoring of environmental background radiations in Azerbaijan, where radiation monitors (external gamma, neutron) will be mounted on fixed stations in Kedabek, Akstafa, Terter, Agdam and Fizuli regions, as well as in the Autonomous Republic of Nakhichevan.

In Uzbekistan environmental monitoring focuses on the impact of nuclear centres. Extensive analysis of samples of soil, plant, precipitation, sewage, ground and surface water showed that the nuclear research activities of altogether six such centres did not contribute to an elevation of the background radiation at the national level.

Turkey has also an extensive environmental monitoring system (RESA) in place, consisting of 67 stations across the country. The system is based on Geiger-Mueller tubes, capable of automatically flagging an abnormal radiation level, thereby functioning as a national early warning radiation monitoring system [8].

5. Nuclear material security

The region is known for being at the cross road of international drug and weapon smuggling [9]. Several regions, such as Nagorny Karabakh, are suspected of becoming lawless and thereby a region suitable as a staging area for terrorist activities. This could pose a serious security threat to the regional nuclear infrastructure, such as the Armenian Nuclear Power Plant.

It is therefore important to ensure that the region does not become subject to illicit trafficking of nuclear and other radioactive materials. Border Guards and Customs Officials carry out the control of import and export. Typically, neither of these organizations have sufficient experience or adequate equipment in order to address these problems in a satisfactory manner. National specialist institutions, such as the Institute of Radiation Protection (Baku, Azerbaijan) with the help of U.S. Department of Energy and other international agencies conduct dedicated training courses in order to provide the Government officials and field officers with the necessary know-how. It should be noted though that further major efforts are needed to ensure that the non-proliferation issue is dealt adequately with in the future, e.g., by intensifying the training of personnel, upgrading the legislation, and by creating an up-to-date technical infrastructure of radiation monitors.

6. Non-ionizing radiation

Despite international efforts to establish the biophysical and biochemical mechanisms of the interaction of electromagnetic fields with biological systems, the current understanding is still limited to the effect of radiofrequency fields above about 100 kHz (primary effect: heating) and at lower frequencies (primary effects: neurostimulation or electric shock). Important observations, such as correlation between power frequency and magnetic field exposure with childhood leukaemia, or the occurrence of brain tumor with the extensive use of handheld wireless phones is still subject of intense scientific debate. In this context it is important that the following results were obtained by Turkish researchers in rats exposed to cellular phones (exposure 20 min/day, 1 month): no histological alteration in the brain, but the fatty acid compositions of phospholipids were found to be affected significantly.

In Azerbaijan an analysis of the environmental impact of a large radar station in the Gabala region (area: 210 ha) showed that the negative impact ranges from changes of free electric charges in the atmosphere, to lowering of the water table due to water requirements for cooling purposes; this in turn impacted negatively on the local forests.

7. Non-governmental organizations

All of the above topics are of great interest not only to the scientific establishment in the region, but also to Non-Governmental Organization (NGO). Frequently lacking the technical and/or financial resources to conduct in-depth studies, they have to rely on pilot surveys or spot sampling. In particular, the following NGOs have discussed their ambitious work plans with the members of the regional scientific community at the NATO ARW:

1. Azerbaijan: “Fovgal”, “Human Environment”, “Ekoil” and “Ruzgar”
2. Georgia: “Radioecolog-21”
3. Russia: “The Creative Person Development”, “Green Don”
4. Kazakhstan: “Nevada-Semipalatinsk” , “Anti-nuclear Movement “Narin”
5. Turkmenistan: Hazards
6. Kyrgyzstan: International Science Centre

8. Conclusions

The workshop demonstrated the significant scientific potential present in the Caspian region. This know-how can be used in the future for the multiple radioecological, radiation safety and radiation security problems waiting to be resolved. However, it is essential to realize that the region needs foremost international support to provide the necessary upgrade of the technical infrastructure, as well as the cross-border willingness of the scientists in the region to engage in international collaboration with each other. International organizations, such as NATO, the European Union, Civilian Research Development Foundation (CRDF), US AID, or the International Atomic Energy Agency (IAEA) can play an important role in facilitating these efforts.

9. Acknowledgement

The NATO ARW Committee provided travel support.

10. References

1. Djahanguiri, F et.al. (1998). Radioactive Pollution on the Apsheron Peninsula and Caspian Sea. http://greenaz.aznet.org/greenaz/issues98/bt7last.htm
2. Gulubov Boris, Sources on Nuclear Pollutants in the Caspian Sea. http://www.isar.org/isar/archive/gt/gt10/golubov.html
3. National Report on the State of Nature in the Republic of Kazakhstan. (1992). UN Conf. on Environ. Development. National Report Summaries, 3, 269-275
4. Semiochkina N, et.al. (2004) Assessment of the Current Internal Dose…Health Phys 86(2) 187-192
5. Simon S.L et.al. (2003). A Summary of Evidence on Radiation Exposure…Health Phys 84(6). 718-725
6. Hamby D.M., Tynybekov. (1999). A Screening Assessment of External Radiation Levels on the Shore of Lake Issyk-kyol in Kyrghyz Republic. Health Phys 77(4). 427-430.
7. Albion Monitor. (1996). Nuclear Pollution Caused by Russians, http://www.monitor.net/monitor
8. Radiation Safety Problems in the Caspian region (2003). NATO ARW Abstract and Program Book, Baku-Azerbaijan. http://hostel.aznet.org/radsafety
9. Zaitseva, L: Illicit Trafficking in the Southern Tier and Turkey Since 1999: A Shift from Europe? Nonproliferation Review, Vol. 9, No. 3 (Fall-Winter 2002).

PUBLIC TRANSPORT POLICY & MEASURES THAT COULD IMPROVE THE AIR QUALITY IN MAJOR CITIES IN SYRIA

A Case Study: Damascus

J. CHAHOUD
World Federation of Scientists and Department of Physics, Bologna University, Bologna 40126, ITALY

Abstract

Energy use in the transportation sector is primarily for passenger travel and freight movements. We shall restrict the present discussion to that part of public passenger transportation in either case of light and high duty vehicles. The focus is to bring about an evolution toward a sustainable public transport system including oil savings, improved energy efficiency and security, and reduced GHG emissions. The baseline of the present case study stemmed, in 2002, from a request of the Syrian Ministry of Transport aiming to introduce the use of Compressed Natural Gas (CNG) and Liquefied Propane Gas (LPG) in public transport vehicles in order to get in the environmental dimension and better air quality in urban areas, within the general framework of reconciling socio-economic development with environmental protection.

1. Introduction

The Mediterranean Sea identifies, according to the E.U. Commission, the southern border of the European Union. The 1995 Barcelona Conference has represented the starting point for the implementation of a new programme of co-operation with the extra-EU countries located at the south and east shores of the Mediterranean Sea.

According to updated figures from Eurostat, by 2020 the population of the South Mediterranean Countries will reach a figure between 300 and 400 millions, with a very young age structure; most of them will be badly looking for labour opportunity. The economy will be concentrated in agriculture, raw materials and manufacturing, as well as in services like transports and tourism. As a result, the urbanization process will likely continue, with negative effects on the environment.

On the Northern side, the European population structure will be much older, averaging around 350/400 millions, with a growth rate close to zero. The economy will be mainly concentrated in the tertiary sector: insurance, banking and financial services, advanced scientific and technological research, etc.

According to this scenario, the E.U. policy is strongly addressed at developing the co-operation and partnership with non-E.U. Mediterranean Countries, in a common

B. Morel and I. Linkov (eds.),
Environmental Security and Environmental Management: The Role of Risk Assessment, 287–296.

effort to identify new forms of socio-economic integration and collaboration in a variety of sectors like, for instance, education, training and technology transfer.

The E.U. programme denominated "**N**egotiation and **A**pplication of **PO**licies for **L**owering **E**missions", better known as "**NAPOLI**" programme, is a programme in view of clear messages to policy decision makers for transport strategies in cities and metropolitan areas. The programme joins the cities of Naples, Athens, and Barcelona with their respective Regions, in the strain to improve mobility and air quality in those urban areas. It included the identification of strategies for the implementation of realistic and successful collaboration amongst the Mediterranean Countries.

The main innovation of the programme is represented by the opportunity for mutual exchange of know-how and of new technological solutions and experiences referring to the energy sector policies. Naples provided its know-how about clean technologies for transport, telematic control for mobility of public transport means in urban areas, new techniques of pollutants monitoring. Barcelona and Athens have provided similar experiences; the former had afforded the 1992 Olympic Games, the latter is organizing the 2004 Olympic Games. Another important result comes from the coordination with other international initiatives and projects as i) the E.U. CENTAUR Project in which the Italian cities involved are Bologna and Naples, ii) the European thematic network EUROTREC-2, in which the group involved is GLOREAM (Global REgional Atmospheric Modelling), iii) the activities of "Remote Sensing and Control of Air Quality in Urban Areas" which is a joint project driven by the Italian National Council of Research and the Italian Ministry of Universities and Scientific Research, with a financial support by E.U.

2. Mediterranean dissemination and co-operation

The first occasion for disseminating the information on the activities of the "NAPOLI" Programme was during the Third Mediterranean Exhibition of Technological Innovation (MEDITERINTEC III) held in Naples on the 10th of December 1998. MEDITERINTEC provides academic and industrial researchers, producers, transformers, end users and market operators with a unique opportunity to mutually share their know-how and innovative technologies in order to identify and explore new perspectives aiming at an eco-sustainable development of the Mediterranean Countries. During the MEDITERINTEC III, each invited country introduced the state of art of transport policy and air quality in their urban areas, as a demonstration of their strong interest in the activities and the results of the "NAPOLI" Programme, and their availability and willingness for a direct co-operation.

The CENTAUR Project has been put forward for financial support from the European Commission Directorate General XVII (Energy) under the JOULE-THERMIE programme in March 1996. It was developed by a European Consortium including organizations from ten cities, namely: Barcelona, Bologna, Bristol/South Gloucestershire, Dublin, Graz, Las Palmas, Leipzig, Krakow, Naples and Toulouse.

The overall aim of the Project is to demonstrate and evaluate "integrated packages" of measures in order to reduce transport related energy consumption and air pollutant emissions. Within this, each of the abovementioned cities will have two main objectives:

- encourage a shift towards the use of energy efficient and low pollution travel modes;
- improve the energy and emission characteristics of public transport.

These objectives will be met through a series of measures (CENTAUR strategies) that can be categorized into three main groups (CENTAUR levels):

- new public transport technologies and use of alternative fuels;
- systems and equipment for a better quality of public transport services;
- supporting planning, policy and infrastructure measures.

Through the use of standardized methodologies for all the ten cities, the following environmental impacts are going to be quantified:

- vehicles energy consumption and comparison for application zones;
- vehicles pollutants emissions and comparison for application zones;
- acoustic pollution;
- users and social acceptance of the innovative measures.

At the Euro-Mediterranean Conference held in Barcelona in November 1995, twelve Southern and South Eastern countries of the Mediterranean area have participated with the fifteen EU countries. All the participants have adopted a Declaration: *establishing a new partnership between the EU and the twelve Southern and South Eastern Mediterranean Partners*. The sustainable development objective and its environmental dimension have been fully integrated in the new partnership. Participants emphasised their interdependency with regard to environment, the need for a regional approach, the need of intense co-operation, and better coordination between existing multilateral programmes.

The European Commission has been entrusted with coordinating the preparation of a Short- and Medium-term-priority environmental Action Plan (SMAP) whose objectives are:

A. to help change of the current trend of environment degradation in the Region;
B. to contribute to the sustainable development of the Region, to the protection of the Mediterranean environment, and to the improvement of the health and the welfare of the population;
C. to contribute to the further integration of environmental concerns in all other policies;
D. to contribute to create opportunities for new employment.

Some of the SMAP specific objectives in this context are:

i) to become the common basis for environmental purposes in the Region;
ii) to ensure a real positive impact through prevention, remedy and rehabilitation;
iii) to offer a better chance of fundraising;
iv) to increase the chances of getting more credit for the environment in the Region.

Pollution and environmental degradation are so far advanced that immediate action must be undertaken. Partners are free, within a broad understanding of the concept, to designate one or several priority hot spots (urgent problems in urban areas).

Syria is engaged to include in this framework its own environmental issues and/or development projects. In fact, one of the priority development programmes of the Syrian Ministry of Transport concerns the internal transport. The programme as proposed by the Ministry includes a project aiming to use CNG and LPG in public transport vehicles. The project requires creating a joint venture with a view of studying, financing, implementing and handling the entire project; specific demands are the building of pipelines, of fuelling stations and of mechanical workshops for conversion of vehicles from gasoline/diesel to natural gas, for maintenance and for repair.

The "Project" goes further foreseeing the set up of modern, well-equipped technical centres for the periodical revision of all motor vehicles and their conditions to fit with the standards. Most importantly is that part which concerns the educational level, namely the creation of technical schools whose scope is the formation of specialized staff to cover all the needs of the transport sector.

3. Air Quality

Limited monitoring indicates that air quality in major cities in Syria is generally poor. Pollution from vehicle traffic is particularly critical in Damascus. Some measured daily values are significantly higher than WHO guideline limits, up to 5 times higher in Damascus. It is estimated that 4.000 people over the whole territory of Syria every year are in danger of dying prematurely due to high particulate concentrations alone.

4. Global Environment

Energy consumption per capita and energy intensity in Syria are about the average for a middle-income country in the Middle East, namely 1,272 toe/capita and 0.530 toe/USD of GDP. However, there are many opportunities to increase energy efficiency and reduce liquid fuel consumption. To mention just one, the current efficiency of thermo-power plants does not reach 50%. Increasing efficiency could reduce CO_2 emissions which is currently around 50,56 Mt/year.

5. Estimated Costs of Environmental Damage in Syria

Environmental degradation and shortage of environmental services affect public health, quality of life, economic productivity and natural and cultural resources. Many such effects impose a direct economic cost on the civil society, whilst the damage caused by most other factors can be estimated in monetary terms using a variety of techniques. Due to air quality degradation, using as indicator the health effect of PM_{10} particulates concentrations and urban SO_2 concentrations, the cost estimated for the year 2005 is around 200 millions USD, at net of mortality costs. The overall estimate of the annual cost to Syria of environmental degradation was between 690-890 million USD in 1997, and is expected to increase to between 1.2 to 1.7 billion USD by 2005 if no action is taken. These estimates exclude the costs of mortality, which, **from air pollution alone,** ranged between 150 and 330 million USD in 1997 and will go up to between 330 and 860 million USD in 2005. In 1995, the GDP of Syria was around 33 billion USD; therefore the losses due to environmental damage represent a significant fraction of national income. The low priority given to environmental issues and challenges in the past has led to lack of investment in infrastructures and innovative technologies and, consequently, to a shortage of technical and managerial resources.

6. The National Environmental Action Plan

The first steps for the NEAP (2001) to take are the building of a good management capability to regulate the environment taking cross-sector linkages into account, the promoting of a campaign in order to educate the population, the improvement of ability of environmental impact assessment to include environmental goals in planning new development projects, and finally compliance with international obligations and treaties.

Within the intervention priorities of the NEAP is improving air quality as well as average vehicle speed in urban areas. The targeted investments of the NEAP represent effective and efficient programme to protect public health, to arrest the rate of degradation and to restore damaged assets. Full implementation of the programme will require approximately doubling the current expense on environment over the next 10 years. Although it is not so easy to relate the costs of each specific investment directly to the value of specific benefits, we might predict that the successful implementation of the ten-year programme, which is strongly dependent on effective management and monitoring, is most likely to bring the damage down to zero. In any case, the damage costs, if no action will be taken, would be much higher than the cost of remedial actions. The example of poor air quality is, in this context, very significant: annual cost of remedial action over the entire territory is around 55 million USD per year while the damage cost reaches a factor of nearly ten times more. The total cost of the NEAP investment programme is estimated at around 5 billion USD over the next ten years, and it is anticipated that some additional funding will be provided through international aid flows.

7. Environmental Management

A Legislative Decree which goes back to 1991 has instituted the "General Commission for Environmental Affairs", but only very recently (2002), a law was enacted that defines the GCEA and contains all comprehensive rules and regulations thereof, besides instituting the "Council for Environmental Protection".

Local Environmental Committees are also instituted in the various Departments for the implementation of environmental policies at local levels. These Committees are expected to identify local environmental problems. However, the support services and the cultural climate needed for effective operation of the environmental institutions are still insufficient.

Moreover, within the many commitments that the Law assigns to the "Commission", two are worthwhile to be mentioned, also for their technological content; one concerns the set up of a network covering the whole territory for the monitoring of the all environmental conditions, and the other concerns the set up of a data base system which manages, in real time, the fluxes of data.

At this point it looks as if Syria is on a good basis, nothing more is needed, to start with the implementation of already existing programmes and to go further on projects of international co-operation, especially in the immediate framework of Euro-Mediterranean Partnership.

8. Status of air quality in major cities in Syria

Measurements of concentrations of air pollutants are limited. However, the results of modelling and limited monitoring indicate that air pollution is generally in excess of the 1988 WHO guidelines for Europe.

1) Suspended particulate matter: Two measures of suspended particulate matter were considered, namely, PM10 and TSP (total suspended particulates in the atmosphere). TSP values in Damascus, Aleppo and Homs are 4-5 times higher than the WHO guidelines. PM10 represents 30-50% of TSP.
2) Sulphur dioxide: limited sampling indicates that SO2 concentrations may reach 0.1 ppm as daily average in Damascus. This is twice as much the proposed Syrian standard. In addition, concentrations up to 100 times the standard have been measured around power plants and industrial sites.
3) Carbon monoxide: the available data indicate concentrations in urban areas between 2-12 ppm as an 8 hour average, occasionally reach 35 ppm in Damascus.
4) Nitrogen oxides: concentrations reach an average daily value of 0.3 ppm in Damascus and 0.5 ppm in Aleppo. The draft standard is 0.1 ppm.
5) Photochemical oxidants: surface ozone concentration could reach 0.07 ppm in Damascus as daily average. Such level value lead to irritation of the eyes often experienced by people who live and/or work in the city centre.
6) Other pollutants: Ammonia, Hydrogen Sulphide and other harmful substances have been found in the atmosphere. Benzene is of particular concern since its concentration could reach values higher than the WHO guidelines. There is also some evidence on acid deposition in Damascus from measurements of pH of rainwater.

8.1 CAUSES OF AIR POLLUTION

The causes of air pollution are a combination of:

- motor vehicle emissions, particularly the old fleet of cars (mainly used in public transport) and the poor traffic management resulting in taxis spending a significant portion of time running, with no passengers;
- industrial emissions mainly in hotspots around industrial sites;

- domestic heaters, which are thought to contribute significantly to CO concentration.

9. Traffic

We restrict the following statistical data to the city of Damascus (intended Big Damascus), updated to year 2000.[1]

1) Motor vehicles by type:
   - automobiles: 76.175. More than 13.000 are public cars used for passengers transport (taxis);
   - buses: 1.720. 600 buses are used for public transport;
   - micro buses: 11.370. 9.600 micros are used for public transport;
   - good vehicles: 31.000. Almost 17.000 are used in public transportation;
   - tankers: 900;
   - pick-ups: 61.500;
   - motorcycles: 19.800.
2) Age of vehicles in Syria:
   - passenger cars: 60% are old more than 13 years, of which 26% are older than 24 years.
   - buses and Microbuses: 40% are older than 13 years, of which 9% are older than 24 years.
   - pick-ups: 57% are older than 13 years, of which 6% are older than 24 years.
   - trucks: 68% are older than 13 years, of which 19% are older than 24 years.

Vehicles ownership is concentrated in the major cities – an estimated 1/3 of the total vehicle fleet operates in Damascus, where, due to poor management of the traffic, the average speed drops to about 4-5 km/h, which is far low with respect to international standards.

A modelling exercise carried out 7 years ago suggested that vehicle emissions were by far the greatest contributors to air pollution. To consider Damascus, the exercise indicates that vehicles emissions contribute by 99% to Nitrogen oxides, 99% to Particulates, 100% to Lead, 81% to SO_2 and 100% to CO.

Data relative to year 1996 indicate that the transport sector consumption of fuel oil derivatives amounts to 2.365 million toe. No use of Natural Gas in this sector, not even of LPG. The 2002 State Budget deserves about 44 million USD for internal transport in the 4 major cities of Syria. 2/3 of this investment programme is dedicated to Damascus.

[1] Source: Ministry of Transport estimates, 1998.

10. Natural Gas Programme in Syria

10.1 NG RESERVES

Proven reserves of NG in Syria amount by now, as confirmed by the Ministry of Oil and Mineral Resources, to about 500 billion cubic meters, to be found mainly in the central and north eastern regions of the Country.

A pipeline brings actually natural gas from those far regions to nearby Damascus (50 Km to the East of Damascus).

At Regional level the four Ministers of Oil and Energy of Syria, Egypt, Jordan and Lebanon have recently agreed upon the full implementation of the regional NG pipeline. They agreed to set up a joint venture company denominated the "Arab Natural Gas Company" based in Damascus and a "Natural Gas Commission" based in Beirut. These two intergovernmental bodies will manage for the thirty years to come the whole business of Natural Gas.

On the basis of what said above and given all good promises we put forward the following

10.2 PROPOSAL

1) 1)To build pipelines to reach the peripheral areas of the city of Damascus. The estimated length of these pipelines is about 40-50 Km.
2) 2)To set up a minimum of five adequate fuelling stations on location.
3) 3)To set up an appropriate number of mechanical workshops in order to convert part of the existing vehicles from gasoline and oil/diesel to NG fuel and to guarantee maintenance and repair.
4) 4)To purchase fifty new buses designed to operate only on CNG.
5) 5)To convert a minimum of 1,000 taxis from gasoline to CNG; taxis converted thanks to public expenditure may be marked as "Green Car" with due acknowledgement to the Ministry of Transport and to the Ministry of Environment; the point of the initiative is to increase public awareness of the issue and to encourage owners of private vehicles to do the same.
6) 6)To set up a joint venture authority to manage the whole project.

10.3 THE SOCIAL, ENVIRONMENTAL AND ECONOMIC BENEFITS OF THE USE OF NATURAL GAS IN VEHICLES AS APPLIED TO THE CASE OF SYRIA

The GDP of Syria has grown in the decade 1990-2000 at a pace of more than 5.5% per year, with a population of more than 16 millions. Given the well-known direct relationship between economic activities and transport demand, the high growth of the economy highlights the importance of identifying mechanisms, which might reduce the negative externalities of transport, and therefore improve the trade off between economic growth and environmental sustainability. This is particularly important in Syria where private transport by cars and lorries has steadily increased over time with a faster rate in recent years (about 9.5% a year between 1992 and 1996).

10.4 PROJECT OBJECTIVES

This project aims at identifying and exploiting solutions based on innovative technologies and strategies that can help the developing countries in the Mediterranean area pursuing an eco-sustainable development and to affect positively the air quality reducing pollution in congested urban areas.

Its feasibility is connected with the assessment of the social (and environmental), economic and financial benefits produced by a policy directed at substituting old vehicles (public transport buses and other public transport means, taxis, private cars, motorcycles, etc.) with vehicles characterized by lower polluting emissions with the objective of:

- exploiting the NG resources, which are already relatively abundant in Syria (about 500 billions cubic meters);
- reducing the environmental impact of urban transport in city areas, disentangling the growth in urban economies from transport demand;
- increasing the efficiency and the quality of public transport services.

The introduction of NG will certainly contribute to improving the air quality, reducing also noise production and exploiting a natural resource largely available in Syria.

In the course of the analysis, the project will provide and measure the economic evaluation of environmental externalities, estimating the costs of the project implementation, on one side, and the benefits of substituting the current means of transport with new vehicles fuelled by NG on the other side.

Given the complexities that a full implementation of the project might bring about at the country level in the Syrian context, it seems quite reasonable to identify a city as a case study. On this regard and for several reasons related to its importance, population, geographical location, etc, Damascus comes out as a good choice.

11. The case of Damascus

In the city of Damascus, as well as in most of the cities in the developing countries, the air and acoustic pollution due to motor vehicles emissions is a very serious issue. The Department of Health, in co-ordination with the Ministries of Health and of Environmental Affairs, is planning to supervise stations for the monitoring of air pollution in many Districts. The stations should identify pollutant types and determine the extent to which they are present in the air. In Syria there are also attempts to curb air pollution through mandatory vehicle inspections, the relocation of parking lots and public transport stops away from residential areas, and the sporadic monitoring of heavy traffic areas at peak hours in order to measure the concentrations of noxious gases.

Nevertheless, Damascus has got in the past few years a dramatic growth in the number of private vehicles with obvious negative effects on traffic congestion. It has also an inadequate bus and tramway fleet, ineffective traffic management, especially in the central business districts and an acute shortage of parking spaces as well as secondary road network. The situation is getting much better in these last 2-3 years. Eventually, an important factor that spurred the growth of private transport is the heavily subsidised cost of fuel.

In this context, the aim of the project is to assess the social, environmental and economic benefits produced by a policy directed at substituting old private cars, taxis and public transport buses with vehicles characterized by lower emissions, with the overall objective of reducing the environmental impact of urban transport.

According to the proposed methodology, the working programme will include:

- the collection of all available data on urban transport;
- the analysis of environmental issues such as, for example, type and amount of fuel consumption, emissions dispersion and concentrations for different types of vehicles and vehicle use in different days and year periods, etc.;
- using the ExternE (a well known European project aiming at evaluating the effects of air pollution due to transport activities in urban areas) results to measure the economic evaluation of environmental externalities, the project will provide an estimate of the costs of the project implementation and the benefits of substituting the old means of transport by new ones; as regards costs, estimates will be provided of the related investment costs; examples of benefits are calculated in terms of the differential between the economic costs of each pollutant before and after the implementation of the substitution procedures or in terms of the reduction of the total amount of fuel consumption due to the use of more efficient transport means. The economic effects of the substitution programme will not only concern air pollution but also acoustic pollution as well as, to the extent that new buses would likely increase the efficiency of public transport versus private one, the congestion costs and the overall polluting emissions related to urban traffic;
- the final outcomes of the cost/benefit analysis will be the Net Present Value (NPV), which is calculated summing up the present values of net future benefits and the Internal Rate of Return (IRR), which is the discount rate that makes the NPV equal to zero (in other words, the rate at which the present value of benefits equals the present value of costs). As it is well known, the project will be acceptable if IRR results greater than the social discount rate adopted;
- for the economic evaluation of benefits the ExternE results will be applied to the exposure-response functions that will emerge from the traffic flows evolution during the implementation of the substitution programme.

The analysis of the results will help the local public authorities to make a choice between different alternative strategies of maximization of the urban quality of life. Such a kind of project will very likely have positive direct and indirect impacts on local urban economy by increasing the attractiveness of the city of Damascus in the tourism sector as well as in the industrial one, boosting thus the creation of new jobs in the local market.

MODELING THE COMBINATION OF IMPACTS CAUSED BY WASTE HEAT DISCHARGE FROM POWER PLANT COOLING SYSTEMS INTO OFFSHORE ENVIRONMENTS

A. ANDRIJIEVSKIJ, A. LUKASHEVICH, A. TRIFONOV
Joint Institute of Power and Nuclear Research, National Academy of Sciences, BELARUS.

Abstract

This paper presents the methodology for spatial modeling of thermal discharge into water systems, including calculations of velocity field evolution, pressure and temperature fields for gas and liquid media, dynamics of temperature fields in solid media, and processes of heat-mass exchange at the interfacial area. The SPACEMORPH_THERMO software package was developed based on the described methodology. Test experiments are discussed for nonstationary three-dimensional modeling of human-induced thermal discharge in both flowing and static water systems. This multifunctional code complex is useful for forecasting the dynamics of ecological systems and risk assessment.

1. Base model of SPACEMORPH_THERMO package

The following system of conservation equations is used for modelling the dynamics of transport flow in a spatial section:

$$\frac{\partial \rho W_i}{\partial x_i} = 0,$$

$$\frac{\partial W_i}{\partial t} + W_j \frac{\partial W_i}{\partial x_j} = -\frac{1}{\rho}\frac{\partial P}{\partial x_i} + \frac{\partial}{\partial x_j}\left(\nu_E \frac{\partial W_i}{\partial x_j} - \overline{W_i' W_j'}\right) + g(T)\delta_{ij},$$

$$\frac{\partial T}{\partial t} + W_j \frac{\partial T}{\partial x_j} = \frac{\partial}{\partial x_j}\left(a_E \frac{\partial T}{\partial x_j}\right),$$

where:

$$\overline{W_i' W_j'} = -\nu_E\left(\frac{\partial W_i}{\partial x_j} + \frac{\partial W_j}{\partial x_j}\right) - \frac{2}{3}\delta_{ij} K \quad ;$$

W_i, W_j, - velocity components of transport flow along axes x_1, x_2, x_3 (in the given model $i, j=1, 2, 3;\ i \neq j$); t - time; P, T - pressure, temperature; ρ - density;

B. Morel and I. Linkov (eds.),
Environmental Security and Environmental Management: The Role of Risk Assessment, 297–305.

$g(T)\delta_{ij}$ - gravity term for the vertical velocity component according to Bussinesk approximation; $\nu_E=\nu_{LM}+\nu_{TB}$; ν_{LM} – laminar factors of kinematics viscosity, ν_{TB} – turbulence factors of kinematics viscosity, $a_E=a_{LM}+a_{TB}$; a_{LM} - laminar factors of thermal conductivity; a_{TB} - turbulence factors of thermal conductivity; and K – turbulent characteristic according to an accepted turbulence model.

Simulation of interphase interaction "an atmosphere - superficial waters"

Kinematics conditions. For incompressible media at determined approximations, the conservation of momentum perpendicular and parallel to the interface surface represents one of the basic boundary conditions:

$$\nabla\left(\sigma + k\nabla\vec{V}_{SF,0}\right) + \mu_{SF}\nabla^2\vec{V}_{SF,0} + \mu^{II}\frac{\partial\vec{V}^{II}}{\partial z} - \mu^{I}\frac{\partial\vec{V}^{I}}{\partial z} = 0$$

where:

σ - factor of a surface tension; k, μ_{SF} - factors of dilatation and shift surface viscosity, correspondingly; μ^{I}, μ^{II} - the appropriate factors of shift viscosity of volumetric media, $\vec{V}_{SF,0}$ - vector of velocity on phase boundary provided that at $z=0, \vec{V}^{I} = \vec{V}^{II} = \vec{V}_{SF,0}$.

The first term in this equation is a condition of momentum transfer parallel to a phase boundary, owing to a gradient of surface tension (Marangony Effect); term $k\nabla\vec{V}_{SF,0}$ characterizes viscous resistance to a local stretching of the phase boundary; term $\mu_{SF}\nabla^2\vec{V}_{SF,0}$ - characterizes viscous resistance to shifting in the same plane as the interface surface. The terms $\mu^{I}\frac{\partial\vec{V}^{I}}{\partial z}$ and $\mu^{II}\frac{\partial\vec{V}^{II}}{\partial z}$ describe momentum transfer in a direction perpendicular to the phase boundary.

Equal values for the module of normal stress and the normal component of the velocity for separate phases should be observed by a deviation of the phase boundary from an initial position:

$$P^{I}{}_{sf} = P^{II}{}_{sf} - \sigma\frac{\partial^2\delta}{\partial x^2}; \quad \vec{V}^{I}{}_{Z} = \vec{V}^{II}{}_{Z} = \frac{\partial\delta}{\partial t};$$

where:

P_{sf} - pressure; δ - disturbance of the boundary along an axis Z; $\frac{\partial^2\delta}{\partial x^2}$ - opposite curvature radius of the disturbance boundary in the given point.

When $Z = \delta$ the disturbance normal stresses are equal:

$$P_{sf}^{I} + g(\rho_2 - \rho_1)\delta = P^{II}{}_{sf} - \sigma\frac{\partial^2\delta}{\partial x^2}.$$

Thermal characteristics. Equal temperatures and thermal flows should be observed on the phase boundary:

$$T_{SF}^{1} = T_{SF}^{2}; \; Q_{SF}^{1} = Q_{SF}^{2} = q_{\Sigma}.$$

Additional decoding is required to establish equal thermal flows:

$$q_{\Sigma} = q_{ev} + q_{tr} + q_{conv},$$

where:

$q_{ev} = \chi\left(c_S'' - c_{SF}''\right)\left(\dfrac{m''}{2\pi K T_{SF}'}\right)$; c_S'' - concentration of aqueous vapor; m - molecular mass; χ - heat of vaporization; K - Boltzman constant; the indexes $s,''$ describes a line of saturation and aqueous vapour. The terms in the right part of this equation represent the transfer of heat due to evaporation (q_{ev}), radiative absorption (q_{tr}) and convective transfer (q_{conv}).

From estimated accounts it is possible to accept:

$$\frac{c_S'' - c_{SF}''}{c_S'' - c_\infty''} << 1,$$

where:

c_∞'' - concentration of aqueous vapor escaping the surface. This term approximates a very small amount of evaporation and requires q_{ev} be determined by the ratio:

$$q_{ev} = \chi\left(I_m''\right)_{SF},$$

where:

I_m'' - vapor mass flow, determined by solving for the convection-diffusion transfer of vapor above the surface.

However, perhaps a discussion regarding the assumption of the quasi-stationary nature of vapor's convection-diffusion transfer is more worthwhile. In this case an approximation of the diffusion layer would be [1-2]:

$$\left(I_m''\right)_{SF} = \frac{D'' Nu}{l^*} \frac{\rho_{mx}}{m_{mx}} \ln\left\{\frac{\left(m_{mx}\left(1-c''\right)\right)_\infty}{\left(m_{mx}\left(1-c''\right)\right)_{SF}}\right\},$$

where:

D - diffusion factor; l^* - characteristic linear scale of surface processes;

$$Nu/l^* = \frac{\left(dT_{mx}/dz\right)_{SF}}{T_{SF} - T_\infty}.$$

A further simplification of the initial system of equations can be accomplished by assuming the vapor's process of convection transfer is quasi-stationary. In this case

$$q_{ev,conv} = \frac{Nu}{l^*}\left\{\overline{\lambda}_{mx}\left(T_{SF} - T_{\infty}\right) + \overline{D''}\rho_{mx}\chi \ln\left\{\frac{\left(m_{mx}\left(1 - c''\right)\right)_{\infty}}{\left(m_{mx}\left(1 - c''\right)\right)_{SF}}\right\}\right\},$$

where:

Nu - Nusselt number is determined for a case of natural convection over a horizontal plane.

2. Method of numerical realization of base model

The numerical depiction of the complex model in a coded PC-package is designed according to the implicit scheme of Patancar S. [3], relying on the method of checking volumes and establishing a set of specialized markers. According to MAC numerical procedure (Markers-and Cells) the values of the pressure and temperatures are placed in the centre of a MAC cell.

The values of velocity components are on the edges of a cell, so that the numerical cell can be interpreted as having a volume in space. Clearly designated cells were used to indicate the position of different phases (gas-liquid-solid). Thus, thermal physical parameters (density, viscosity and so on) were assigned to each cell depending on the number of the marker.

Special markers were used to characterize the boundary conditions, the integrated calculation areas, and the allocation of volumetric and surface sources of heat and mass.

Effective simulation of the boundary surface between interacting phases is the crucial element of the **SPACEMORPH_THERMO** package. This includes the modelling of heat and mass transfer, the effect of wave formation, etc.

The package code **SPACEMORPH_THERMO** is developed as a multi-window application for PC by using FORTRAN program language.

Software of the package code **SPACEMORPH_THERMO** uses Windows 95-XP incorporated mechanisms of preemptive multitasking and synchronization of processes and threads.

A sample of the graphic interface of the software package is presented on Fig.1.

3. Simulation of heat release in static water system

The example below was completed by the 3D fluid dynamics model, SPACEMORPH_THERMO. Fig. 2 depicts a shape of water surface and distribution of inlet and outlet pipes.

Initial data:

- inlet and outlet pipes were located at the depth of 7.5 m ;
- the inlet pipe was oriented 90 degrees in a horizontal plane, and 45 degrees in a vertical plane
- distance between pipes was 40 m;

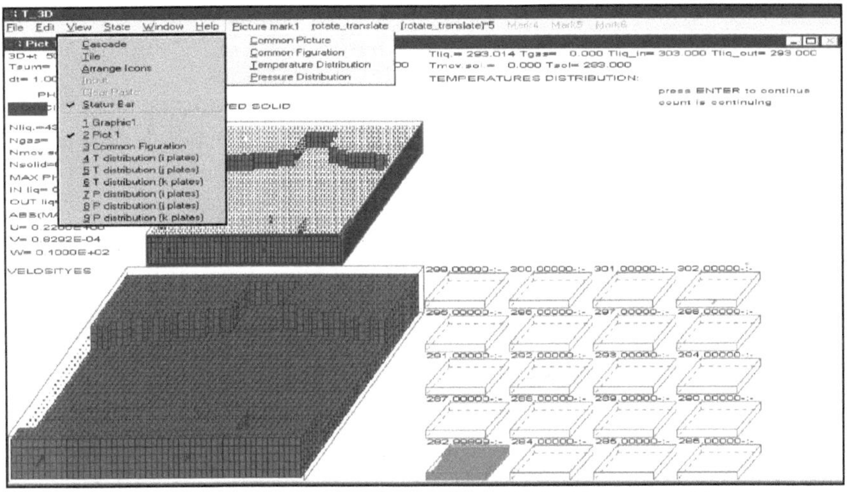

Fig. 1 Element of graphic interface of 3d fluid dynamics model

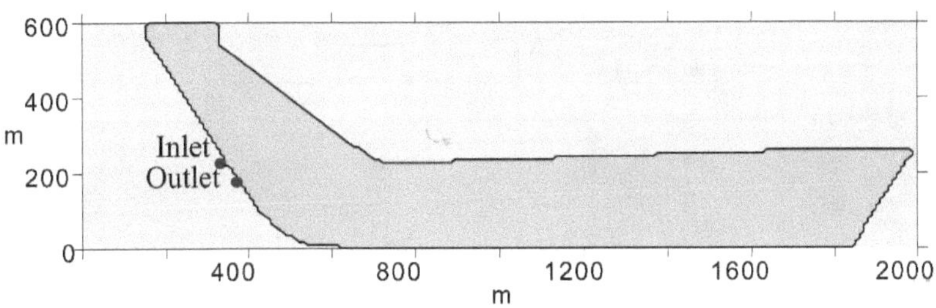

Fig. 2 Shape of water surface

- water discharges for the inlet and outlet pipes were 2.7 m^3/s and –2.7 m^3/s respectively;
- water temperature in the inlet pipe was set to 38°C;
- initial temperature of the water was the same as ambient temperature, equal to 23°C.

The amplitude and the direction of the wind were ignored for these simulations. The numerical grid of this experiment was (3m)x(3m)x(3m). Five horizontal layers in the vertical plane describe the body of water.

The results of the described simulations are presented in Figs. 3 - 6. The size of the simulated area was 320x300x10m. Formation of the discharged heat plume is illustrated by Fig. 7, whereby the inlet pipe is given a horizontal orientation.

As follows from Figs. 3-7, stratification of the primary thermal release is observed by:

- convective flow (directed to near bottom layers).
- thermo-convective flow (directed to near surface layers).

The similar stratification decreases the thermal pollution of near bottom layers, resulting in the expansion of a thermal spot on the surface of the harbor and, subsequently the increase of surface layer temperature.

Fig. 3. Vertical distribution of temperature at initial time of heat discharge (vertical cross-section of calculated area along outlet pipe, size of presented area 10x300 m)

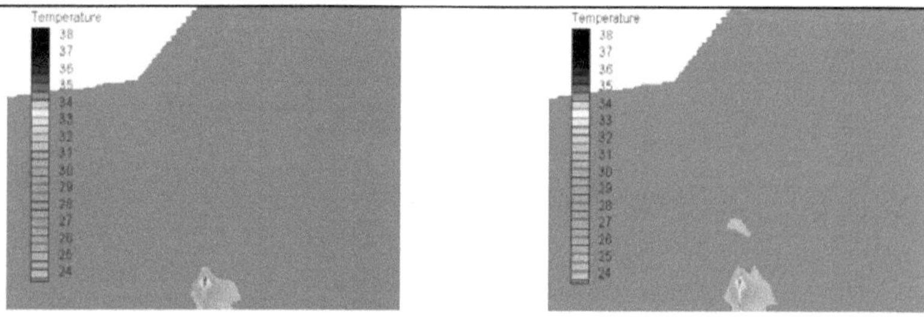

Fig. 4. Horizontal distribution of temperature at initial time of heat discharge. Layers at depths of 9 m and 7 m (size of presented area 300x320 m)

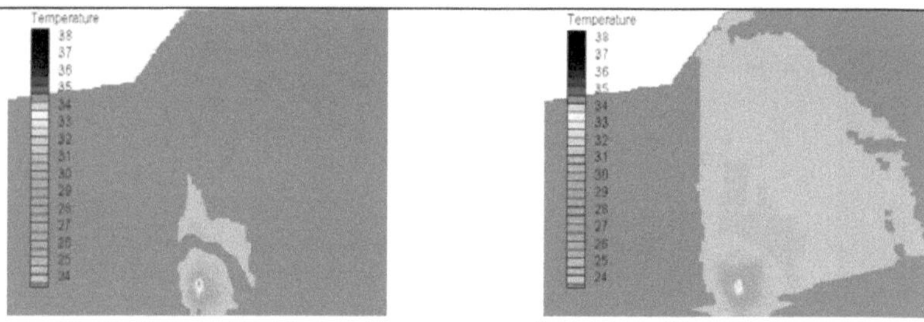

Fig. 5. Horizontal distribution of temperature at initial time of heat discharge. Layers at depth of 5 m (layer of the discharge) and 3 m (size of presented area 300x320 m)

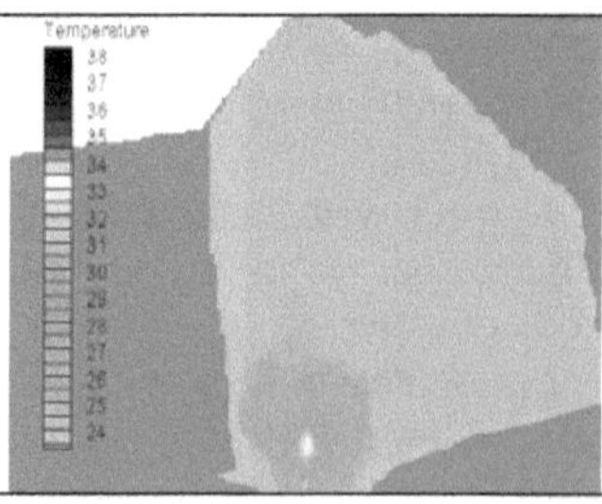

Fig. 6. Horizontal distribution of temperature at initial time of heat discharge for the near surface layer (size of presented area 300x320 m)

4. Results

As a result of the SPACEMORPH package simulations, the following basic tendencies are revealed:

- a thermo-convective effect and integrated reconstruction of pressure and velocity fields are predominant as applied to simulation of heat release dispersion in water media,
- to a lesser degree, the velocity of mass discharge has an influence on similar modelling processes.

Due to these relationships it is possible to conclude that in order to reliably simulate this phenomenon, it is necessary to use the base model simultaneously with the 3D conservation equations describing:

- the thermo-convective process,
- the spatial fields of pressure and velocity,
- the interaction between phases.

In order to provide basic technological recommendations regarding the location of inlet and outlet pipes in static water system, it is necessary to consider:

- separating pipes located at the same level by constructing dividing bars between them,
- locating pipes at different levels, such that the outlet pipe is located near the surface and the inlet pipe near the bottom layers.

In the last case there are positive effects of thermo-convective buoyancy of the heat plume and its tendency to spread along the surface.

5. Discussion

The discharge of the heated effluent can appreciably influence the ecological health of the surrounding marine environment, including native and migratory fish populations as well as benthic organisms. It can cause or exacerbate eutrophication, a process in which warm nutrient-rich water can cause excessive growth of algae resulting in extreme oxygen depletion. Eutrophication is often characterized by an increased green and brown color in the water.

The effects of thermal discharges on the marine environment can be subdivided into direct effects (organisms directly impacted by changes in the temperature regime) and secondary effects arising in the ecosystem as a result of the changes in the organisms directly affected (/4-5/).

The direct effects of thermal discharges on the marine environment include:

- changes in the temperature regime of the water column, and perhaps the sediment, of the receiving environment;
- lethal and sub-lethal responses of marine organisms to the change in temperature regime;
- increased productivity in a range of organisms; reduction in dissolved oxygen capacity.

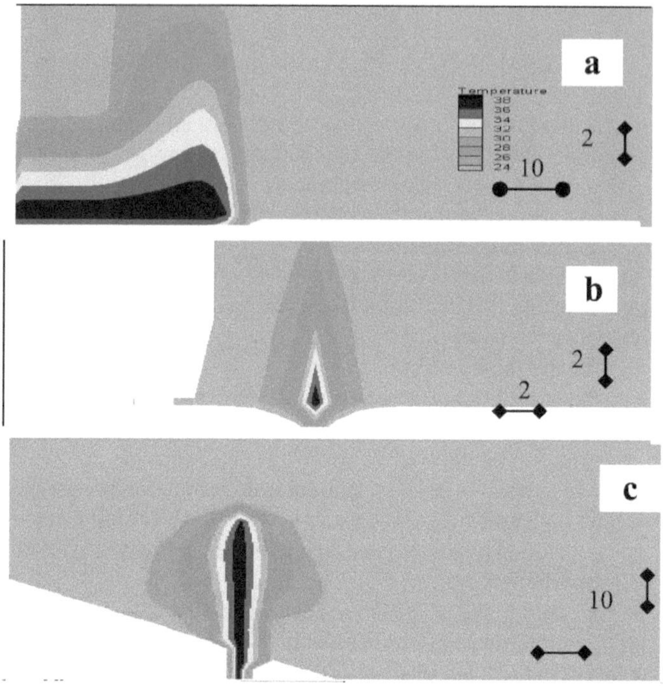

Fig. 7 Formation of the heat plume at initial time of discharge (the outlet pipe at a depth of 7.5m): a – vertical longitude section of heat plume at the zone of outlet pipe, b – vertical cross section of heat plume at the zone of outlet pipe, c - horizontal layer at depth 7.5m

The indirect effects of thermal discharges on the marine environment include:

- changes in the distribution and composition of communities of marine organisms (particularly estuaries);
- localized variations in bird distributions usually in response to increased macro-invertebrate or fish food supplies close to thermal discharges.

The primary sources of thermal discharges in marine/estuarine environment are from power station cooling water discharges; although cooling-water from other industrial processes could be responsible for more localized temperature changes. Even though there has been a significant increase in marine environment research over recent years, further scientific data is required to realize concrete policy decisions. The management will only be effective and successful, if an integrated, multidisciplinary approach is adopted combining both scientific disciplines and institutions.

The model described shows a positive effect of thermo-convective buoyancy of a discharged heat plume spreading along the water's surface. In order to minimize disruptive impacts of thermal discharges outlet pipes should be placed near the surface and inlet pipes should be located near the bottom of water body. The thermal plume is vertically stratified, with a relatively large surface area and a relatively small bottom contact area. The advantage of this design is that it maximizes heat exchange with the atmosphere and prevents benthic organisms from undergoing extreme temperature increase. Further research must be accumulated on marine organism's sensitivity to the thermal regime variation. One application of research could be exploring the effect of cooling-water discharge on the development of concrete industrial objects.

6. References

1. Andrijievskij A., Mikhalevich A., Trifonov A. (1997) "Modelling of mobile boundaries and interface surface in the problems of description of transfer processes in real aqueous systems." *Journal of Engineering Physics and Thermophysics*. V 69, No 6.
2. Andrijievskij A., Lukashevich, Mikhalevich A., and Trifonov A. (1997) "RIVMORPH Model." *Radiation Protection Dosimetry*, V. 73, No 1-4.
3. Patancar S. "Numerical Heat Transfer and Fluid Flow." Hemisphere Publishing Corp., New York, 1980
4. Langford, T.E. 1990. Ecological Effects of Thermal Discharges. Elsevier Applied Science, 468pp.
5. Langford, T.E., Hawkins, S.J., Bray, S., Hill, C., Wels, N. and Yang, Z. 1998. Pembroke Power Station: impact of cooling water discharge on the marine biology of Milford Haven. Report No. UC285 by the Aquatic and Coastal Ecology Group, GeoData Institute, University of Southampton for the Countryside Council for Wales

ROLE OF SYNERGY IN BIOLOGICAL RISK ASSESSMENT

N. GONCHAROVA
Laboratory of Environmental Monitoring, International Sakharov Environmental University, 23 Dolgobrodskaya Str., 220009 MINSK, BELARUS.

Abstract

A substantial effort has been made over the past few decades to label toxicological interaction outcomes as synergistic, antagonistic or additive. The mathematical characterizations of "synergism" and "antagonism" are most closely related to the characterization of no interaction, rather than being some intrinsic toxicological property. For now, labels such as "synergism" are useful to regulatory agencies, both for qualitative indications of public health "risk" as well as numerical decision tools for mixture "risk" characterization. Efforts to quantify interaction designations for use in "risk assessment" formulas, however, are highly simplified and carry large uncertainties. Several research directions, such as pharmacokinetic measurements and models as well as toxicogenomics should promote significant improvements; by providing multi-component data to allow these pair-wise interaction labels to be replaced by biologically based mathematical models of joint toxicity in "risk assessment" procedures

Synergism is not always dangerous or even significantly more dangerous than the individual toxicities. Statistical analyses do not necessarily indicate any significant impacts on public health. Similarly, antagonism does not necessarily mean the mixture is safe, only that it is less toxic than the no-interaction model would predict [1].

1. Combined effect in live meristems of plants from ionizing radiation and heavy metals

The results of experiments investigating the induction of genetic effects in live meristems of winter rye of exposure to a combination of ionizing radiation and heavy metals revealed that the induction of aberrant cells depends equally on the levels of both studied factors; however the extent of injury is determined mainly by the presence of heavy metals. Under the combined influence of ionizing radiation and soil pollution with cadmium salts, the type of non-linear effect is determined by the levels of heavy metal (*e.g.* synergistic effects take place at low concentrations and antagonistic at high concentrations). During the analysis of the combined influence of ionizing radiation and lead soil pollution, another irregularity appears. The decisive role in their formation is determined by the dose of ionizing radiation: low doses combined with various lead

B. Morel and I. Linkov (eds.),
Environmental Security and Environmental Management: The Role of Risk Assessment, 307–314.

concentrations show antagonistic effects, while high doses in the same conditions show synergistic effects. It is particularly complicated to provide an accurate quantitative assessment of a biological system response to a combined effect of factors from very different sources. Nevertheless, in order to present adequate solutions for the protection of the environment demands the creation of a solid scientific basis to evaluate the ecological risk of combined effects and to develop the corresponding universal quantitative criteria and methods where possible [2-4]. The complicated, unequally directed processes are induced in a cell by the combined factors of different pollutants and include, in particular, both the induction of reparation systems and their suppression. They are able – depending on their interaction – to form various responses: antagonistic or synergistic. This phenomenon is confirmed via non-linear changes in a biological system's value and character of response, in particular, the order, level, and terms of effects caused by the exposure to injuring agents. Therefore, the resulting response of a biological system to combined influences cannot be predicted solely from data on the separate effects of each factor.

In order to study the genetic effects of the combined effect of gamma-irradiation and heavy metal soil pollution, two 2-factor vegetation experiments were performed by a full factor experimental design. The data on frequency of aberrant cells in live winter rye meristems are adduced in Table 1, the data on the load of aberrant cells with injuries and their distribution for all studied combinations of factors is described in Table 2. The maximum values for the frequency of aberrant cells and the average number of injuries per cell were observed after the combination of the maximum dose of ionizing radiation paired with a number of cadmium concentrations, other than the control, and than the variant with lead. It is interesting to note that another important index – the load of aberrant cells with injuries – reaches maximum values in other combinations of acting factors: 6 cGy dose with cadmium in any concentration is different from the control and the maximum concentration of lead with the maximum dose or absence of irradiation.

2. Ionizing radiation and cadmium.

The combined influence of irradiation and cadmium soil pollution leads to a significant increase in cell injury (the experimentally determined values of this index for all studied combinations of factors consistently exceed the corresponding values of injury in the absence of these agents.) However, the combination of both tested concentrations of cadmium with a maximum irradiation dose (15 cGy) leads to the decrease in the number of cells with injuries. We must note, though, that the decrease of this index takes place with a considerable growth in the number of aberrant cells as well as the number of injuries per cell as a background (Tables 1, 2).

The interaction of disturbances induced by mutagens of different natures in the cell conceals one enigma surrounding the theory of mutations. It seems the interaction of mutagens causing various primary injuries in a DNA molecule is additive, due to the independence of induction processes. The total output of mutations must be the sum of effects of every factor. However, this rarely occurs. The appearance of non-linear effects is possible mainly due to the following reasons:

TABLE 1.Change Of Frequency Of Aberrant Cells And The Coefficient Of Interaction

Variant	Dose, cGy	Concentration, mg/kg	Frequency of aberrant cells,%	Coefficient of interaction
γ	control	control	0.00±2.04	
	2	control	6.05±2.46	
	6	control	4.31±2.55	
	8	control	9.12±2.65	
	15	control	9.89±2.64	
	50	control	15.25±2.68	
	500	control	25.01±3.06	
Cd+γ	control	2	1.42±2.05	
	2	2	5.86±2.34	0.81
	6	2	13.66±2.83	2.39
	15	2	25.23±3.14	2,19
	control	5	10.84±2,45	
	8	5	19.01±2.67	0.98
	50	5	16.45±267	0.65
	500	5	2815±3.21	0.78
	control	10	10.58±2.54	
	2	10	11.34±2.65	0.66
	6	10	13.54±2.67	0.89
	15	10	26.65±3.06	1.23
	control	50	3.78±2.45	
	8	50	5.57±2.76	0.46
	50	50	19.54±2.65	1.05
	500	50	24.89±3.15	0.89
Pb+γ	control	20	4.35±2.29	
	2	20	2.18±2.09	0.22
	6	20	12.54±2.76	1.46
	15	20	22.01±2.67	1.52
	control	25	6.40±2.56	
	8	25	14.00±2.76	0.89
	50	25	15.89±2.76	0.75
	500	25	24.98±3.07	0.82
	control	100	9.56±2.45	
	2	100	7.56±2.43	0.45
	6	100	16.31±2.51	1.20
	15	100	28.17±2.67	1.48
	control	250	10.24±2.56	
	8	250	19.21±2.45	0.97
	50	250	24.45±2.98	0.89
	500	250	30.79±2.86	0.92

Means followed by the same letter are not significantly different ($P<0.05$)
±:standard deviation

Table 2. Load Of Aberrant Cells With Injuries And Their Distribution By Cells

Variant	Dose, cGy	Concentration mg/kg	Total number of cells	Aberrant cells	Number of frequency per cell*10^{-2}	Load of injury
γ	control	control	500	45	9.0	1.2
	2	control	500	73	14.3	1.2
	4	control	500	82	16.4	1.2
	15	control	500	93	18.6	1.4
Cd+γ	control	2	500	63	12.6	1.2
	2	2	500	87	17.4	1.4
	6	2	500	108	21.6	1.5
	15	2	500	160	32.0	1.6
	control	10	500	84	12.2	1.2
	2	10	500	94	18.8	1.4
	6	10	500	103	20.6	1.5
	15	10	500	158	31.6	1.6
Pb+γ	control	20	500	61	12.2	1.2
	2	20	500	68	13.6	1.2
	6	20	500	110	22.0	1.4
	15	20	500	153	30.6	1.6
	control	100	500	92	18.4	1.3
	2	100	500	104	20.8	1.4
	6	100	500	127	25.4	1.4
	15	100	500	234	46,8	1.8

Means followed by the same letter are not significantly different ($P<0.05$)

1) the long-term existence of an injuring agent in a cell and the possibility of the modification of potential injuries as a link between the injury of DNA's primary structure and its mutations;
2) the possibility of altering the efficiency of a cell's reparation system due to exogenic factors.

Experiments studying the genetic effects of the combined (chemistry + irradiation) exposure on seeds of higher plants [5] showed that the interaction of mutagens occurs on the level of potential injuries. Therefore, the long-term existence of potential injuries induced by one of these factors may lead to an authentic mutation. Some agents have the ability to influence the efficiency of a reparation system – increasing the output of mutations for spontaneous or induced injuries in DNA molecules. The results of experiments [6, 8, 17] reveal the last mechanism describing the formation of non-linear effects on different objects (*E. coli*, *D. radiodurans*, *S. cerevisiae*). These experiments showed that, unlike the wild type, the mutant, created by reparation stocks, displayed a less obvious synergism. There are at least three molecular mechanisms that decrease the efficiency of a reparation system in conditions of combined influence [7]:

- alteration of the second factor of chromatin structure leading to the decreased accessibility of DNA, partially injured by the first factor for repairing enzymes;
- direct inhibition of reparation systems;

- competition between potential injuries induced by both factors, either mutation or reparation.

One more mechanism of synergistic effects involves the interaction between DNA damages and cell membranes. A cell membrane is a "switching panel" for multiple ambient signals. Their relative intensity is expressed on the membrane where they are summated and transferred to secondary messengers, thereby determining the cell's response to its surroundings. Injuries in membranes can lead to local changes of ionic interactions of intracellular medium. These changes – taking into consideration the operative character of chromatin reaction and its connection with the membrane – can change substantially the formation of the DNA molecule, causing a change of gene expression. In particular, changes to the nuclear matrix or membrane induced by outside factors, are able to change the efficiency of a cell's reparation systems [5, 9]. Therefore, independently of the type of studied cells and their mechanisms of adjusting to injuries in their genome or membranes, these two structures may be considered components of one of a cell's co-operative reactions to injuries caused by a combination of factors.

The expressiveness of non-linear effects depends also on the value and the dose (concentration) of acting agents. At relatively low dose levels, the contribution of non-linear effects does not decrease but, generally increases. In the area of low values of exposure, the form of the "dose-effect" curve is determined by the characteristics of the biological system being investigated and often has a non-linear character [10-12]. Therefore the combined effect of different factors at low doses and concentrations tends to lead to the non-linear response of a biological system rather than the case of strong effects, when the ability of a biological system to adapt is exhausted and its response curve follows the increase in intensity of present acting agents.

Formally the response of a biological system to the combined action of many agents is assessed [6, 13] by the comparison of a theoretical situation in space of biological surfaces exhibiting real and/or additive responses from a combination of tension factors. The location of real response points about the additive surface will indicate the character of interaction. The coefficient of interaction is used as a quantitative measure to describe the deviation of a real situation from the additive as well as for classifying effects of the combined influence by groups – additively, synergistic, antagonistic. It is determined as a ratio of response reaction to combined action and sums the effects of the separate actions of each factor.

The values for the coefficient of interaction and an assessment of its variability from 1 for all combinations of studied factors are described in Table 1.

3. Ionizing radiation and lead.

The coefficient of interaction (Table 1) is reliably smaller than 1 at low irradiation doses, i. e. when the non-linear effects are antagonistic. On the contrary, the combination of a maximum irradiation dose (15 cGy) with both tested levels of soil pollution from lead leads to verifiable synergistic effects. The weakening of the mutagenic effect of lead with low doses of ionizing radiation corresponds to the total output of aberrant cells as well as to the degree of their injury. So, the combination of low doses of ionizing radiation (2 and 6 cGy) with maximum tested level of soil pollution (250 mg/kg) leads (Table 1) to considerable decrease in the load of cells with injuries. And in the absence of irradiation or in the combination with the maximum dose, the load of cells with injuries grows along with the growth of lead concentration. According to [14], the assessment of contributing factors into observed changeability corresponds to the share of the mean square of dispersion of this factor in comparison to the total dispersion of the studied index. What are the causes and molecular mechanisms of non-linear effects in our experiment? The unusual genetic effects that occurred are predominantly determined by the biochemical properties of heavy metals. It is known that lead ions are able to influence the cell simultaneously through several mechanisms:

1) immediate interaction with DNA and induction of mutations [19, 20];
2) partial inhibition of reparation systems, decrease of their efficiency [8, 20] and increase in occurrence of mutations for spontaneous and induced injuries of DNA molecules under normal conditions;
3) general toxic influence as a result of the formation of stable complexes with amino acids [17, 22].

Obviously, the ability of lead to inhibit the reparation systems explains the picture of non-linear effects observed in our study and in works of other authors [16-18, 21]. Really, if the presence of an excessive number of ions in a cell is able to decrease the efficiency of their mutagenic reparation systems [8] induced by low ionizing radiation doses [12, 23], then the combined action of lead ions and ionizing radiation must lead to antagonistic effects under low doses (due to the inhibition of mutagenic reparations creating an increased output of genetic injuries in the area of low doses), and simultaneously – to synergistic effects from medium-level doses (due to the inhibition of mutagenic reparations in this dose range causing the efficient removal of multiple genetic disturbances).

Unlike the lead, the cadmium toxicity and mutagenicity are based on its denaturing influence on metabolically important proteins [6]. Apparently, this determines the observed picture of non-linear effects. If low, close to maximum admissible concentrations of cadmium exert a predominantly mutagenic effect combined with ionizing radiation leading to synergism, the high cadmium concentrations cause grave disturbances and lethal cell injuries. They do not increase the frequency of aberrations, revealing the mostly additive and perhaps antagonistic character ionizing radiation and cadmium salts in toxic concentrations on genetic material. The absence of frequent aberrations in chromosomes from experimental exposure to cadmium salts in toxic concentrations was noted in experiments on cells of mammalia and *Crepis capillaris* by other authors [24]. Therefore the qualitative difference of observed non-linear effects of cadmium and lead interaction with ionizing radiation is due to the difference between their mechanisms of biological interaction.

4. Conclusions

It is worth paying attention to fact that synergistic and antagonistic effects are not observable at high irradiation doses, confirming our hypothesis that non-linear effects are induced mainly by the combination of low exposure to injuring agents.

Under the combined influence of ionizing radiation and soil pollution with cadmium salts, the type of non-linear effect is determined by the heavy metal (synergistic effects take place at low concentration, antagonistic – at high concentration). From the analysis of the interaction between ionizing radiation and soil pollution with lead, other irregularities appear, with the cause of their formation being due to the dose of ionizing radiation: low doses combined with various lead concentrations cause antagonistic effects, high doses in the same conditions lead to synergistic effects. The obtained results testify to the necessity to count non-linear effects of combined interaction of different factors when fixing rates of technologic load on agrocoenoses.

5. References

1. R. C.Hertzberg and M. M.MacDonell.Synergy and other ineffective mixture risk definitions//The Science of The Total Environment 2002,V288, issue.1-2, P 31-42
2. N.V.Goncharova Separate and combined effect of radiation and nonradiation factors on pea shoots XXX ESNA Annual Meeting 2000, Hungary Keszthely, p. 86 Abstract
3. N.V.Goncharova Effect of ionizing irradiation and heavy metals on the photosynthetic apparatus of plants XXXI ESNA Annual Meeting Chania, Crete, 2001Greece p.53 Abstract
4. N.V.Goncharova Effect of Cu^{2+} ions on photosynthetic apparatus of pea plants in the conditions of gamma-irradiation// Revue de Cytologie et Biologie végétales , Le Botaniste 2000 V.23 issue 1-2 P.29-34
5. M. M. Vilenchik. Radiobiological effects and environment//Ehnergoatomizdat. Moscow, 1991. 160 p. (In Russian)
6. T.I. Evseeva, S.A. Geras'kin, E.S. Khramova (2003) The comparative estimation of plant cell early and long-term responses on ^{232}Th and Cd combined short-time or chronic action. *Tsitol. Genet*, **37** (3) 61-66.
7. S. A .Geraskin V. G Dikarev , N. S Dikareva, A. A Udalova Induction of genetical effects in leaf meristem of barley by ionizing radiation and heavy metals. I. Separate effect//Genetika, 1995, V. 31. (In Russian)
8. A. A Petin., V. P. Komarov .Quantitative description of radiosensitivity modification//Ehnergoatomizdat, Moscow, 1989. 190 p. (In Russian)
9. C.W Theodorakis, B.G. Blaylock, L.R. Shugart (1997) Genetic ecotoxicology I: DNA integrity and reproduction in mosquitofish exposed *in situ* to radionuclides. *Ecotoxicology*, **6**, 205-218.
10. D. M Spitkovskij., A. A Ermakov., A. I Gorin. et al. Dependency of reparation of DNA induced by genetically dangerous influences on the ion power of medium in which are situated the cells//Tsitologiya, 1992, V. 34, no. 7 . P. 76-85 (In Russian)
11. A. V Sevankaev. Actual state of problem of quantitative assessment of cytogenetical effects in the area of low radiation doses//Radiobiologiya, 1991, V. 31, issue 4. P. 600-605 (In Russian)
12. S. I.Zaichkina., G. F., Aptikaeva., A. Kh. Akhmadieva. et al. Peculiarities of performance of cytogenetical disturbance at the influence of low irradiation doses on cells of mammalia and plants//Radiobiologiya, 1992, V. 32, issue 1. P. 38-41 (In Russian)
13. Ionizing Radiation: Sources And Biological Effects. UN Scientific Committee on Effects of Atomic Radiation. Annual report to General Assembly. New-York, 1982, V. 2. 780 p.
14. S. A. Ajvazyan , I.S.Enyukov ,L.D. Meshalkin Applied statistics. Investigation of dependencies//Finances and Statistics, Moscow, 1985. 487 p. (In Russian)

15. S.A.Geraskin,V.D. Dikarev ,N.S. Dikareva et al. Induction of chromosome aberrations in leaf meristem under the combined influence of irradiation and heavy metals. Study of the surface of response of multifactor experiment//Radiation Biology. Radioecology, 1993, V. 33, issue 3 (6). P. 890-899 (In Russian)
16. V.I.Shershunova , E.R.Vilenskij , A.I.Voronin ., I.G.Kogan Genetical effects of combined action of chronic gamma-irradiation and lead nitrate on barley waxy-locus//Radiobiologiya, 1990, V. 30, issue 2. P. 224-227 (In Russian)
17. S.B.Dineva , V.I.Abramov ,V.A. Shevchenko Genetical consequences of effect of lead nitrate on seed of chronically irradiated populations of *Arabidopsis thaliana*//Genetika, 1993, V. 29, no.11. P. 1914-1919 (In Russian)
18. T.P. Reddy , K.Vaidyanath Synergistic interaction of gammarays and some metallic salts in the induction of chlorophyle mutation in rice//Mutat.Res. 1978. V.52. N.3. P.361-365.
19. A.M.Reichmayr-Lais , M. Lead. Kurchgessner //In: Biochemistry of the essential ultratrace elements / Ed. Frieden E. N.Y., L.:Plenum Press, 1984. P.367-387.
20. N.V.Reutova ,V.A. Shevchenko Mutagenic effect of inorganic silver and lead compounds on spiderwort// Genetika, 1992, V. 28, no. 9. P. 89-96 (In Russian)
21. A. Hartwig , R. Schlepegrell , D. Beyersman Indirect mechanism of lead-induced genotoxicity in cultured mammalian cells//Mutat.Res. 1990. V.241. N.1. P.75-82.
22. B.Venugopal , T.D.Luckey Metal toxicity in mammals. 2.Chemical toxicity of metals and metalloids. N.Y.: Plenum Press,1978.
23. D.M.Spitkovskij ,A.V. Ermakov ,A.I. Gorin . et al. Peculiarities of non-planned DNA synthesis and changes of structural parameters of nuclei of human lymphocytes after the influence of X-rays in low doses and in combination with UV-irradiation//Radiation Biology. Radioecology, 1994, V. 34, issue 1. P. 23-30 (In Russian)
24. .Eh. N.Vaulina , I.D.Anikeeva , I.G.Kogan Effect of cadmium ions on cell partition of root meristem of Crepis capillaris (L.) Wallr.//Tsitologiya i Genetika, 1978, V. 12, no. 6. P. 497-503 (In Russian)

COOPERATIVE NETWORK FOR ENVIRONMENTAL RISK ANALYSIS STUDIES: THE CASE OF THE MIDDLE EAST REGION

M. K. ZAIDI
Radiological and Environmental Sciences Laboratory (RESL)
1955 Fremont Avenue, Idaho Falls, Idaho, USA.

J. GANOULIS
UNESCO Chair for Water,
Laboratory of Hydraulics, Department of Civil Engineering,
Aristotle University of Thessaloniki, 54124 Thessaloniki, GREECE

Abstract

The role of regional networks for applying Environmental Risk Analysis (ERA) in order to promote environmental and human security is analysed in this paper. These networks should take national and regional approaches and have a multidisciplinary view, and implement a financing strategy, a training policy for specialists and technicians according to specific needs and a social policy for informing and sensitising citizens, and encouraging policies of prevention. The case of a specific network to be developed in the Middle East is further discussed.

1. Introduction

In public health, Risk Assessment (RA) may be defined as a process aiming to analyse different hazards that may induce risks to human health following exposure to a particular substance [1]. RA of health effects depends on the purpose and scope of available information (data) used in assessing the situation. Some assessments examine the impact after an accident has happened whereas others precede any incident and identify hazards, estimate exposure and potential health dangers and characterize or describe the risk. A more general definition for RA may be applied not only to health effects but also to environmental, ecological and technical risk problems based on a multidisciplinary approach [2,3,4]. According to this definition, RA is a complex scientific domain involving different disciplines (e.g. physics, chemistry, toxicology, engineering, law, economics, sociology and political sciences) and inter-disciplines (e.g. ecology and environmental sciences) and addresses difficult technical, economic, environmental and social problems.

In order to effectively apply ERA methods and tools to resolve complex regional and national problems regional networks of institutions and specialists may play a major role. A strategy for developing sustainable cooperative networks should include: a multidisciplinary view for environmental protection, financing, a training

B. Morel and I. Linkov (eds.),
Environmental Security and Environmental Management: The Role of Risk Assessment, 315–320.

policy for specialists and technicians, preventive action for environmental protection and sensitisation of the public, starting at primary school level, according to the specific needs and social conventions for information dissemination. Methodologies to evaluate and manage environmental and health issues should be adapted to particular regional problems. A suggested first step is to create an open information network to develop and sustain cooperation in the region. Modern Information and Communication Technologies (ICTs) and the Internet should be intensively used in order to increase effectiveness and save available resources [5]. A scale to measure the usefulness of such a regional network should also be agreed upon. Selected topics of interest as well as the tools and methods to be used are further discussed in this paper.

It is important that such a network should develop, maintain and share an objective information database. Given the regional nature of the network, scientific and financial support from international organisations, such as the United Nations Educational, Scientific and Cultural Organisation (UNESCO), the United Nations Environment Programme (UNEP), the Global Environment Facility (GEF), the World Bank, the International Waters Exchange and Resource Network (IW-LEARN), the European Union (EU), the United States Agency for International Development (USAID) and the North Atlantic Treaty Organisation (NATO) could be sought. Methods and tools for developing such a network, selected topics of interest and ways to plan future action are further discussed below.

2. Methodology

On the issue of environmental security in the Middle East, it should be pointed out that while there is a large surplus of water in some places in the region, such as Syria and Turkey, in others, such as Jordan and the Palestinian Occupied Territories there is a great shortage. [6,7]. Independent of climate change there will be a great stress on water supplies by the year 2025 [8].

To face this situation, it was suggested [9] that the challenges of water resource management quantity and quality should be addressed by considering not only technical and economic objectives (Fig. 1) but also environmental and social objectives (Fig.2). Sustainable management of water resources may be achieved by applying Dublin principles [10], i.e. the:

- *Ecological Principle.* This is based upon river basin management, multi-sectoral management of agricultural, industry, and households, and the idea that land and water must be managed together.
- *Institutional Principle.* This encompasses all stakeholders, the state and private sector and civilians, and requires them to be involved in management using the principle of subsidiarity or action at the lowest level.
- *Instrument Principle.* This uses economic mechanisms to improve allocation, enhance quality and adopt pricing policies.

Figure 1: Technical and economic objectives.

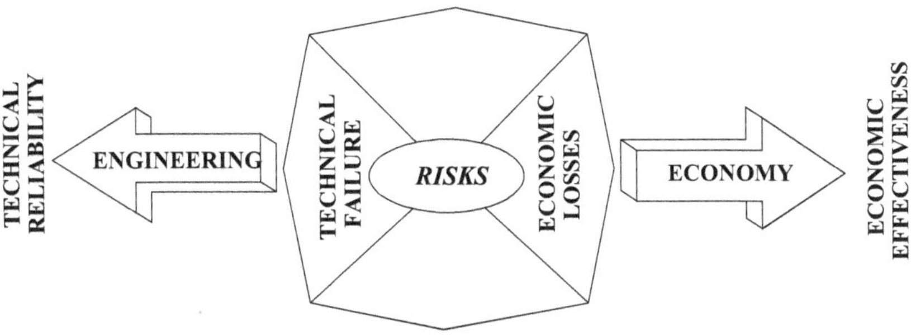

Figure 2: Technical, economic, environmental and social objectives.

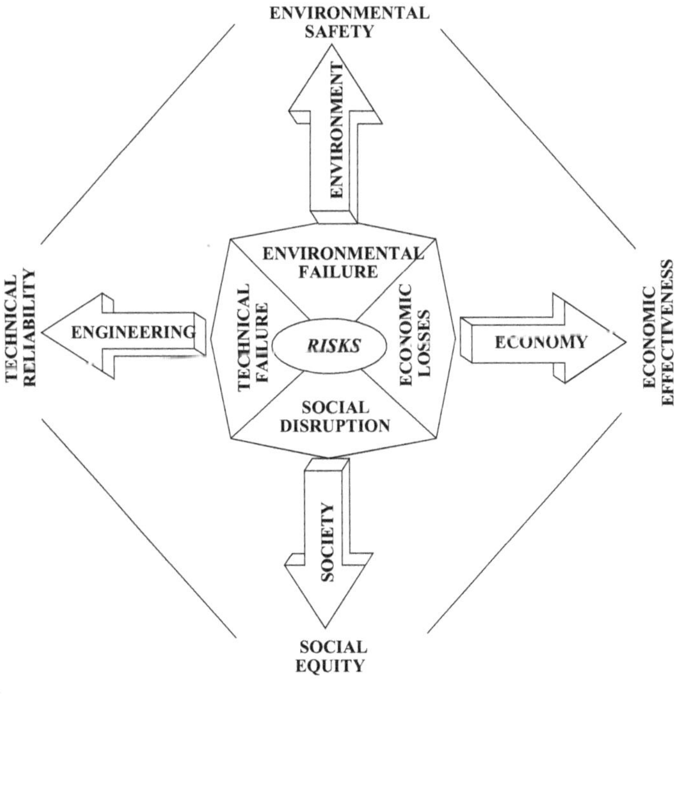

Regardless of the approaches agreed upon, the public needs to be well informed so that the age-old scenario of "us against them" can be overcome by making people realize that both sides of an issue will always have both pros and cons. It has been stated that most of the time the public does not share/agree upon common goals, nor does it collectively have the goodwill and spirit to get a job done [11]. Therefore politicians, elected representatives and policy makers should take the role of leaders so that professionals can conduct research and provide background knowledge, technical data and pertinent information to administrators, who in turn will provide solutions to create a lasting outcome for the recipient public.

The proposed network for the Middle East will involve the countries of Cyprus, Egypt, Israel, Jordan, Lebanon, Palestine, Syria and Turkey. This network will be only one in the wider Mediterranean region, and later undertake inter-regional cooperation with similar networks from Middle East and North Africa (MENA) countries, and South East Europe (SEE) or the Balkans. As the network will be a multi-national group, different languages, customs and ways of life play should be taken into consideration. Wars in the region have created mistrust and bitterness, and there is potential conflict over the use of international waters. The existing institutional structure is also very weak and depends on or is dominated by external international politics and pressure put on participating countries when making decisions.

When considering water management and environmental protection, issues of water supply alternatives, demand, sharing international waters, water recycling and use, wastewater treatment and disposal, desalination, coastal waters, ecology and water quality should be taken into account. The Middle East is an ecologically very sensitive area, being both semi-arid and arid. Rapid changes in the demand for water and food may occur in this region, which may be among the first victims of global climate change [12]

Air pollution from industry and transporting of soil contaminants are major factors affecting water quality and may put public health at risk [13], as was the case with the accident at the Chernobyl nuclear plant in April 1986. The risks posed by the remains of the destroyed reactor (and its surrounding shelters, the radioactive waste storage and disposal sites were assessed, as was environmental contamination in the region. Other risks posed – such as collapse of the shelter, radionuclides migration from storage and disposal facilities and transfer from soil to vegetation and its potential regional impact were also investigated [14].

The proposed sustainable cooperative network for the Middle East will apply theories of ERA. The network will create a website, managed by scientists, and open to local Non-government Organizations (NGOs), administrators, end users, research institutions, representatives of the general public, law makers and ministers. The network will consist of three major components:

1. *Information Network* (IN): This will be responsible for collecting all necessary information, mapping existing information/data, analysis of data reliability, monitoring networks, guidelines/standards, sharing data, internet e-networking, education, training, capacity building, exchange of publications, paper, reports etc.
2. *Knowledge Network* (KN): This will formulate the problem, prepare the model, analyse case studies using past experience, develop tool boxes, conduct

workshops and seminars, exchange experience and learn from traditional techniques, identify gaps in knowledge, perform comparative analyses and apply best practices.

3. *Practice Network*: This will seek financing and sponsors, involve decision makers, address public opinion, communicate knowledge/information to the media and the public through publications, handouts and brochures etc.

3. Conclusion

Any national RA related policy should be based on national and regional approaches. This may increase transboundary cooperation at the local level. Effective institutional arrangements and networks for environmental management, will support sustainable use of water resources, encourage development of local economies, promote stability and peace, harmonize legislation, guidelines, and methodologies, and attract international support and funding. The technical, economic, environmental and social objectives of region will be effectively addressed.

4. Acknowledgement

This work was developed at the NATO-ARW – "The role of risk assessment in environmental security and emergency preparedness in the Mediterranean Region" held in Eilat, Israel from 15-18 April 2004. The authors wish to thank the organising committee for the financial support provided to attend the meeting.

5. References

1. U.S. Environmental Protection Agency (USEPA) (1984) Risk Assessment and Management: Framework for Decision Making, EPA 600/9-85-002, Washington, DC: U.S. December 1984.
2. U.S. Environmental Protection Agency (USEPA) (1998) Guidelines for Ecological Risk Assessment. Federal Register 63: 26846-26924, 14 May 1998.
3. Duckstein, L. and E.Plate (eds.) (1987) Engineering Reliability and Risk in Water Resources, E.M. Nijhoff, Dordrecht, The Netherlands, 565 pp.
4. Ganoulis, J. (1994) Engineering Risk Analysis of Water Pollution: Probabilities and Fuzzy Sets. WILEY-VCH, Weinheim, Oxford, NY, 306 pp.
5. Ganoulis, J., I. L. Murphy and M. Brilly (eds.) (2000) Transboundary Water Resources in the Balkans: Initiating a Sustainable Co-operative Network, NATO ASI SERIES, Partnership Sub-Series 2: Environmental Security, Vol. 47, Kluwer Academic, Dodrecht, Boston, London, 254 pp.
6. Amery H.A, Wolf A.T. (2000). Water in Middle East: geography of peace, Austin, TX.
7. Bellissari A. (1994). Public health and water crisis in occupied Palestinian Territories. J.Palestinian Studies 23 (2). 52-63
8. Robert T. Watson RT. (1996). Implications of Climate Variability and Human-Induced Climate Change for Water Resource Management, ESSD, World Bank. Inter-governmental Panel on Climate Change.
9. Ganoulis, J. (2004) Integrated Risk Analysis for Sustainable Water Resources Management In: Comparative Risk Assessment and Environmental Decision Making, 275-285, NATO Science Series, Kluwer Academic, The Netherlands.
10. Water Engineering and Research in a Learning Society. (2003). Proceeding -30th Biennial World Congress of the IAHR, Thessaloniki, Greece.

11. Homer-Dixon HF. (1991). On the threshold: environmental changes as causes of acute conflict. Int. Security 16 (2). 76-116.
12. Tal A, Linkov I. (2004). The role of comparative risk assessment in addressing environmental security in the Middle East. *Risk Analysis,* In Press
13. Israel Ministry of Environment. (2003). Risk assessment for air pollution in Tel Aviv – Ashdod area, 1995-99, Jerusalem
14. Vargo GJ. (2000). The Chernobyl Accident: A comprehensive risk assessment. 2004 Books from Battelle.

LIST OF PARTICIPANTS

Abu al Taher, Mohammed	Amman, JORDAN	E-mail: hyper2022@yahoo.com Ph: 962-6-5353036
Abu Dwaih, Dawlah Mosa	Ziyad bin Abeeh - Amman 11181, JORDAN	E-mail: dawlah@hotmail.com 962-79-6735909
Acma, Bulent	Anadolu University Department of Economics Yunwemre Campus 26470 Eskisehir, TURKEY	E-mail: bacma@anadolu.edu.tr Phone: +90(222) 3350580ext.6171 FAX: +90(222) 3353616
Al-Sairafi, Radi Daoud	P.O.Box 7273, Amman 11118, JORDAN	E-mail: radi1973@hotmail.com Ph: 962-6-4915783
Arlosoroff, Saul	Mekorut National Water Corp. 2/15, Menora Street Tel Aviv 69416, ISRAEL	E-mail: Sarlo@inter.net.il Phone: +972 3 648 3279 FAX: +972 577 977 720 Cell : +972.505.507 74220
Brenner, Shmuel	Director Green Enterprises 16 Israel Yeshayahu Street Herzelia 46803, ISRAEL	E-mail: shmubr@netvision.net.il Phone: +972 9 9554232 FAX: +972 9 9589582 Cell: +972 55 774373
Darwish, Abdel Raouf	ASEZA P.O. Box 2509, Aqaba, 77110, JORDAN	E-mail: darwish1jo@yahoo.com Ph: 962-79-5515971
Figovskiy, Oleg	Journal Scientific Israel-Technological Advantages, ISRAEL	E-mail: figovsky@netvision.net.il
Ganoulis, Jacques	Aristotle University of Thessaloniki Department of Civil Engineering Thessaloniki 54124, Macedonia GREECE	E-mail: iganouli@civil.auth.gr Phone: +30 (2310) 995682 FAX: +30 (2310) 995681

Garb, Yakov	Arava Institute for Environmental Studies, D.N. Hevel Eilot 88840 ISRAEL	E-mail: ygarb@cc.huji.ac.il Phone: 972 8 6356694 FAX: 972 8 6356634
Girgin, Serkan	Research Assistant, Middle East Technical University Department of Environmental Engineering, Inonu Bulvari, 06531 Ankara, TURKEY	E-mail: girgin@metu.edu.tr Phone: +90-312-2102641 FAX: +90-312-2101260
Goncharova, Nadezhda	International Sakharov Environmental University, 23 Dolgobrodskaya Str., Minsk, 220009, BELARUS	Email: nadya_goncharova@yahoo.com Phone: +375 172 24 43 11 Fax: +375 172 30 68 97 or 30 68 88
Grebenkov, Alexandre	Head of Laboratory #15 Joint Institute of Power and Nuclear Researach 99 Akadeonik Krasin Street Minsk, 220109, BELARUS	E-mail: greb@sosny.bas-net.by Phone: 011-203-5010572 FAX: 011-4285792
Green, Manfred	Director, Israel Center for Disease Control Gertner Institute Tel Hashomer 52621, ISRAEL	E-mail: mgreen@icdc.health.gov.il Phone: 972 3 7371500 FAX: 972 3 5349881 Cell: 972 56 242012
Haruvy, Nava	Netanya Academic College 1 University Rd Netanya, ISRAEL	E-mail: navaharu@netvision.net.il Phone: 972 8 9463189 FAX: 972 8 9365345 Cell: 972 53 611260
Hijazin, Lubna Samir	Amman Dabouk, Amman 1182, JORDAN	E-mail: lubna_hijazeen@hotmail.com Ph: 962-6-541-1210
Jubier, Alain	NATO, BELGIUM	E-mail: a.jubier@hq.nato.int
Letnik, David	Israel Academy of Engineering, Haifa, ISRAEL	E-mail: daile@actcom.co.il Fax: 04-8626570

Levner, Eugene	Lab. of Strategic Environmental Studies, Holon Academic Institute of Technology 52, Golomb St. Holon 58102, ISRAEL	E-mail: levner@hait.ac.il Phone: 972 3 5026741 FAX: 972 3 5026733 Cell: 972 6 6246757
Linkov, Igor	Cambridge Environmental Inc., 58 Charles Street, Cambridge, MA, 02141, USA	E-mail: Linkov@CambridgeEnvironmental.com Ph: 617-225-0812 FAX: 617-225-0813
Lipchin, Clive	Arava Institute for Environmental Studies, D.N. Hevel Eilot 88840 ISRAEL	E-mail: clive@arava.org Phone: 972 8 6356694 FAX: 972 8 6356634
Lukashevich, Alexey	Joint Institute of Power and Nuclear Research 99 Krasin STY. Minsk, 220109, BELARUS	E-mail: lag@sosny.bas-net.by Phone: 375 172 99 44 75 FAX: 375 172 99 43 55
Marquina, Antonio	Faculty of Political Science Campus Somosaguas 28223 Madrid, SPAIN	E-mail: marioant@cps.ucm.es Phone: 34 91 3942924 FAX: 34 91 3942655
Maxwell, Carl	USAID 16648 S.W. 88th Pl., Tigard, Oregon 97224, USA	E-mail: carlmaxwell@hotmail.com Phone/fax: (503) 684-7838
Maxwell, Magdalena	USAID Maxwell, Magdalena 8250 Dili Place, Dulles, Va 20189-8250, USA	E-mail: mmaxwell@usaid.gov Phone: (3741) 270-262
Morel, Alexandra	World Council on Renewable Energy, UK	E-mail: xelatine@hotmail.com
Morel, Benoit	Engineering and Public Policy, Carnegie Mellon University Pittsburgh, PA 15213, USA	E-mail: bm1v@andrew.cmu.edu Phone: 412-268-3758 FAX: 412-268-3757
Nsheiwat, Zein Bassam	Al-Hamam St., Amman, JORDAN	E-mail: zein_nsheiwat@hotmail.com Ph: 962-6-5517808

Ramadan, Abou Bakr	Egyptian Radiation and Environmental Network,Egyptian Atomic Energy, 3 Ahmed El Zomor St., P.O. Box 7551, Cairo, EGYPT	E- Mail: ramadan58@yahoo.com Telephone: 202 2740 238 FAX: 202 2740 238,
Shtern, Yuri	Committee on Environment, Knesset, ISRAEL	
Simpson, Lori	Massachusetts Institute of Technology 7901 Anna Pl, Lincoln, NE 68507, USA	Lorial@rocketmail.com
Tal, Alon	Arava Institute for Environmental Studies, D.N. Hevel Eilot 88840 ISRAEL	E-mail: alon@arava.org Tel. 972-58-629368
Tsikerman, Alexander	Director of Centre of Ecological System and Technologies Emer Beit Shaan Street 50/6 Modiin, 71700 ISRAEL	E-mail ecostisi@hotmail.com Phone: 972 8 9728896 Cell: 972 54 705723
Zaidi, Mohammed	US Department of Energy, Radiological Environmental Sciences Laboratory, 1955 Fremont Avenue, MS-4149, Idaho Falls, ID 83401 USA.	E-mail: zaidimk@id.doe.gov Ph: 208-526-2132 Fax 208-526-2548

AUTHOR INDEX

MIX
Papier | Fördert gute Waldnutzung
FSC® C083411

Kolibri 360 GmbH
Ferdinand-Jühlke-Straße 7
99095 Erfurt, Deutschland
produktsicherheit@kolibri360.de